William Watson Cheyne
and the Advancement
of Bacteriology

William Watson Cheyne and the Advancement of Bacteriology

CHARLES DEPAOLO

McFarland & Company, Inc., Publishers
Jefferson, North Carolina

LIBRARY OF CONGRESS CATALOGUING-IN-PUBLICATION DATA

Names: DePaolo, Charles, 1950– author.
Title: William Watson Cheyne and the advancement of bacteriology /
    Charles DePaolo.
Description: Jefferson, North Carolina : McFarland & Company, Inc.,
    Publishers, 2016. | Includes bibliographical references and index.
Identifiers: LCCN 2016038394 | ISBN 9781476666518 (softcover :
    acid free paper) ∞
Subjects: LCSH: Cheyne, William Watson, Sir, 1852–1932. | Bacteriology—
    History.
Classification: LCC QR21 .D47 2016 | DDC 616.9/201—dc23
LC record available at https://lccn.loc.gov/2016038394

BRITISH LIBRARY CATALOGUING DATA ARE AVAILABLE

ISBN (print) 978-1-4766-6651-8
ISBN (ebook) 978-1-4766-2641-3

Front cover image of Sir William Watson Cheyne Photograph by Elliott &
Fry (Wellcome Library, London); petri dish © 2016 Rudi Gobbo/iStock

Printed in the United States of America

McFarland & Company, Inc., Publishers
  Box 611, Jefferson, North Carolina 28640
    www.mcfarlandpub.com

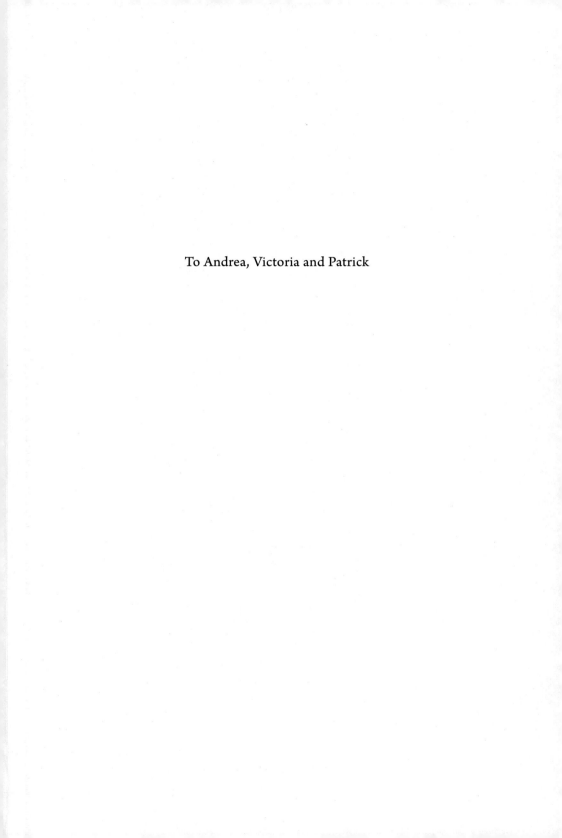

To Andrea, Victoria and Patrick

Acknowledgments

An earlier version of chapter 1 appeared in *The Victorian Web: Literature, History, and Culture in the Age of Victoria,* http://www.victorianweb.org, as "Paster and Lister: A Chronicle of Scientific Influence." I would like to thank Dr. George P. Landow, Editor-in-Chief and Webmaster, for his constructive criticism and editorial attention. To the anonymous reviewers of this book who offered indispensable advice and correction, I extend my gratitude. My colleagues at Borough of Manhattan Community College, City University of New York, offered support and votes of confidence. Through the Faculty Publications Program, sponsored by the College, I received a reduced teaching schedule for the Spring 2016 semester, providing me with time to complete the project. Professor Dorothea J. Coiffe-Chin, Faculty Research and Scholarly Communications Librarian, expedited my requests for documents that were difficult to find and essential to the research. For having advocated my work through the years and before College Personnel and Budget Committees, I owe a debt to Dr. Joyce Harte, Professor and Chairperson, Department of Writing and Literature, to Dr. Robert M. Zweig, Professor of English, and to Dr. Erwin J. Wong, Dean of Instruction. I wish also to thank Dr. Antonio Pérez, President of the College, for supporting traditional research, and especially for recognizing its importance in the professional lives of senior faculty. Finally, without the patience and technological expertise of my wife, Andrea, the manuscript would still be on my desk.

Table of Contents

Introduction

Sir William Watson Cheyne (1852–1932), a surgeon by training, has been called "one of the most prominent bacteriologists in England" (Bulloch, *History*: 358).[1] In addition to publications on surgery, he produced more than 60 papers and 13 monographs, from 1879 to 1927, either entirely or in part on medical bacteriology. The large number of writings and the complexity of his thought in this area warrant a sustained reading. Michael Worboys is correct to say that Cheyne's work, along with that of his contemporaries, "deserves to be better known and understood" (*Spreading Germs*: 15). The objectives, arrangement, and content of this book are delineated below.

Historical Perspectives: 1889 and 1897

The object of this study is threefold: (1) to characterize the complex development of emergent bacteriology (a term coined in 1884) through a survey of primary texts; (2) against this intellectual background, to describe Cheyne's professional activities as laboratory investigator; as clinician and surgeon; as translator, editor, biographer, international liaison, and educator; and (3) to emphasize Cheyne's idea of medicine and microbiology (a term coined in 1888) as interdependent branches of science ("bacteriology," *OED*. I:155; "microbiology," *OED*. I:1788).

Keenly aware of the rapid development of laboratory science and of its implications for health care, Cheyne periodized his milieu into five stages, identifying each with a representative figure(s) whose contribution or Landmark was a turning point in the history of medical bacteriology. Constructed according to preconceived ideas, periodic history often displays a pattern of phasic development, a prevailing intellectual, cultural, or moral state, and prominent individuals who are believed to embody its ethos (R. G. Collingwood: 264;

R. Stover: 280; D. Gerhard: 476). All periodical historians, no matter what the doctrinal or intellectual orientation, organize historical time into discernible and meaningful patterns (C. DePaolo, "Periodization": 447).

As a surveyor and periodic historian, Cheyne efficiently employed chronology, phasic development, the representative figure, and the Landmark metaphor to convey his understanding of medical bacteriology as an emergent discipline. I have adopted his periodization scheme as the organizing principle of this study primarily to preserve his perspective and secondarily to provide the presentation with chronological unity, intellectual depth, and topical coherence. I have also made two adjustments: the first to integrate at Landmark I the complementary research of Pasteur and Lister, and the second to designate Cheyne's contribution, five chapters in all, as representative of Landmark VI.

Cheyne outlined the recent history of biomedical discovery briefly in his Preface to the 1889 *Suppuration and Septic Diseases* and in an 1897 paper on pathology. He suggested, in the 1889 Preface, that a review of scientific knowledge, from the recent past to the present, would make it possible to "realise what we have gained, what are the points which most urgently require elucidation, and what direction future investigations should take" (vi). The great scientific achievements of the medical microbiologists of his time were the results of the individual initiative and the collaborative work of those who founded the new discipline. The retrospective scheme of 1889 begins with Lister's bacteriology in the 1870s: "The first landmark of importance to our present views on this matter is the first papers by Sir Joseph Lister on Antiseptic Surgery" (vi). Cheyne's scope at Landmark I encompasses the period from 1867 to 1889, the years at Edinburgh, Glasgow, and London, when Lister produced 23 papers on antiseptic surgery, six on bacteriology, and practiced medicine and surgery. I have paired Lister with Pasteur at this Landmark to describe the interactivities of biochemistry and medicine in an age when the two disciplines were considered independent of one another.

Three achievements qualified Alexander Ogston's (1844–1929) research as the "second great landmark" (Preface: ix). Indebted to the findings of German researchers such as Theodor Billroth (1829–1894), Frederich D. von Recklinghausen (1833–1910), and Theodore Albrecht Klebs (1834–1913), Ogston elucidated the relation between micrococci and suppuration in abscess pathology, described two types of cocci (*Staphylococcus* and *Streptococcus*), and concluded that cocci caused abscesses and other infections (ix-x).

Cheyne affixed "the third landmark" to the work of A. J. F. Rosenbach (1842–1923), whose 1884 publication on the infectious diseases of wounds "still forms the chief basis of our present knowledge of the subject" (Preface: x). Cheyne recognized the significance of Rosenbach's contribution: not only had he identified varieties of "pyogenic organisms," but he had also studied their characteristics and the diseases associated with them (Preface: x).

A second text containing Cheyne's historical perspective is the 1897 paper "On the Progress and Results of Pathological Work," delivered at the Annual Meeting of the British Medical Association. In this paper, he delineates contemporary lines of research and updates the 1889 model with references to physiology, pathology, and immunology. As for bacteriology, a field integral to all branches of medicine, Cheyne observes that it has led to "most important practical results," has illuminated "the processes which go on in the body as a whole, and has stimulated research in other directions not immediately associated with it" ("On the Progress": 586). The development of antiseptic surgery and in-depth studies of bacterial pathogenesis, above all, have demonstrated the interrelatedness of medical research and praxis. Cheyne understood that locating these trends in a historical framework emphasizing the individuation of each discipline and their unity of purpose was needed, and he was dismayed that, in the public mind and in professional circles, the recent history of bacteriology was obscure. From 1872 to 1897, the record of the emergent discipline of medical bacteriology had amounted to an "absolute blank" (586). Resistance to the new science and to its cross-disciplinary implications waned as experimental evidence continued to be compiled and corroborated. According to Cheyne, fundamental misunderstandings about the nature of micro-organisms had to be corrected. Some scientists, for example, did not accept the idea of bacterial pathogenesis, subscribing instead to spontaneous generation theory or to some variant of it. Others continued to deny the existence of bacteria, while a less extreme position was that, instead of invading wounds, "organisms were always present in the healthy tissues." The most obstinate critics of Lister had refused to believe that micro-organisms had anything to do with disease at all, and that the reputed success of the antiseptic system did not depend on "the exclusion of micro-organisms from wounds" (586–587). These objections, Cheyne writes, had inspired him to convince practitioners that antiseptic surgery depended upon bacteriology. He saw as his initial objective, in the mid–1870s, "to ascertain whether or not, as a result of antiseptic treatment, organisms were absent from the discharges from the wounds" (587).

In the 1897 model, Cheyne reaffirmed Lister's research as, unquestionably, the first historical landmark. Through fractional cultivation, Lister had demonstrated that the micro-organism *Bacterium lactis* soured milk, and in 1877–1878 his investigation revealed that *B. lactis* was, morphologically and physiologically, a unique species ("On the Nature of Fermentation," *CP*. **I**: 349–350; "The Lactic Fermentation," *CP*. **I**: 352, 373, 376, 381). This discovery had much greater implications. For Lister not only had obtained the first pure bacterial colony and recorded its physiological reactions, he also inferentially compared *B. lactis'* fermentation of milk to the pathogenic properties of some bacteria, thereby suggesting that microbes could be implicated in disease (M. Santer: 59–65).

Associating Koch, once again, with Landmark II of the 1897 model, Cheyne called his work "the foundation of all modern bacteriological research." This praise was deserved: Koch's major contributions were to the etiology of wound infection, to the development of staining methods, and to the cultivation of bacteria on solid media (587). Instead of moving directly to Ogston at Landmark III, Cheyne modified the 1897 model in the light of the latest research. To accommodate the latest developments, he described Landmark III as a bifurcation. One branch moved in the direction of parasitology (587). In Cheyne's milieu, c. 1870 to 1894, great strides were indeed being made in the understanding of parasitic disorders, such as Elephantiasis, Schistosomiasis, Trichinosis, Liver and Lung Fluke Diseases (F. E. G. Cox). The other research branch of the 1897 model combined into a single period the discoveries of Landmarks III and IV of the 1889 outline. In 1889, Cheyne had positioned Ogston's research at Landmark III and Rosenbach's at Landmark IV. In the 1897 model, however, Ogston, Rosenbach, and their contemporaries were collectively assigned to Landmark III, since their research focused predominantly on morphology, physiology, and pathogenesis. This revision was not arbitrary. It represented a broadening of Cheyne's perspective: the emphasis of the 1897 model, represented by the conflation of Landmarks III and IV, was on the process of discovery and, particularly, on the collaborative work of a nascent scientific community. Cheyne made this adjustment, for example, to reflect the interactions of Ogston and the German bacteriologists. The medical and laboratory scientists at Landmark III (in the 1897 model) explored the natural history, taxonomy, physiology and disease-causing properties of microorganisms, and, in so doing, established unequivocally the interdependence of bacteriology and medicine (587).

Élie Metchnikoff and the birth of immunology, in the 1897 order of discovery, are identified as the next "landmark" (unnumbered by Cheyne) (587). Phagocytosis, in Cheyne's perspective, was one aspect of a profound area of enquiry on cellular reactions to parasites, on the differentiation of cells, on the chemical stimuli that attract phagocytes, and on the development of protective agencies (antibacterial substances, antitoxins, and immunity) (587).

The metaphoric Landmarks of 1889 and of 1897, numbered I to VI below, provide this study with chronological order and conceptual unity. The resulting, six-part chronology extends from Pasteur's 1857 "Report on the Lactic Acid Fermentation" to Cheyne's multi-faceted contribution to the development medical bacteriology into the 1920s: Landmark I (1889/1897): French chemist/bacteriologist Louis Pasteur (1882–1895) and British surgeon/physiologist Joseph Lister (1827–1912); Landmark II (1889/1897): German physician/bacteriologist Robert Koch (1843–1910); Landmark III (1889): British surgeon/bacteriologist Alexander Ogston (1844–1929); Landmark IV (1889): German surgeon/bacteriologist A. J. F. Rosenbach (1842–1943); Landmark V (1897): Russian zoologist Élie Metchnikoff (1845–1916); and Landmark VI: British physician and bacteriologist William Watson Cheyne (1852–1932).

Content Preview

The book's 11 chapters, delineated in greater detail below, are subdivided into two parts. Chapters 1 to 6 survey the development of nascent bacteriology, from 1860 to the 1890s. Chapter 1 (Landmark I) chronicles Pasteur's and Lister's scientific relationship as it is presented in published texts. Included in this chapter is background on the genesis of the germ theories, on the chief writings of Pasteur's predecessors (Redi, Spallanzani, Schwann, Cagniard-Latour, and Henle), and on two proponents of spontaneous generation, Bastian and Pouchet. The chapter traces Lister's transition from physiology to antisepsis (1852–1864), through his experimental replication, verification, and adaptation of Pasteur's findings (1864–1870), and it documents their 1871-to-1877 exchange of ideas. Lister and Pasteur pursued interdisciplinary work on microorganisms, and Pasteur, who cultured bacterial pathogens, conducted hospital-based experiments and developed vaccines (1878–1888). Pasteur and Lister who corresponded with each other, and who met on three occasions, discussed the nature of pathogenic micro-organisms. Lister's 1896 "On the Interdependence of Science and the Healing Art," an encomium to Pasteur, revealed the importance of collaboration (Pasteur's firm belief), of the cross-fertilization of allied fields, and of the advancement of laboratory technology.

Chapter 2 (Landmark I) briefly recounts Cheyne's association with Lister, the former's medical school days and the beginning of his bacteriological career in antiseptic surgery, under Lister's mentorship at the Edinburgh Royal Infirmary and at King's College Hospital, London. German visitors, among other medical professionals, toured Lister's wards, and two of them translated the seminal Listerian paper on war wounds, intended for use during the Franco-Prussian War (1870–1871). Just as Lister had visited Munich, Leipzig, and Halle in 1875–1876, Cheyne inspected German (along with Austrian) hospitals. Both gathered data on the use of antisepsis and evaluated conditions wherever antisepsis had appeared to have failed. Cheyne's visits to Vienna (winter 1875) and to Strasbourg (spring 1876), for example, revealed mixed results on antisepsis, attributed to the failure to change dressings regularly and to strengthen carbolic-acid concentrations sufficiently.

Cheyne's earliest bacteriological work leading up to the publication of his first paper in 1879 is contextualized in chapter 3 (Landmark II). For background, I discuss three areas of early bacteriology: Ferdinand Cohn's work on bacterial systematics and the problems he faced regarding instrumentation and nomenclatures; Robert Koch's 1878 "Investigations" and crucial experiments on bacterial pathogenesis using various micro-organisms and animal models; and the convergence of laboratory technologies and methods involving precision instruments, advanced microscopy, staining methods, and slide preparation (e.g., the contributions of Abbé, Zeiss, Weigert, and Ewart).

Cheyne's experimental findings on micrococcal virulence in abscesses and on the efficacy of carbolic acid as a germicide are set forth in the 1879 paper "On the Relation of Organisms to Surgical Dressings." He began his bacteriological work in January 1877 at the Edinburgh Royal Infirmary and continued with his investigation of bacteriology and antiseptic surgery, from 1878 to 1879, at King's College Hospital, London, where he had been appointed House Surgeon under Lister's direction. Even though Cheyne was thoroughly familiar with current scientific research, inadequate equipment, shortcomings in procedural methodology, and editing weakened the paper. Nevertheless, in this context, he was still able to make an original contribution in regard to the bactericidal potency of carbolic acid: when diluted below 1–20 (acid-to-water) strength, exposed micrococcal species survived. Following the inaugural paper was Cheyne's 1882 monograph *Antiseptic Surgery* and the 1884 paper "Micrococci in Relations to Wounds" which contain early reflections on antiseptic surgery, on the latest laboratory methodology, on bacterial systematics, and on the intellectual tradition of Pasteur, Lister, and Koch.

Alexander Ogston's discovery of *Staphylococcus* is the focus of chapter 4 (Landmark III). Here, I outline the relationship between Lister, Cheyne, and Ogston, beginning with Lister's enigmatic microbe *Granuligera* and the influence of Koch on all of them. With new technology, Ogston synthesized the doctrines of Lister and Koch to reveal the virulent nature of micrococcus species in acute abscesses. The Lister-Cheyne debate with Ogston on microbial virulence was primarily concerned with the causes of bacterial infection, manifested in acute and chronic abscesses. Whereas Lister and Cheyne ascribed abscesses to neurological inflammation, Ogston correctly associated infection with invasive pathogens; implicitly, this debate involved suppuration, inflammation, and assumptions about innate immunity. Cheyne and Lister eventually acknowledged the validity of Ogston's findings. Furthermore, in light of Ogston's success, Cheyne recognized and then emended the methodological errors of his 1879 paper, and Lister followed suit, abandoning a non-bacterial hypothesis on abscess formation; in turn, once Cheyne had access to more up-to-date equipment at King's College Hospital and had gained facility with Koch's method, he was able to correct Ogston on several original points pertaining to bacterial physiology.

Cheyne's periodical historiography located the bacteriology of A. J. F. Rosenbach, of Friedrich Fehleisen (1854–1924), and of contemporaries at Landmark IV, and I present their contributions in chapter 5. Rosenbach whose erysipelas paper Cheyne had translated from German to English described *Streptococcus pyogenes* and two species of *Staphylococcus* (*albus* and *aureus*), all of which are pathogenic. As Cheyne had with Koch's "Investigations" in 1880, in 1890 he published his translation of Carl Flügge's comprehensive *Micro-Organisms: with Special reference to the Etiology of the Infective Diseases* (1886), both of which were projects under the auspices of the New Sydenham Society. Cheyne assessed Flügge's taxonomy of the Genus *Micrococcus,* according to pathogenicity and ecology; 14 species were found to be pathogenic to human beings, seven to nonhuman vertebrates, and 27 were saprophytic. Flügge's work on *S. pyogenes aureus* supplemented Rosenbach's.

Chapter 5 also includes a section on Cheyne's multiple roles as translator of important German-language texts, as editor of the collected essays *Recent Researches on Micro-Organisms in Relation to Suppuration and Infectious Diseases* (an original project), and as instructor and liaison between bacteriologists and the public-health sector. He contributed the essay "Biological Laboratory" to the 1884 edition of *Public Health Laboratory Work,* created a replica of a laboratory for visitors to the London International Health Exhibition, delivered lectures to the general public, and his work heralded the appearance of the first bacteriological textbooks and manuals. Originally offshoots of chemically-orientated public health manuals, these texts were designed for Medical Officers. In 1885, Cheyne published his own *Manual of the Antiseptic Treatment of Wounds.*

Élie Metchnikoff's work which Cheyne considered representative of immunological history in the 1880s–1890s is associated with Landmark V (chapter 6). Cheyne held an intermediate position with respect to the argument over natural immunity in the so-called cellular-humoral debate. He thought that phagocytosis (the cellular uptake of microbes) was valid. But he also suspected that immunity in human beings was more dependent on the complex defensive properties of the tissues (the humoral theory), which vaccines stimulated. The researches of Bordet (the complement system), of Wright and Douglas (opsonins), and of Ehrlich (antibodies) confirmed Cheyne's dualistic view.

Chapter 7 (Landmark VI), "Tuberculosis and Anti-Vivisection," makes a case for Cheyne's importance as a tuberculosis researcher, clinician, and surgeon. Faced with two adversaries, a crippling endemic disease and a political movement obstructing animal experimentation, he tried to maintain legal and ethical standards while conducting important research. But despite his judicious and professional activities, he was, for a time, unjustly stripped of a license to use animal models. Nevertheless, through painstaking labor and as political activists criticized him publicly, in 1882–1883 he tested the experimental findings of Honoré Toussaint (1847–1890) and of Koch who had claimed to have isolated the cause of tuberculosis. Disproving through experiments a number of alleged causes, Cheyne determined that Toussaint's and Koch's claim that the tubercle bacillus caused tuberculosis was correct. Cheyne developed surgical procedures to treat the disease, established that it was a metastatic affliction, and, despite heated opposition, averred that animal research was absolutely essential to its management.

Cheyne's research on cholera is surveyed in chapter 8 (Landmark VI) through six "Reports" to *The British Medical Journal*, published *seriatim* in the spring of 1885. His work was an urgent response to the sixth Cholera Pandemic which had erupted in British India, and his findings supported those of Koch who had claimed to have isolated the bacillary cause. Cheyne studied the putative organism and carefully reviewed Koch's morphological, physiological, ecological, and epidemiological data. On the strength of Koch's work, the validity of which he confirmed, Cheyne refuted the charge of histologist Emanuel E. Klein (1844–1925) that Koch had failed to isolate the cause of cholera. Klein who, in 1884, had been delegated by the British Cholera Commission to investigate the outbreak in the Lower Bengal region, India, had subscribed to von Pettenkofer's soil-based theory of cholera pathogenesis. The limitations of this theory made it impossible for him to disprove Koch's experimental and histological evidence that *Vibrio cholerae* was responsible for the disease.

Chapters 9 and 10 (Landmark VI) review Cheyne's innovations in military medicine during the Second Boer War (1870–1902) and during World War I (1914–1915), respectively. The ninth chapter covers Cheyne's early efforts to devise a practical antiseptic system for battlefield use in South Africa. Lister had proposed a system, in 1870, that had been translated into the German language and communicated to the German medical corps, but the system was not widely employed during the Franco-Prussian War. I survey variants of Lister's system, calling for the use of carbolic acid and other chemical germicides on the battlefield and in field hospitals, and collateral systems developed by European battlefield surgeons, namely J. F. von Esmarch (1823–1908), Carl von Reyher (1846–1890), and Cheyne's British colleagues (MacCormac,

Crookshank, Moriarity, and Ogston). Cheyne who was a combat surgeon during the Second Anglo-Boer War experienced firsthand the poor logistics, the difficult environment, and the devastation of infection after traumatic injury. He recommended that field hospitals be moved closer to the Front and that effective, portable disinfectants and pre-treated dressings be routinely used prior to and during evacuation.

During World War I (chapter 10), Cheyne experimented with antiseptic formulae for battlefield use, but the Medical Service was reluctant or unable to apply them systematically, despite the devastating injuries and infections that struck infantrymen. Environmental conditions on the Western Front, notably the fertilized soil of Flanders, were conducive to wound invasion by myriads of bacterial species, the worst of which caused tetanus and gangrene. In the "Treatment of Wounds" lectures of 1914–1915, Cheyne assessed the disinfection of gunshot wounds, using a variant of Lister's 1870 plan. He formed a committee at the Royal Naval Hospital at Chatham to devise a protocol for war wounded, based on experiments with disinfectants (i.e., Tricresol, Lysol, corrosive sublimate). One aim was the invention of life-saving antiseptic kits to inhibit the growth of pathogens prior to evacuation from the war zone. Front-line surgeons, however, either ignored Cheyne's instructions or used the antiseptic compounds incorrectly. In that period, he engaged in an acrimonious debate with the physician Colonel Almroth Wright (1861–1947). Contrary to Cheyne's subscription to temporary, pre-evacuative disinfectants, Wright argued for the complicated use of hypertonic saline lavage to stimulate innate immunity and against the use of chemical antiseptics, except in an auxiliary role. On physiological grounds, critics debunked Wright's theory. Cheyne's efforts and the fieldwork of his assistant, Dr. Arthur Edmunds (1847–1952), were ignored. Hypochlorite sodium and other compounds which Alexis Carrell (1873–1944), Henry Drysdale Dakin (1880–1952), and others developed as germicides eclipsed Cheyne's campaign. The toxicity of Cheyne's tar-derivative antiseptics and of Dakin's solution has today been established.

In chapter 11 (Landmark VI), "Bacteriology and Medicine: 1886–1899," I review Cheyne's conviction that bacteriology and allied branches of medicine are interdependent, an idea unifying his writings. In 1884, for example, he emphasized the enduring importance of Koch's Postulates and how essential bacteriology was to public health, even though, at the time, the emphasis in that domain was on chemistry. Epidemiology which deals with the incidence, distribution, and management of disease in a population depended upon information uncovered in the lab, and knowing the identity of a pathogen allowed for timely countermeasures against outbreaks of typhoid, diphtheria, cholera, or tuberculosis. Bacteriology was integral to surgery in several respects: antisepsis provided the surgeon with greater technical latitude, along with more time to perform complex operations and greater ability to control post-operative sepsis. The association of pathogen with pathology endowed the surgeon with predictive knowledge about a disease and an opportunity to improve treatment modalities. For these and other reasons, Cheyne argued for the inclusion of bacteriology in British medical school curricula and training. Bacteriology and chemistry were also allied fields, a fact exhibited physiologically through analyses of bacterial metabolism and of toxicity. The corpus of Cheyne's work, along with symposia and the venues in which he participated, helped to institutionalize bacteriology in the clinic, in the hospitals, and in the public-health sector. Cheyne's role as humanistic scholar and humanitarian are highlighted throughout. He contributed to the nascent field as clinician, surgeon, instructor, *litterateur*, international liaison, naval commander, and political advocate.

As a participant in, and historiographer of, a new science, Cheyne had the unique opportunity to contribute to bacteriology and, at one remove, to survey its cooperative development as an emerging discipline. He understood that a unifying concept such as the interdependence of all branches of medicine could reveal how trenchant hypotheses were formed, tested, abandoned, modified, or accepted. In the mid-nineteenth century, according to Cheyne, investigators and clinicians perceived their respective endeavors as being compartmentalized, neither community recognizing the complementariness of the other's specialization. On the other hand, as a worker in both areas, Cheyne perceived the interconnectedness of bacteriology and medicine, and he articulated a balance between them. A trained clinician who had studied the subject "from both points of view," he hoped "to reconcile the apparently conflicting evidence" (Preface, *Suppuration and Septic Diseases*, Preface: v).

1

1857–1896

Pasteur and Lister—Landmark I

Pasteur and Lister cooperated with each other in several investigative areas.[1] In 1864 Pasteur's research on fermentation and putrefaction began to influence Lister's medical and surgical practice. From 1864 to 1896, Lister meticulously replicated, verified, and lectured on Pasteur's findings and endorsed his work before the British medical community; in turn, Pasteur promoted Lister's medical applications of antiseptic surgery in French circles and personally encouraged Lister at every opportunity. The investigations of Lister, Ogston, Koch, and others, were based in large part on the microbiological and etiological work of Pasteur; and their work, in its entirety, formed "the foundation of the science of bacteriology" (Caird: 872). As student, liaison, and researcher, Cheyne would become a member of this young scientific community.

1857–1863: Pasteur and Spontaneous Generation

Five of Pasteur's papers, published between 1857 and 1863, directly influenced Lister's work on micro-organisms. In 1854, while Professor of Chemistry and Dean of the School of Science at the University of Lille, Pasteur experimented on beet sugar in alcohol fermentation and witnessed the process taking place in the presence of living organisms, confirming what predecessors, such as Francisco Redi (1688), Lazaro Spallanzani (1799), Theodore Schwann (1837), Charles Cagniard-Latour (1838), and Jacob Henle (1839), had reported (Dubos, *Pasteur*: 40); Redi, for example, had proven that worms in decaying meat were not generated *de novo* but had emerged from fly eggs (Redi: 189). Using hermetically-sealed vessels containing seeds and legumes, Spallanzani

detected "animalcula" or bacteria, the origin of which he traced to the deposi-
tion of airborne organisms onto infusion surfaces, and the growth of which he
attributed to oxygen (13–15). Schwann determined that fermentation produced alcohol; that immersion in boiling water or exposure to hot air prevented
putrefaction, fermentation, and the growth of micro-organisms (18); and that,
if yeast were immersed in sugar solution, the fermentation process rapidly commenced (19). Charles Cagnaird-Latour, in his 1838 "Memoir on Alcoholic Fermentation," conjectured that the "simple vesicles" in fresh beer yeast, each less
than ten micrometers wide, seemed to reproduce by budding or lengthening
to form globular chains; these organisms were likely related to plants (20–21)
      On the basis of the above-cited experiments, Jacob Henle claimed that a
reproductive, autonomous, and parasitic micro-organism could cause disease
(76). As a way of explaining how pathogenic organisms grew and multiplied
in blood and tissue, he compared them to fermentative bacteria and fungi. He
constructed this parallel based on the findings of Cagnaird-Latour and Schwann
that, during fermentation, lower fungi feeding on nitrogenous and sugary nutrients develop, multiply, and convert substances into alcohol (Schwann: 19;
Henle: 77). Henle, as noted, anticipated one of Koch's 1884 Postulates: to
prove that parasitic microbes cause disease, one had, first, to isolate the contagious organism or substance (Henle: 78–79; Brock, *Comment*: 79). The germ
theory of fermentation was eventually connected to medicine, and Lister's 1878
paper "On the Lactic Fermentation and Its Bearings on Pathology" heralded
this transition (*CP*. I: 353–386).[2]
      The theory that microbial life developed from prior life-forms of the same
kind challenged the long-standing idea of *spontaneous generation*: that life sprang
from inanimate matter (Wainwright and Lederberg: 421–423). Farley, Lawrence,
Dixey, Strick, and Worboys are among the modern scholars who have studied
the history of the idea in depth and in context.[3] Basic definitions and a historical
sketch, provided here, will help to orientate the reader.
      Attributed, in 1655, to the British scientist and inventor Robert Hooke
(1635–1703), the concept of spontaneous generation holds that living organisms develop "without the agency of pre-existing living matter," and result
instead from changes "in some inorganic substance" ("spontaneous generation,"
OED. I: 2977). In the nineteenth century, several related terms entered the
life-science lexicon. One, having a threefold meaning, was heterogenesis: the
abnormal or irregular development of fetuses (1854); "the birth or origination
of a living being otherwise than from a parent of the same kind" (as per Thomas
Huxley, in 1870); and "the generation of animals or vegetables of low organization from *inorganic* matter" (*OED*. I: 1298; italics added). Henry Charlton
Bastian (1837–1915), a neurologist, subscribed to the latter definition of

heterogenesis (Worboys, *Spreading Germs*: 87–88; Strick, *Sparks of Life*: 90). In the 1870 Presidential Address to the British Association for the Advancement of Science, Huxley redefined these categories: biogenesis, for him, was the principle that living matter always arose by "the agency of pre-existing living matter," whereas abiogenesis was equivalent to Bastian's heterogenesis: that living matter could be "produced by not living matter" ("biogenesis," *OED.* **I**: 218; "abiogenesis," *OED.* **I**: 8; Huxley, "*Biogenesis and Abiogenesis*": 229–271).

Although a Darwinian, Bastian rejected Pasteur's biogenic theory of fermentation (*Beginnings*: 344–345, 374, 384, 403–404). Bastian's support for spontaneous generation had ironically developed "from his commitment to the new evolutionary science of Darwin, Spencer, Huxley and Tyndall" (Strick, "Darwinism": 51). Bastian's avid interest in the morphology of bacteria and fungi seemed incongruent with the idea that microscopic life could originate *de novo* from inanimate matter (*Beginnings*: 267, 271–75; "Epidemic": 407). But it is interesting to observe that, in the late 1830s, both Bastian and Darwin felt that abiogenesis and evolution were compatible doctrines. In *Notebook* entry B228 (c. July 1837–January 1838), though intrigued by the diversity of species in the Galápagos Archipelago, Darwin had retrospectively imagined the molecular genesis of primordial life from nonliving matter. He believed that the study of biogeography could account for the forces of adaptation, for instinct, and for the organization of living beings; and that knowledge of these principles could lead to an understanding of the "first germ" and of how life on Earth began. In one note, for example, Darwin reflected on the mystery of life's origin—on "what is added to the composition of the atom, to make it alive, and how the laws of generation were impressed on it" (B228, *CDN*: 227–228). Huxley, too, envisioned a Genesis germ. In "On the Physical Basis of Life" (1868), having in mind a "protoplasm, simple or nucleated, [to be] the formal basis of all life" (104), he imagined the existence of a primordial cell, but one devoid of hereditary material and of an organic progenitor. Furthermore, in Huxley's philosophical construct, a triadic unity of power or faculty, of form, and of "substantial composition," informed the natural world ("On the Physical Basis of Life": 96–97). Bastian's interest in micro-organisms and his affinity to Darwin and Huxley suggest that, in the 1870s, abiogenesis was not considered to be outside the mainstream of biological thought. Eventually, in the light of Pasteurian research, the idea of abiogenesis in its early nineteenth-century usage became untenable.

In April 1864, the biogenesis-abiogenesis debate reached a climactic point. On the one hand, Pasteur's laboratory work, from 1857 to the early 1860s, indicated that fermentation depended on microbial life, that these living organisms were external to, and independent of, the fermentative process they produced, and that they were not generated from inanimate matter. These ideas resonated through the early papers. Pasteur's 1857 "Report on the Lactic Acid Fermentation," which describes the discovery of the microbe responsible for fermentation in beer yeast, inaugurated his microbiological studies: both lactic and alcoholic fermentation involved "a nitrogenous material" having "the properties of an organized body of the mycodermal [i.e., fungal] type … probably related to the yeast of beer" (29; Brock, "Comment": 31). After being appointed Director of Scientific Studies at the *École Normale Supérieure*, he inquired further into the origin and role of micro-organisms. In the 1860 "Memoir on Alcoholic Fermentation," he presented further evidence that living organisms in yeast (fungi such as *Saccharomyces cerevisiae*) were responsible for the effervescent change ("Memoir": 31–38).

A closely related aspect of Pasteur's experimentation, seminal to Lister's earliest understanding of bacterial sepsis, was presented in the 1861 paper "Animal Infusoria Living in the Absence of Free Oxygen, and the Fermentations They Bring About" (39–41). Pasteur proposed in this paper that the ferment producing butyric acid was an anaërobe, a microbe living in the complete absence of free oxygen. He wondered how these micro-organisms related to fermentation ("Animal Infusoria": 40). Pasteur had intuited, by this time, that the fermentative organisms, rather than being the direct offspring of chemical processes, had a life independent of the sugar and related compounds that they metabolized. Experiments described in the 1861 monograph "On the Organized Bodies Which Exist in the Atmosphere; Examination of the Doctrine of Spontaneous Generation" explicitly tested the hypothesis that life in boiled infusions arose from spores, solid particles suspended in the air. Additionally, he proved that airborne microbes could be cultivated: thus, if inseminated into a liquid boiled to sterility, the very same organisms, upon exposure to the air, reappeared and multiplied ("On the Organized Bodies": 43–48). The fifth influential paper in the series, the 1863 "Investigation into the Role Attributed to Atmospheric Oxygen" established that fermentation, putrefaction, and slow combustion destroyed organic substances and that these processes were necessary for life. He learned that the slow combustion of organic material, in most cases, was connected to anaërobes, and his experiments with blood and urine were important contributions to contemporary debates over spontaneous generation and fermentation ("Investigation" [1863]; Bulloch, *History*: 67–145).

In 1864, a simple experiment and an unforeseen outcome discredited the doctrine of spontaneous generation considerably. The French naturalist Félix Archimède Pouchet (1800–1872), Director of the Natural History Museum in Rouen, had hoped to disprove Pasteur's 1863 claim that micro-organisms suspended in the air were responsible for fermentation and putrefaction. Pouchet believed, conversely, that life could spring directly, not from pre-existent life, but spontaneously from organic molecules of living matter.

Pasteur, in "An Address Delivered ... at the 'Sorbonne Scientific Soirée' of April 7, 1864," recounted the details of the experiment. Pouchet had hoped to show that a mold could arise in a hermetically sealed flask containing boiled water and sterilized hay. If microbes were to appear in an environment reputedly devoid of life, he would have disproven Pasteur's claim that micro-organisms could only arise, biologically, from pre-existent forms of the same kind.

When mold was detected in the flask, Pouchet felt vindicated. Pasteur, however, was able to demonstrate that the microbial growth had not arisen spontaneously. On the contrary, the colony had actually originated from external spores that had collected on the surface of a vat of mercury coolant. When Pouchet had inverted and then immersed the hot flask in the coolant, he unknowingly picked up spores that adhered to the seal, penetrated it, and, once inside the flask, germinated in the moist hay. Consequently, Pouchet's experiment had failed: life had sprung from pre-existent life ("Sorbonne"; B. Barnett).

During this period of experimentation, Pasteur established six points that would influence Lister's antiseptic system: (1) microbial metabolism, in most instances, causes fermentation and putrefaction (the Italian chemist Adamo Fabbroni had hypothesized as much in 1787; and Edouard Buchner (1860–1907), in 1897, would show that enzymes also caused yeast to ferment) ([Bulloch, *History*: 42–3, 62–3, respectively]); (2) fermentative and putrefactive microbes were abundant in the environment, vulnerable to heat and chemicals, and filterable; (3) each organism produced a specific fermentation; (4) some organisms derived oxygen from the air; others, called anaërobes, absorbed the gas metabolically; (5) organic or vegetable substances, collected under sterile conditions, neither fermented nor putrefied; and (6) spontaneous generation was invalid.

1852–1864: Lister's Transition from Physiology to Antisepsis

Joseph Lister earned a degree in medicine from the University of London, 1844 to 1852, and, after passing the examinations, was made a Fellow of the Royal College of Surgeons (Guthrie: 34–35). In September 1853, he joined the medical and teaching staff at the Edinburgh Royal Infirmary, became an Assistant Surgeon there in 1856, and was appointed Professor of Surgery at Glasgow in 1860 and, in 1861, Surgeon at its Royal Infirmary. In the period of

1853–1863, while Pasteur was producing the papers described above, Lister was publishing 11 papers on physiology, dealing with the mechanisms of arterial contraction, of blood coagulation and circulation, of anesthesia, and of inflammation. For Pasteur and Lister, this was a prolific period, and their theoretical interests would soon converge.

Soon after Lister's 1869 appointment to the Edinburgh Royal Infirmary, he conducted animal experiments on inflammation, on the causes and effects of blood coagulation, and on the production of pus (Godlee: 43–57). He had suspected, by 1867 while at Glasgow, that the physiological work he had undertaken from 1853 to 1863, along with that on antisepsis beginning in 1864, were related areas of inquiry, possibly involving ways of controlling the growth of pathogens. Before the Pasteur-Pouchet debate had been settled, the germ theory of disease, of which many variants existed between 1865 and 1900, was in its nascent stage (Worboys, *Spreading Germs*: 1–3). But even in this early period Lister had been interrelating vertebrate physiology to micro-organisms, and some of his intuitions would be borne out by future research; one was that blood clots, suppuration, and antiseptics all had anti-microbial properties. He was correct to assume as much: clots (which trap and dispose of germs) and pus (the leukocytes and other cells of which destroy bacteria) are both elements of innate immunity (Tortora and Derrickson: 354–7, 425–7). In 1881, Lister asserted, in "An Address on the Treatment of Wounds," that a blood-clot could inhibit the development of septic bacteria (*CP.* **II**: 280–1). In 1907, he interconnected the physiological research of 1858–1863, along with unpublished essays, notes, and drawings, directly to the antiseptic and bacteriological work that had begun in 1864. While at Walmer in Kent, he planned but never completed an essay synthesizing the early research on physiology and antisepsis under the title "On the Suppuration of Blood Clot." Lister's purpose, in 1907, might have been to clarify, not only the linkage between physiology, bacteriology, and surgery, but also the trajectory of his thinking just prior to having read Pasteur's 1857–1863 papers. In hindsight, Lister might have believed that the early physiological inquiries had prepared him, beginning in 1864, to adapt Pasteur's findings to medicine (Godlee, Appendix 3: 639–651; Cheyne, *Lister*: 9; Fisher: 121–122).

1864–1870: Lister Replicates, Verifies and Adapts Pasteur's Findings

In 1864, while Lister was at the Glasgow Royal Infirmary, a chemistry professor and colleague, Dr. Thomas Anderson, drew his attention to the latest work of Pasteur, specifically to the papers "On the Organized Bodies Which Exist in the Atmosphere" (1861) and to "Investigation into the Role Attributable to Atmospheric Gas" (1863) (Fisher: 121). The advice was serendipitous since Lister had begun to read Pasteur's papers at exactly the moment when struggling to control post-surgical infections. At the time of his appointment as head of the Glasgow Royal Infirmary in 1861, Lister had observed that *hospitalism*, a term James Y. Simpson (1811–1870) had coined, was a prevalent danger in surgical wards. In *Hospitalism: Its Effects on the Results of Surgical Operations*, Simpson wrote about hospital cross-infections, dreadful complications to surgery (Cameron: 50–52; Guthrie: 16–19). Lister, in an 1875 paper, alluded to the problem and to possible solutions, as evidenced by Dr. Saxtorph of Copenhagen's successful application of the antiseptic treatment in the centuries-old Frederick's Hospital, which he oversaw. The three major postoperative infections, gangrene, pyemia, and erysipelas, had nearly vanished from this institution once Lister's regimen had been employed (cited in "An Address on the Effect of the Antiseptic Treatment," *CP*. **II**: 247). Rickman John Godlee (1849–1925), Lister's nephew, noted that Saxtorph who frequently visited Edinburgh and Glasgow had become acquainted with the method as early as 1869 (as mentioned in a letter to Lister) and, from 1870 on, acquired positive results (351–352). After having departed Glasgow for Edinburgh, Lister published an article on the problem in *The Lancet* of 1870, "On the Effects of the Antiseptic System of Treatment upon the Salubrity of a Surgical Hospital" (*CP*. **II**: 123–136). The article outlined the unhealthy space in which Lister routinely operated, and he grouped three conditions under the *hospitalism* rubric: pyemia (purulent septicemia), erysipelas (inflammatory bacterial infection), and hospital gangrene (necrosis from interrupted blood supply or bacterial toxicity) (*CP*. **II**: 123–124).

Also in 1864, Lister began to investigate whether Pasteur's work on microorganisms could illuminate the mechanisms of wound infection and whether antiseptics could be used to control infection. He investigated chemical compounds to prevent or to combat *hospitalism*; and, while replicating Pasteur's bacteriological experiments, he hoped to verify the former's discoveries and to gain a clearer understanding of these micro-organisms. Lister had come to realize that the study of bacteria and the practice of surgery were potentially interdependent sciences, a principle to which Cheyne would later subscribe. In 1891, though still speaking of bacterial infection in terms of fermentation and putrefaction, Lister strongly suspected that the problem of post-operative infection was approaching a solution. Pasteur had proven that micro-organisms were responsible for putrefaction, that spontaneous generation was invalid, that microbes did not emerge *de novo* in the blood, and that the prevention or destruction of these organisms, notably in the case of simple fracture, was essential (*CP.* **II**: 340). The extraordinary revelation was that, since bacteria were thought not to arise from within the body, theoretically they could be barred, chemically or by other means, from entering open wounds; and, as a routine part of surgery and rehabilitation, infections could be managed and even prevented. A landmark in the history of bacteriology had been reached.

Lister composed a series of interdisciplinary papers exploring the relation of bacteria to surgery. In "On a New Method of Treating Compound Fracture, Abscess, Etc." (1867), he tried to determine how atmosphere related to the decomposition of organic substances (*CP.* **II**: 1–36). Pasteur's research and Lister's everyday practice strongly suggested the "essential cause" of putrescence to be micro-organisms (*CP.* **II**: 2). Lister then applied the idea of the fermentative germ directly to the case of a patient suffering from a punctured lung, caused by a fractured rib. An injury such as this one was extremely dangerous because of the risk of infection. Lister surmised that, if an external wound were to penetrate the chest, and if it exposed the pleura to bacteria-laden surroundings, germs invariably entered the body. For Lister, these circumstances raised questions to which he would give considerable thought for decades to come: how did harmful organisms enter and develop in the wound? What were the effects of their presence? And was it correct to consider inflammation, suppuration, and associated phenomena reactions to or as concomitants of infection?

In other 1867 papers, we find that Lister was acutely aware of how much he owed to Pasteur's research of 1857–1863. "On the Antiseptic Principle in the Practice of Surgery" (9 August 1867), which is one example, emphasized the theoretical importance of Pasteur's research to medicine (*CP.* **II**: 37–45). Once it had been proven that micro-organisms exist freely in the environment and that some are pathogenic, Lister realized that preventing germs from entering wounds and destroying those that had already done so would benefit postoperative patients immeasurably. Pasteur had shown that minute organisms suspended in the atmosphere were responsible for infection. This could be avoided either by excluding the air or by applying a dressing imbued with an agent capable of destroying external microbes (*CP.* **II**: 37).

The 1867 papers cited above represent the earliest stage of Lister's application of the fermentative germ theory to medicine and of the development of his antiseptic methods. Its surgical application progressed from possibility to certainty: hence, the "New *Method*" (or unprecedented technique), in the 9 August 1867 *Lancet* paper, had become a "*Principle*" or fundamental doctrine; and, ultimately, after sufficient clinical and experimental evidence had been compiled, it emerged as a bona fide "Antiseptic *System*"—that is, as an organized and established procedure.

Pasteur's systematic activities in the laboratory guided Lister as he described and afterwards replicated the former's work. In "Illustrations of the Antiseptic System of Treatment in Surgery," for example, Lister outlined one instance of the chemist's study of bacteria, designed to prove that the atmosphere contained the spores of minute vegetation and infusoria, believed to be densely concentrated in cities and forests where life abounded (*CP.* **II**: 46–85). Pasteur's experimentation led him to conclude that in the Alpine glacial regions that he had visited airborne pathogens, because of environmental factors, were less concentrated there than in rural or urban regions. The experiment Pasteur devised to replicate the Alpine findings was ingeniously simple. According to Lister, Pasteur suspected that if he passed a stream of air through a narrow but contorted glass tube, microscopic particles would be deposited in the twists and bends. The inflowing, aseptic air (devoid of spores and microorganisms) would then contact the organic material at the bottom of the flask. Since the organic infusions in the experiment did not putrefy, this suggested that aseptic air, devoid of mold spores and of other dust-borne organisms, was not responsible for the decomposition of organic material (*CP.* **II**: 47–48).

At this point in the essay, Lister recalled an analogy that Pasteur had drawn, in the March and April 1865 edition of the *Annales des Sciences Naturelle*, between fermentation and putrescence, on the one hand, and infection, on the other. The analogy had led Lister to suspect that the microbial processes involved in fermentation and putrefaction might be similar to those responsible for infections. Further, after viewing Vibrio (motile, **S**-shaped bacteria) moving on the surface of bodily fluids, Lister suspected that pathological germs might also be motile and, as a result, have the capacity to invade and destroy living tissues (*CP.* **II**: 47n.). In the light of Pasteur's papers, it became clear that, wherever there was an open wound, the well-being of the patient depended either on the eradication of germs or on the maintenance of an aseptic condition. Lister emphasized this conclusion in the above-cited 1867 *Lancet* paper "Illustrations"; but he also reminded practitioners that carbolic acid, if used in excess or in the wrong concentration, could be irritative and even toxic.

After having replicated an 1861 Pasteurian experiment, Lister presented his findings. The results of both experiments further undermined the argument for spontaneous generation. In "On the Causation of Putrefaction and Fermentation" (delivered at the University of Edinburgh, 8 November 1869), Lister explained how Pasteur had prepared a flask similar to that used in the experiment described above but with a significant difference (*CP.* **II**: 477–88). After the yeast had been introduced, the neck of the flask had been heated to the point of malleability, drawn into the shape of a narrow tube, and then bent into an **S**-curve (Dubos, *Pasteur*: 58). The fluid was boiled, but the end of the neck was left open to allow air to pass into the flask as the heating lamp was withdrawn. This done, the container was left untouched to determine what effects, if any, diurnal changes of temperature had on the decoction. Once again, the alleged contaminative property of atmospheric gas was being tested. The gases in the test flask expanded diurnally and condensed nocturnally. Even though atmospheric air had entered and exited the vessel, over an indefinite period of time the decoction did not putrefy. Although this experiment did not explicitly account for the disposition of airborne life, it supported the view that the gases of themselves did not generate microbes spontaneously.

Pasteur and others had the strong suspicion that airborne, microbial spores had settled in the angles of the contorted neck ("On the Causation," *CP.* **II**: 486). The Committee of the French Academy which adjudicated the proceedings tested this hypothesis in a simple way. Re-sealing the neck and once satisfied that the flask was sterile, the examiners inverted and shook it, in this way forcing liquid into the angles of the re-shaped tube, to flush any microbial spores down into the nutritive solution. Once this had been done, they then let the liquid settle. In a short period of time, micro-organisms that had indeed adhered to the angles and recesses of the bent tube germinated in the fluid. As a result of this test, it became even more difficult to dismiss the possibility that microscopic growth originated from germs carried by the diurnal influx of air onto the glass surfaces of the angular neck. This intriguing supposition encouraged Lister to think about its clinical importance, but first he needed to convince colleagues, one of whom, James Young Simpson, had assumed incorrectly, in 1867, that Lister had claimed precedence for the antiseptic use of carbolic acid (Godlee: 159–61, 199–205). Lister's forerunner in carbolic-acid research was the pharmaceutical chemist Jules Lemaire who, from 1860 to 1863, demonstrated the antimicrobial properties of carbolic acid and published his findings in the book *De L'Acide Phénique* (H. A. Kelly: 183–189; Godlee: 159; D. C. Schechter and H. Swan: 817–826). Despite the professional rivalry and confusion surrounding Lister's claims and intentions, he steadfastly subscribed to Pasteur's work. For this reason, he was obliged to duplicate the chemist's experiments, to survey the latter's work in his own publications, to allude to collateral investigators, and to consider what these findings could mean for his patients.

Lister had to show, beyond doubt, that bacteria were ubiquitous in the operating theaters and recovery rooms, as well as on clothing, instruments, dressings, sponges, and hands. He decided to replicate Pasteur's experimental methods and to present the results publicly so as to persuade detractors. Thus, on 8 November 1869, at the University of Edinburgh, he did just that. Instead of the yeast decoction that Pasteur had used, Lister, in 1867, used urine as a medium, introducing it into four flasks, each being one-third full. Three of four were heated to the point of malleability. The glass necks were then elongated; and the spout openings, reduced to transverse, diametrical slits. Before the tubes had cooled and hardened, Lister, emulating Pasteur, had bent the necks of three vessels at acute angles. The fourth vessel, its shorter neck elongated and diameter narrowed more than those of the other three, remained vertical. Lister boiled the flasks for five minutes until steam issued from the narrow necks; the boiling killed microscopic life. When the burner was removed, air passed into the vessels, displacing the condensed aqueous vapor.

After having recounted his version of the Pasteurian experiment, Lister then demonstrated to the University of Edinburgh audience that, after two years, more than one cubic inch of fresh air had entered into each vessel, contacting the liquid ("Causation," *CP.* **II**: 486–8). In the bent flasks, the air had evaporated the moisture in the tubes. Despite the entrance of air, for two years the urine had remained uncontaminated (*CP.* **II**: 487). Fluid in the vertical-tube flask, however, had changed considerably, exposure to airborne microorganisms having transformed the urine's transparent straw color to a muddy brown, clogging the vessel with sediment and dead fungi; in addition, a strong odor of ammonia was present (*CP.* **II**: 487). Lister, therefore, had experimentally demonstrated that "the gases of the air, however abundantly supplied" were of themselves unable to produce the *torula* and other minute organisms that appeared in atmospherically-exposed decoctions of yeast. The essential source of such development, he concludes, had to be "suspended particles of germs" (*CP.* **II**: 485–6).

At the six-month mark in the experiment and after examining the contents of the bent vessels microscopically, Lister recalled having not found organisms. He attributed the aseptic state to his having filtered out the airborne spores: because the necks had been bent, airborne micro-organisms had collected in the glass bends and curvatures. The Listerian version of the experiment successfully confirmed Pasteur's threefold discovery: atmospheric gases were not responsible for putrefaction; organisms did not arise spontaneously from infusions; and microbes germinating in the flasks originated from airborne germs and spores. Once exposed to the air, the uncontaminated urine samples from the six-month-old flasks became odorous and full of microscopic life ("Causation," *CP.* **II**: 487).

1871–1877: Communication and Exchanges

Visiting London, in 1871, Pasteur met the British physicist John Tyndall (1820–1893), who had recently toured Lister's wards. Tyndall had embraced

germ and contagion theory and, in June of 1871, asserted that parasitic organisms responsible for epidemics arose, not from spontaneous generation, but
"from ancestral stock whose habitat is the human body" ("Dust and Disease":
661–662). A surveyor of contemporary germ theories, he informed Pasteur of
Lister's successes; and this was the first time Lister's work "had come to Pasteur's attention" (Fisher: 198; Worboys, *Spreading Germs*: 125–126). As we
have seen, from 1864 on Lister had become well-acquainted with Pasteur's
thinking, was methodically testing and communicating the theory, and was
exploring ways of applying this knowledge to medicine. Part of this overall
strategy, at this time, was to persuade his colleagues as to its merits.

Besides carefully reproducing Pasteur's experiments, whenever the opportunity arose for Lister to present the antiseptic system of surgery at a professional forum, he acknowledged his indebtedness to the former's groundbreaking
research of 1857–1863. "The Address in Surgery" (1 August 1871), read in Plymouth, was typical of this tendency. In it, Lister reviewed the theoretical basis
of antiseptic treatment, his replication of Pasteur's experiment on putrefaction
(that is, of having boiled organically-infused liquid in bent flasks), and stressed
the fact that airborne germs cause decomposition. Once again, he related that
three of the four containers of originally sterilized urine, presented at the University of Edinburgh in November 1869, had not putrefied, even though they
had been continually exposed to atmospheric air—that is, to air from which
dust and suspended organisms had been filtered. He also reminded the Edinburgh audience that, at the six-month mark in the experiment, he had decanted
small amounts of urine from the purportedly-sterile flasks, exposing the samples directly to unfiltered, atmospheric air. As had been predicted, in a short
time the samples teemed with micro-organisms and exhibited fungal growth;
thus, the urine proved to be a medium chemically favorable to the germination
of life (*CP.* II: 175–78).

Once the bacterial hypothesis of putrefaction had been proven to Lister's
satisfaction, he turned to surgery, armed with carbolic acid, crude aerosolizing
sprayers, and ingenious methods of dressing wounds and of setting broken
bones, using saturated putty, rubber, tin, and other materials. An inventive
stage in Lister's career had begun, as he devised practical ways of applying antiseptic compounds before, during, and after an operation, the intention always
being to promote natural healing. In these interventions, Lister consulted colleagues, such as John Tyndall, who had been experimenting successfully with
cotton-wool filters as a way of trapping germ-laden dust. Lister presented what
he had learned from Tyndall in the 10 August 1871 "Address on Surgery" (*CP.*
II: 175–78). In 1871, Lister had succeeded his father-in-law, James Syme (1795–
1890), as Surgeon-in-Ordinary to the Queen in Scotland; prior to having held

that appointment, Syme had been Chair of Clinical Surgery at Edinburgh University (1933–1863). The most famous surgery that Lister performed at this time was to drain an infected abscess in Queen Victoria's shoulder, using lint strips soaked in an oily solution of carbolic acid, supplemented by antiseptic spray (which, unfortunately, irritated the Queen's eyes) (Godlee: 305, 473). The operation was a success: the lint worked beautifully to drain the pus while the carbolic acid, however awkwardly applied, prevented infection (305).

Lister's respect for Pasteur's work is further evident in published correspondence of this period. While in London, performing dissections and conducting lectures, Lister forwarded a 13 February 1874 letter to Pasteur that included a copy of the October 1873 paper "A Further Contribution to the Natural History of Bacteria and the Germ Theory of Fermentative Changes" (Letter, Godlee: 274–275; *Joseph, Baron Lister: Centenary Volume*: 163; "Natural History," *CP.* **II**: 309–34). Furthermore, Lister invited Pasteur to review the British journals on bacteria and antisepsis and, should he visit Edinburgh, to tour the hospital. Responding in a cordial 27 February 1874 letter, Pasteur explained that he had learned of Lister's antiseptic system through conversations with Tyndall and with the chief surgeon at the Val de Grâce Hospital in France (Letter, Godlee: 275–277). Although Pasteur had been unacquainted with the details of the research, he was nevertheless genuinely impressed by Lister's thoroughness and invited him to send along a detailed account of the system.

The exchange of views described above produced important publicity. Just as Lister had found in Pasteur's biochemical research a promising new approach to the management of hospital infections, Pasteur felt obliged to transmit Lister's surgical applications to the French medical community. In the interest of humanity, both men felt that the system needed to be widely disseminated. To promote greater cooperation, Pasteur requested from Lister copies of his principal scientific publications on the germ theory. To support Lister's laboratory research, Pasteur, in turn, offered professional suggestions on safeguarding cultures from contamination (Letter [27 February 1874], Godlee: 275–277).

In a letter of 29 June 1876, Pasteur gratefully accepted two of Lister's papers and informed him that, although the antiseptic system had not yet been universally accepted in French medical circles, interest in it was increasing. Unknown to Pasteur at that time, a prominent French physician had already been instrumental in bringing Lister's system to the French medical community. When visiting Glasgow in 1868, Dr. Just Lucas-Championnière (1843–1914) had been impressed by Lister's method, and, in 1876, he wrote the first complete account of it (Letter, Godlee: 307–310, 353; Vallery-Radot: 239).

Following Lister's lead, Pasteur investigated the possibility of bacterial infection in surgery. In June 1876, he had witnessed an operation at Necker Hospital, from which no post-surgical sepsis arose (Letter, Godlee: 307–10). The satisfactory outcome, Pasteur averred, was attributable to the surgeons having followed Lister's instructions on the use of carbolic acid. Citing a remark by his colleague, Dr. Charles-Emmanuel Sédillot (1804–1883) (coiner of the word *microbe*), Pasteur wrote that recent successes and failures in surgery could be logically explained in terms of the germ theory. Upon this theory, a new approach to surgery had been made possible, and Lister understood its significance ("microbe," *OED*. **I**: 1788; "The Germ Theory": 116–117).

The effectiveness of boric acid in surgical-wound treatment using Lister's method encouraged Pasteur to consider its antiseptic properties in conjunction with those of carbolic acid (Letter, Godlee: 307–310). Pasteur corroborated Lister's notion that boric acid was, indeed, a promising germicidal agent, for it, too, had been used in the Necker Hospital operation, and he suggested that, to acquire more data, Lister should experiment with the agent further. Thus, both scientists began to work with boric acid, and the interchange of ideas and findings continued. Pasteur even included in the dispatch a copy of his latest book on the fermentation of beer, in which was reprinted Lister's 10 February 1874 correspondence to him.

From 1875 to 1878, Lister replicated Pasteur's bacterial experiments in order to learn more about microbes and, through this knowledge, to improve wound management. The 1875 article "A Contribution to the Germ Theory of Putrefaction and other Fermentative Changes" exemplified Lister's strategy (*CP.* **I**: 275–308, esp. 276). He began, as expected, with enthusiasm for the "philosophical investigations" of Pasteur that, since 1867, had completely converted him to the germ theory and that, to date, had guided his practice unwaveringly (*CP.* **I**: 276). In addition, this principle meant that he had to expand his surgical methodology to include chemical and bacteriological content, as evidenced by the 1878 paper "On the Lactic Fermentation and Its Bearings on Pathology" (*CP.* **I**: 353–85).

1878–1888: Interdisciplinary Research

In the 29 April 1878 paper "The Germ Theory and Its Applications to Medicine and Surgery," Pasteur declared before the French Academy of Sciences that chemistry, the emerging discipline of bacteriology, and medicine were interdependent sciences that "gain[ed] by mutual support." I shall call this his declaration of interdependence. Just as Lister recurrently acknowledged his debt to Pasteur, Pasteur acknowledged what he owed to his predecessors, one of whom, in particular, was the French pathologist and parasitologist Casimir-Joseph Davaine (1812–1882) ("The Germ Theory": 110).

To confirm that microscopic organisms caused disease, Pasteur described in this paper how he cultivated pathogens. Through 12 successive cultivations, each ten cubic centimeters in volume, an original drop of the anthrax bacillus in culture had become progressively more attenuated. Pasteur followed this procedure in sterile liquid a number of times, at each stage in the experiment seeding cultures with minute drops. A significant finding was that the final culture in the series was still virulent. Augmenting the work of Davaine, of Robert Koch, and of others, Pasteur and his colleagues, through this method, acquired further evidence that anthrax was a bacterial disease ("The Germ Theory": 111). Pasteur reviewed the progress of the bacteriological research that he and his co-workers had undertaken in 1877, as they searched for definitive proof of the bacterial causation of certain diseases. Research in 1876, he recalled, had shed light on the etiology of anthrax, but the cause of septicemia remained obscure. He and his colleagues suspected, however, that septicemia could be traced to a micro-organism. To prove this hypothesis, he realized that it was necessary to cultivate the putative microbe out of the body ("Germ Theory and Its Applications": 111).

This was a fecund time for bacteriological research, generally, and for the Pasteur-Lister interchange of ideas. Along with Pasteur's laboratory work on anthrax, in 1877–1878, Lister delivered an Introductory Address at King's College, London, on 1 October 1877, entitled "On the Nature of Fermentation" (published in the *Quarterly Journal of Microscopical Science*, April 1878 [*CP*. **I**: 335–52]). Koch was also linking bacteria to medicine, as his concurrent research and publications plainly showed. His paper "The Etiology of Anthrax, founded on the Course of Development of the *Bacillus Anthracis*" had appeared in 1876; and "Investigations of the Etiology of Wound Infections" in 1878 (*Essays*: 1–17, 20–56, respectively).

Joseph and Agnes Lister visited Paris in June 1878, for the Prince of Wales had appointed him President of the British Commission to the Jury on Medicine, Hygiene and Public Relief (Fisher: 248). Lister had a wonderful opportunity, at this time, to meet with Pasteur at the home of Dr. Gueneau de Mussy, close friend of Pasteur. After the dinner at de Mussy's home on 20 June, Pasteur and Lister met again for lunch on 21 June (248). Unfortunately, no record of what transpired is publicly available. One could reasonably infer that the discussions included pathogenesis and surgical antisepsis.

Pasteur congratulated Lister, in a letter of 2 January 1880, on his presentation of surgical methods at the Amsterdam Medical Congress, apprised him of his present work on contagious diseases and bacteria, and hoped that his investigations would uncover "indisputable proof" that microscopic organisms played a part in infectious disease (Letter, Godlee: 436–437). He was convinced that research of this kind, ranging as it did across disciplines, required a cooperative effort.

In the 3 May 1880 paper "On the Extension of the Germ Theory to the Etiology of Certain Common Diseases," read before the French Academy of Sciences, Pasteur wrote in the spirit of professional cooperation (118–130). "On the Extension" surveyed efforts to connect the germ theory to infectious disease. Its content arose from a fortuitous occurrence. In May 1879, one of Pasteur's laboratory co-workers had been diagnosed with furuncles, a bacterial infection of hair follicles. With the cooperation of the volunteer, research was conducted from 2 June to 21 July 1879 ("On the Extension": 118–122). From suppurating lesions, Pasteur took samples that, hours after being cultured, produced bacterial growth; the same organism was identified in each case. Repeated cultivations in the presence of atmospheric oxygen led him to conclude that furuncles contain "an aerobic microscopic parasite" (121). Thus, the disease had been traced to its probable cause.

Pasteur thought that knowledge about the furuncle germ could illuminate the etiologies of osteomyelitis and of other infectious diseases. He had also been accumulating observations on puerperal fever, a disorder that Dr. Ignaz Philip Semmelweis (1818–1865) had worked so diligently to prevent by requiring obstetric surgeons to wash their hands in chlorinated lime ("Lecture": 80–82). Pasteur's initial entry on the subject was based on a visit to a maternity wing under the supervision of Dr. Hervieux ("On the Extension": 123). Pasteur concluded that "puerperal fever" comprised a group of distinct pathologies having in common a single pathogen (128). Infectious organisms spread from traumatized areas to remote sites. Undoubtedly, he deduced from this experience that inhibiting the production of these organisms could improve the chances of recovery. Confident that Lister's antiseptic method was the key to treatment, Pasteur suggested that either carbolic or boric acid could be satisfactorily applied to the patient (128–129).

These clinical experiences of Pasteur dramatized that understanding the mechanisms of disease required a concerted enquiry. Three disciplines were interdependent in this effort: bacteriology (established in Germany in 1884), chemistry, and medicine. Acutely aware that a cooperative effort was involved in this work, Pasteur invited opinions from every circle. As a chemist, Pasteur realized that, to study the role of germs in post-surgical infections, he needed the guidance of medical professionals. He reiterated that research in this area must be both open-minded and collaborative ("On the Extension": 129–130).

1880–1896: Fowl-Cholera, the Jubilee and "Interdependence"

Pasteur and Lister met for the second time in London at the August 1881 International Medical Congress. At this symposium, and with Pasteur in the audience, Lister delivered the 5 August 1881 "Address on the Relations of Minute Organisms to Inflammation" (*CP.* **I**: 399–410). Lister began the "Address" with a reference to an earlier paper of 12 August 1880, read during the meeting of the British Medical Association, at Cambridge. In the earlier paper, he described how he and Pasteur had tried to find a suitable culture medium for the fowl-cholera organism. Because Pasteur valued Lister's opinion, he had sent to him specimens of chicken-broth cultures. Once they were received, Lister located an unidentified bacterium developing in chicken broth and, as expected, conveyed his observations to colleagues at the Cambridge meeting (*CP.* **I**: 390–391).

The interchange between Pasteur and Lister on the fowl-cholera issue is found in four consecutive papers. The first was Pasteur's 26 October 1880 paper "The Attenuation of the Causal Agent of Fowl Cholera" (126–131). Lister's "Address" followed. This summary was read on 12 August 1880 to the Pathological Section of the British Medical Association (B.M.A.) at Cambridge and was titled "On the Relations of Micro-Organisms to Disease" (published April 1881) (*CP.* **I**: 387–398). In June 1881, Pasteur's efforts to develop a vaccine by attenuating the pathogen were publicized in "On a Vaccine for Fowl Cholera and Anthrax" ("Vaccine": 131–132). And an English translation of his "Address on the Germ Theory," read at the International Medical Conference in London, 12 August 1881, was published *verbatim* in *The British Medical Journal*, on 13 August 1881, as "An Address on Vaccination in Relation to Chicken Cholera and Splenic Fever" (283–284).

Pasteur's experiments on fowl-cholera impressed Lister so much that the latter described the entire experiment in the 12 August paper "On the Relations of Micro-Organisms to Disease" (*CP.* **I**: 387–398). Rightly attributing the discovery of the pathogen to Toussaint, professor at the Lyons' Veterinary School, Lister concentrated on its etiology. Lymphatic swelling, pericarditis and duodenal ulcerations were typical symptoms of this highly-infectious blood disease of chickens. If the blood were mixed with chicken feed, contagion occurred in two-thirds of the animals. Lister was interested in knowing how the bacterial pathogen (misidentified both as "cholera" and as "virus") entered the circulatory system. According to Toussaint, an abrasion of the oral epithelium was the probable route of infection because the throat and lymph nodes were so affected, and because similar symptoms were produced if chickens were inoculated orally with infected blood or serum. Lister attributed to Toussaint the observation that resistant chickens were not immune to the bacteria when inoculated, suggesting that the previous immunity to the tainted feed was accidental (*CP*. **I**: 390).

Lister had firsthand knowledge of how Pasteur cultivated the microorganism in chicken broth. As mentioned above, Pasteur had shipped to Lister fowl-cholera cultures in tubes. At the 12 August 1880 British Medical Association meeting, and as his Cambridge University audience looked on, Lister took a drop of Pasteur's specimens for microscopic examination. He described a small ovular bacterium that multiplied through transverse constriction, frequently appeared in pairs, and occasionally formed chains (*CP*. **I**: 390–391). It resembled the *Bacterium lactis* in diameter, both microbes ranging from 1/50,000th to 1/25,000th inch in length. Pasteur had also been kind enough to send Lister a representation of the organism, a Woodcut of a camera lucida sketch, drawn to scale. To Lister's knowledge, this was the first time anyone in Britain had ever seen the organism. After a number of successive cultivations, Pasteur claimed that the original virulence remained undiminished, inoculation of the culture or of the blood proving lethal (*CP*. **I**: 390). These findings, Lister believed, increased the probability that the isolated organism was responsible for the disease.

Lister underscored several of Pasteur's observations. One was that the pathogen did not putrefy chicken broth. Another was that the organism reached a threshold of development and then ceased to multiply. Pasteur reasonably conjectured that the microbes had subsisted on the broth, the volume of which diminished over time; furthermore, if another sample of fowl-cholera fluid were introduced into the very same dish, the new specimen did not flourish, which seemed to imply that the nutritional content of the medium had been depleted. This led Pasteur, logically although incorrectly, to suspect that the broth had been rendered unfit for the further development of the organism. Lister, on the other hand, surmised that the answer lay not in the medium, but in the organism's constitution: Pasteur had unknowingly "enfeeble[d] the organism," altering it in some way so that, when the culture was injected into a healthy fowl, only a mild form of the disease occurred (*CP*. **I**: 391–392). If this were true, and perhaps Lister had the physician Edward Jenner (1749–1823) in mind, then, through this procedure, the bird had been rendered secure, or (to use a later term) had become immune to, the ordinary form of the disease. The chicken, in effect, had been vaccinated, having received an attenuated, immunogenic form of the bacillus. Pasteur had taken the trouble of sending an immunized hen along to Lister. The chicken, none the worse for the overseas trip, was doubtless a source of humor (*CP*. **I**: 392).

Because Pasteur's method and findings were so significant, Lister outlined both carefully. Pasteur had begun by cultivating a pure, unmixed microbial strain; the culture was then exposed to air but shielded from dust to avoid contamination (*CP*. **I**: 396–398). If the period of cultivation were two months or less, the organism remained unchanged. But, by the third or fourth month, virulence began to decline, and by the eighth month, the organism, though it could cause a mild attack if inoculated into a chicken, was no longer lethal; if the period was sufficiently prolonged, the bacillus was rendered harmless. Although neither Pasteur nor Lister was immediately aware of the fact that the inoculated chickens had developed antibodies to the disease itself, their concerted efforts contributed significantly to the nascent science of immunology.

At that time, the reason for the organism's attenuation was unknown. Lister recounted how Pasteur had tested hypotheses on the subject. Speculating that oxygen exposure had degraded the germ, he conducted a physiological experiment. In closed tubes with a minimal oxygen volume, the organism lived until free oxygen was used up, then it sank to the bottom, but when retrieved, it was reputed to have retained its vitality and virulence if inoculated into healthy chickens (*CP*. **I**: 397). A prominent researcher, whom Lister mentioned, subscribed to the theory that the medium attenuated the germ. William Smith Greenfield (1846–1919), Professor of Pathology at the University of London's Brown Animal Institution, was known for his pioneering work on anthrax and immunity. His theory was that the colony grew either until nutrients were exhausted or until the broth had undergone chemical changes, enfeebling the organism's progeny. As for the sealed tubes with limited oxygen, Greenfield (according to Lister) suspected that with the exhaustion of oxygen the colony ceased reproduction, becoming inert; and each, "like a vigorous seed" (or spore), remained inherently fecund. Whether the cause of the organism's diminution lay in the oxygen or in the medium remained a topic of interest and debate.

Eventually, Pasteur, Toussaint, and Greenfield learned that the source of resistance was immunogenic, involving the micro-organism and the host's reaction to it (*CP*. **I**: 398). Lister was certain that Pasteur's study of chicken cholera, begun in 1878, was leading to extraordinary innovations in the science of vaccination. More definitive explanations were on the horizon. In 1890, Emil von Behring (1854–1917) developed a vaccine against tetanus and diphtheria and introduced the concepts of passive immunization and of antitoxins. The science of serology was inaugurated by the paper "The Mechanism of Immunity in Animals to Diphtheria and Tetanus" (1890), co-authored by von Behring and the Japanese bacteriologist Shibasaburo Kitasato (1852–1931), known for his later work on influenza and plague (Brock, *Milestones*: 138–140).

Lister met Pasteur for the third and final time at the latter's Jubilee cele-
bration, which took place on 27 December 1892. As a representative of the
Royal Society of London and of the Royal Society of Edinburgh, Lister delivered an eloquent "Address" in French (Godlee: 518–522). Despite the occasional figurative cliché or superlative phrase, the text reads as a précis of
Pasteur's genuine achievements. Lister writes that Pasteur's research had transformed surgical wound-treatment from an uncertain and sometimes disastrous
undertaking to a scientific art. Not only had the chemist's pioneering efforts
helped to dispel terrors associated with post-surgical infection, but they had
enlarged the capabilities of surgeons who no longer had to hurry through a
procedure in order to preclude bacterial infection.

Uncovering the nature of microbes, Lister states, had made it possible to
diagnose the plagues of the human race and also to design prophylactic and
curative treatments. He alluded to Pasteur's method of attenuation to create
protective vaccines, no more dramatically illustrated than in the campaign
against rabies. Because his originality was so striking, many doctors, at the
time, were suspicious of Pasteur. How could a man who was neither a medical
doctor nor a biologist instruct the greatest medical minds about a disease they
have for centuries tried but failed to cure? As far as Pasteur's scientific method
was concerned, Lister applauded his colleague's clarity of thought. Events justified Lister's confidence, as the entire world recognized Pasteur's victory over
rabies. The vaccine, painstakingly developed by 1885, had dispelled the anguish
and uncertainty that every person exposed to a rabid animal experiences.

One year after Pasteur's death, Lister presented "On the Interdependence
of Science and the Healing Art," the 1896 Presidential Address to the British
Association for the Advancement of Science (*CP*. **II**: 489–514). In this important document, he observed that Pasteur's fowl-cholera experiments revealed
much about immunity. Chickens that had received the attenuated microbe had
developed immunity to the disease it caused (*CP*. **II**: 504). Lister was certain
that Pasteur's knowledge of the history of medicine and the latter's conviction
that chemistry, biology, and surgery were mutually dependent, had brought
him to a decisive phase in the experiment: the inoculation of vaccinated chickens with a virulent strain of the organism. Lister rightly emphasized the Jenner-
Pasteur connection: Pasteur had recognized that the immunity to fowl-cholera
produced in chickens inoculated with the attenuated bacillus was, in principle,
comparable to the protection vaccination afforded against the smallpox virus.
Convinced of this, Pasteur proceeded to weaken a variety of microbes for prophylactic vaccination against anthrax, swine erysipelas, and rabies (*CP*. **II**:
504).

Lister included in the 1896 "Interdependence" Address an encomium to

Pasteur who had died the year before, and this public expression of gratitude and indebtedness helped to define the latter's position in the history of biochemistry and medicine. Spanning more than three decades, Lister's Pasteurian commentary, in many ways a professional tribute, included critical evaluations, replicated experiments, important communications, conferrals with a multidisciplinary scientific community, and adaptations of theory to practice. Pasteur's statement, in 1878, that the natural sciences and medicine "gain by mutual support" was indisputable ("The Germ Theory and Its Application": 110).

2

1871–1878

Cheyne and Lister—Landmark I

The previous chapter described the biomedical tradition within which William Watson Cheyne would work. In 1871, the year that Lister and Pasteur had met for the first time in London, Cheyne commenced his medical studies, in May, at the University of Edinburgh (*Lister*: 23; "Lister, the Investigator": 923; Coutts, *Microbes*: 132–155). In a biography and several papers, 1925–1927, Cheyne recounted his years in medical school and his long association with Lister.[1]

During the 18-month period after of matriculation, Cheyne studied botany, natural history, and chemistry. In October 1872, while attending classes in surgery, physiology, and anatomy, he became acquainted with Lister's research when the latter was at work on the antiseptic system, on bacteriology, and on related surgical and physiological topics. During the winter session at Edinburgh, Lister became Cheyne's advisor and would remain his close associate for 40 years.

Their association dated back to a rainy afternoon in October 1872 when nineteen-year-old Cheyne, a second-year student, sought shelter during a storm. Since no Students' Club at Edinburgh was available at that time, he thought about spending the midday, one-hour break in the library, but the building was small and crowded. Entering the hospital for shelter, he casually thought about auditing a lecture and thus using his free-time productively. So, at noon, he crossed the road to the Infirmary amid students hurriedly crowding in the lobby. Joining them out of curiosity, he followed them to Lister's lecture room. The large operating theater, the gallery, the back of the room, and the gangways, were crammed with more than 200 fourth-year surgery students. The number and enthusiasm of the students surprised Cheyne ("On the Opposition"; *Lister*: 23–25).

The clarity of Lister's delivery surprised Cheyne even more and stimulated his interest in surgery. He continued to audit the lectures, taking notes and reflecting on what Lister said (this notebook is currently archived among Lister's records at the Royal College of Surgeons of England). In an ampler notebook, he copied the lecture notes and looked up unfamiliar technical terms in his surgery textbook. He followed this routine assiduously after every lecture during the winter session of 1872–1873 ("Lister, the Investigator": 924). Cheyne was particularly struck by Lister's unique pedagogical method and by how much it differed from that to which he had grown accustomed. Unlike typical lecturers at Edinburgh, Lister avoided esoteric terms and abstruse theory; instead, the presentations were organized, clearly presented, and the content understandable. Cheyne's notes reflect Lister's acuity. At the end of the day, Cheyne reviewed the notes and contemplated "the wonderful future that was opening up for surgery." An "apparently trivial occurrence"—a visit to Lister's lecture on a rainy afternoon in October 1872—had "shaped the whole course" of his life ("On the Opposition": *Centenary Oration. II.*).

Obligatory class examinations at Edinburgh, in the 1870s, were held twice during winter session. When the sessions were over, all those who received 75 percent and higher on the medical examinations were announced in order of merit. The examinations were highly competitive. For the two or three medical students whose scores qualified at the higher levels, bronze medals were awarded. Cheyne was tempted to take the upper-level exam, even though it was risky. At first, being a second-year student, Cheyne thought it unwise to compete in the clinical-surgery class since he would have been up against better-prepared fourth- and fifth-year medical students. When the time approached for the examinations, however, he decided to give it a try, with the aim of acquiring a sense of the test questions and the format, so as to be better prepared for his own tests in the following year.

Recollecting that the notes taken while auditing Lister's lectures had aided him during the examinations, Cheyne was surprised to find that he had earned an exceptional score of 93 percent, qualifying him for the silver award. Lister's humorous custom was to award an outstanding scorer with a case of silver catheters. Impressed by Cheyne's score, Lister interviewed him, bestowed the award, and offered him an appointment as a dresser during the summer of 1873. The summer internship was followed, during the winter of 1874, by a clerkship. Cheyne's progress was thereafter quite remarkable: he was appointed House Surgeon at the Edinburgh Royal Infirmary in 1876–1877, to work "with great interest and unanimity" with Lister and with the latter's nephew, Rickman J. Godlee; and then, under Lister's direction, to a surgical post, in 1877, at King's College Hospital.

Cheyne reflected on how uncharacteristic for him the 1872 impromptu visit to Lister's lecture had been since he was a shy person (Cheyne, *Lister*: 31, 34; "On the Opposition": *Centenary Oration*. II). Cheyne found Lister to be a humble, generous, and open-minded person, one who was sympathetic to patients, respected by students and co-workers, and in so many ways ideally suited to teaching science. Lister's analytical mind, acute observations, inferences, and generalizations impressed Cheyne. Instead of "curious theory, almost impossible to understand, and very difficult to memorize," Lister maintained students' interest "by reason of the new and intelligent statement of the subject" ("On the Opposition": Oration II). John Rudd Leeson wrote that "Lister's mind was a researching one" (40). In his biography of Lister, Cheyne portrayed his mentor as one who was unafraid to test a hypothesis rigorously, despite contrary opinions and practices, and as a tenacious problem-solver who devoted his entire life to surgical antisepsis (*Lister*: 14). Not only did Lister have "a very logical brain," he was never peremptorily blinded by his own ideas and was prepared "to modify his views and plans at once if proper evidence demanded it" (*Lister*: 15). An example of this flexibility in the light of Alexander Ogston's research would be his revised thinking on inflammation and on the role of bacteria in abscess formation.

German medical institutions were an important testing ground for Listerism in the 1870s. Although German physicians were interested in Listerism and visited Lister's wards in the 1870s, their application of his method in hospitals was, at times, not always in accord with the latest instructions. Where aseptic results were sporadic, Lister's and Cheyne's on-site inspections showed that the practitioners, and not the system, were often to blame (Appendix II). Cheyne's fact-finding visit in the winter of 1875 to Vienna and in the spring of 1876 to Strasbourg, of which more will be said in chapter 5, exemplified his role as Lister's inspector and medical liaison. During these visits, Cheyne who was fluent in the German language fielded questions from colleagues who were eager to learn about antisepsis and septic micro-organisms. Before his continental mission, he had studied the writings of eminent German surgical pathologists, such as Billroth, von Recklinghausen, and Klebs. On this tour, Cheyne confirmed Lister's mixed appraisal of how Listerism was being practiced in German hospitals during the latter's summer 1875 visit (Godlee: 311–312, 366–371; Cheyne, *Lister*: 74).

In Vienna, Cheyne attended lectures and in Strasbourg, in the spring of 1876, worked for three months in the laboratory of the anatomist von Recklinghausen. This was a year rich in theoretical and practical experience for the young medical bacteriologist (*Lister*: 74). In Austria and Germany, despite lapses in wound management in hospitals, students and professors alike continued to express enthusiasm about Lister's antiseptic system. From 1867 to 1875, the method had been articulated in 12 papers, with more to come. German scientists and students, however, lacked access to Lister's papers, and many could not overcome the language barrier; but, in 1870, the latter problem was partially solved. At the onset of the Franco-Prussian War, 1870–1871, for example, Lister's paper on antiseptics in warfare was recognized for its humanitarian value. The German medical researchers, A. W. Schultze and L. Lesser, while visiting Lister, translated it and other salient texts into the German language (Schlich: 3). The English-to-German translations adumbrated Cheyne's reciprocal efforts, in the 1880s, to make the German and French research accessible to English speakers.

Cheyne apprised Lister on his continental experiences and asked if he might continue with the work, using specimens from clinical cases at the Edinburgh Royal Infirmary. Unlike the enthusiastic Germans and Austrians whose results were mixed and uncertain he could study cases that Lister had personally treated. There would be no question about the consistency with which the system was applied since Lister personally oversaw the wards. Cheyne declared in the biography that "the actual bacteriological facts" were manifest; and the clinical evidence could be used to vindicate antiseptic surgery, and to make the system more efficient (*Lister*: 75). At Edinburgh, Lister and Cheyne perceived that bacteriological research and medical practice could function interdependently. Lister gave Cheyne freedom to perform lab work, there and later at King's College, in conjunction with surgery. Cheyne's enthusiasm and energy, no doubt, added an important dimension to Lister's work (75–76). Inadequate laboratory technology in the Edinburgh Infirmary, however, was a deficit no degree of enthusiasm could overcome. Even though Cheyne worked for Lister and in a prestigious academic setting, his laboratory was cramped and under-equipped, lacking up-to-date incubators, staining techniques, growth medium, and microscopes (*Lister*: 76).[2] While the Germans had the lead in hardware development and the advantage in funding and facilities, they needed Lister's instruction and guidance if inconsistencies and errors in wound care were to be rectified (77). So there was an obvious need for cooperation on many levels.

In the spring of 1877, Lister who had been appointed Professor of Clinical Surgery at King's College Hospital, London, invited Cheyne, along with several colleagues, to join him. Cheyne was given the prestigious appointment of House Surgeon. At King's College, Lister would continue to apply bacteriology to surgery, with the dual incentive of demonstrating the efficacy of the antiseptic system and, hopefully, of changing the minds of British colleagues, many of whom had "strenuously opposed his views" (Cheyne, "First Oration": *The Lister Centenary*, 1927).

Cheyne contributed significantly to the bacteriological dimension of medical practice in these early years, and Lister acknowledged his hard work. In a 10 February 1880 correspondence, for example, Lister evaluated his performance from 1877 to 1880, observing his "extraordinary excellence" and diligence as a medical student at Edinburgh, in pathology at Vienna and Strasbourg, and as House Surgeon, demonstrator, registrar, and published researcher ("Letters from Joseph Lister"; *The Royal College of Surgeons of Edinburgh*). As further evidence of his great achievements was his reception, in 1877, of the *Syme Surgical Fellowship* ("Medical News": 187). Rickman J. Godlee, in 1917, seconded his uncle's favorable opinion of Cheyne, who was his friend and past co-worker: "An ardent admirer of Lister as a man, and an out-and-out believer in his principles and methods, he assisted him in the care of patients, carried out for him many complicated experiments, and conducted very important bacteriological enquiries of his own, bearing on the infection of wounds. His writings have done much to familiarize the world with Lister's teaching, [along with] his example and precept to bring it home to many generations of students" (607).

3

1878–1882
Robert Koch—Landmark II

This chapter, designated Landmark II, covers the early years of Cheyne's career as medical bacteriologist, leading up to the publication of his first paper in 1879. Background sections treat Ferdinand Cohn's bacterial taxonomies of 1872 and 1875, Robert Koch's 1878 paper "Investigations of the Etiology of Traumatic Infective Diseases," and related discoveries. The research developments described here and in chapter 1 shaped Cheyne's early thinking on systematics, pathophysiology, and antisepsis.

Cohn's Taxonomy

The classification of bacteria with respect to their natural relationships is called *systematics* (Campbell, Reece, et al.: 536, 565–570). In the late eighteenth century, naturalists, such as Otto Friedrich Müller (1730–1784), Christian Gottfried Ehrenberg (1795–1876), and Felix Dujardin (1801–1860), attempted to assign bacterial forms to Genera based exclusively on shape; but the results were inconsistent (N. A. Logan: 1). Ehrenberg (1838) who expanded Müller's nomenclature constructed the formal taxonomy reproduced below. Four of the five generic members of Ehrenberg's *Vibrionia* Family below are rod-shaped; and a fifth is a flattened spiral:

    bb. Family *Vibrionia*
        c. Cells not flattened.
            d. Cells not flexuous
                e. Forming straight rods ... ***Bacterium***
                ee. Forming spiral rods ... ***Spirillum***
            dd. Cells flexuous

   e. Straight rods ... *Vibrio*
  ee. Spiral rods ... *Spirochaeta*
 cc. Cells in form of a flattened spiral ... *Spirodiscus*

[Cohn, *Bacteria*: 121; Buchanan: 18].

Under each Genus, Ehrenberg included species; hence, **Bacterium** had three; **Spirillum** had four; **Vibrio** had nine; **Spirochaeta** had one; and **Spirodiscus** had two (Buchanan: 18). Ehrenberg has been credited for taking an important first step in microbial systematics (Cohn, "Studies": 212).

At the time that Cheyne was visiting Vienna in 1875 and Strasbourg in 1876, the German botanist Ferdinand Cohn (1882–1898) had been attempting to improve the Müller-Ehrenberg system, particularly by solving the problem of arbitrary and inexact nomenclatures (Cohn, "Studies on Bacteria" [1875]: 210; Bulloch, *History*: 192–195; G. Drews). Consistency in bacterial taxonomy, he understood, could only be acquired through the use of high-quality instrumentation. In 1872, even though Cohn had a suitable Hartnack microscope, he complained that visual acuity was still limited (*Bacteria* [1872]: 15; "Studies" [1875]: 210). Thus, tiny bacteria were sometimes beyond "the limits of resolution"; and the form and interior structures of a cell, along with the stages of microbial growth, were not always determinable (210–211).

In the early 1870s, while advanced microscopy was being developed, Cohn realized that biological criteria other than form had to be found if a microorganism was to be properly classified. Whereas morphology involved the bacterium's visible structure, physiology took into account biological functions, such as metabolism and reproduction (Cohn, "Studies" [1875]: 213). Cohn implied, in the 1875 paper, that physiology, more than morphology, was a surer determinant of Genus and species: for example, bacteria found exclusively in a specific environment or that were constant in their fermentative properties had more in common with each other, systematically, than did bacteria only externally similar to one another (213). In fact, bacteria that looked alike could actually be physiologically dissimilar from each other (213–214).

Relying on form, on biological properties, and on other distinguishing features, Cohn concluded that bacteria were devoid of chlorophyll, had specific shapes (spherical, oblong, cylindrical, twisted, or curved), reproduced by transverse fission, and existed either individually or in groups ("Studies" [1875]: 214). With the Müller-Ehrenberg model as a prototype, he formulated a four-part classification system: Tribe I. **Sphaerobacteria** (sphere bacteria) had one Genus (Micrococcus); Tribe II. **Microbacteria** (Rod bacteria) had one (Bacterium); Tribe III. **Desmobacteria** (Filament bacteria) had two (Bacillus and Vibrio); and Tribe. IV. **Spirobacteria** (Cork-screw bacteria) had two (Spirillum and Spirochaete). In the 1870s, a bacteriologist interested in the medical aspects of the field would find in Cohn's work four methodological conditions upon which further progress would depend; advances were contingent on precedent, on improved optical instrumentation, on consistent experimental method, and on knowledge of bacterial physiology.

Koch's 1878 "Investigations" and Micrococcus

In the paper "Investigations of the Etiology of Traumatic Infective Diseases" (1878), translated by Cheyne from German to English, Robert Koch applied Cohn's ideas to the study of pathogenic bacteria. A sense of bacterial taxonomy's incremental progress can be gathered from the history of micrococcus research in the 1870s. A neologism combining the Greek words, μικρὸ / mikros (small) and κόκκος / kokkus (berry), micrococcus ("small berry") was a general nomenclature based on form. The botanist Ernst Hallier (1831–1904), of Jena, was the first to use the term, micrococcus; but, since he was referring to a mold rather than to a bacterium, he was not credited with discovering the Genus (Breed and Hucker: 113). In 1872, Cohn who was on the right track had used the word in reference to spherical, chain-forming, achromatic, and immotile micro-organisms (Bacteria: 113). Ehrenberg had been the first to assign the name to a specific micro-organism, Micrococcus prodigiosus, but its shape did not correspond to Cohn's description (113–114). Cohn's assistant, Joseph Schroeter (1835–1894), isolated Micrococcus luteus, the second bacterium of this kind to be discovered; and it has been recognized as being the type species (113–114). In 1885, Wilhelm Zopf (1846–1909), botanist and bacteriologist, subdivided Cohn's cocci into five Genera. In the Zopf model, 14 species occupied the Genus **Micrococcus** (114).

As Koch explored the relation between micrococcus and disease, he initially attempted to identify the species that Cohn had classified under the general epithet **Sphaerobacteria**; all spheroidal microbes, in Cohn's system, had been assigned to the Genus **Micrococcus**. A second aim was to find out which of these spheroids could cause disease. Inducing wound infections in animals by injecting them with putrid material and human blood, Koch was then able to isolate from the test animals a number of spheroidal micro-organisms having distinct features. Equipped with a state-of-the-art Zeiss of Jena microscope, an oil-immersion system, Ernst Abbé's optical condensers, and Carl Weigert's staining techniques, he conducted six artificial wound experiments to learn if specific pathogens in this group existed, and if they could be linked to particular diseases (Koch, "Investigations": 23–32).

The results of Experiments II–V, described in the 1878 paper, were especially relevant to the Genus **Micrococcus**. A nonspecific micrococcus identified in mice that had been injected with putrid blood in Experiment II multiplied rapidly and formed chains. From the inoculation site, these micrococcal chains spread outwardly, pressed together, and then congealed into "thick masses" (41). The ensuing "gangrene" (as Koch called it) was attributed to the bacterial excretion of soluble substances, diffusing throughout and destroying murine tissue (42–43). Further along in Experiment II, he inoculated contaminated blood from an index mouse into a series of healthy ones, successively inducing in each a gangrenous condition in which chain-like micrococci were detected (43–44). Koch, at this point in the paper, stressed the importance in an experiment of this kind of adequate ocular instrumentation, especially since widely-dispersed micrococcal chains were not visible at low magnification (42).

Laboratory results demonstrated that the micrococci had discrete variants, and these pathogenic micro-organisms affected healthy tissues in different ways. Putrid blood, in Experiment III, for example, caused abscesses in rabbits (Koch, "Investigations": 44). On the walls of the abscesses, Koch found thin layers of micrococci in zoöglœal or gelatinous masses ("zoöglœa," *Webster's*, 1040; *OED*. **II**: 3471). Microbes approximately 0.15 micrometer in diameter had grown in enclosed dead tissue, forming spore-like encapsulations. Koch conjectured that, though inactive, they remained pathogenic (45–47).

In Experiment IV, Koch subcutaneously injected one square centimeter of macerated murine skin into the back of a rabbit (47). Microscopic examination after two days revealed the widespread presence of micrococci, individual and paired cells, throughout the body. The individual cells, averaging 0.25 micrometer in width, were in size midway between the chain-like micrococci of "gangrene" in Experiment II and the zoöglœa-forming micrococci of rabbit abscesses (50). Morphological differences permitted Koch to tell these cells apart. There were more prominent distinctions, as well. The rabbit in which pyemia was induced in Experiment IV, for example, reacted differently to red blood cells than did the gangrenous or the abscess-forming microbes of Experiments II and III, respectively; when the pyemic organism contacted red blood cells, they adhered to one another and a clot formed (51).

Injections of putrid meat into rabbits caused septicemia in Experiment V. Large oval micrococci turned up in the first rabbit, as well as in another inoculated with the former's blood (53, 54). The second rabbit that had been infected with micrococci only experienced venous hemorrhage and edema (54). The microbes in Experiment V, when congregated, appeared as slightly curled, "trough-like pellicles" (55). Some colonies were loosely populated while others were densely-obstructive masses (55); the largest of the Experiment V cells, in diameter, measured from 0.8 to 1.0 micrometer (55).

Since the term *micrococcus* in the scientific vernacular of the time included all small, ovular bacterium, Koch was making extraordinary differential progress as he characterized subtypes in terms of size, of shape, and of pathology. In his analysis, he began conventionally with morphology. Between the various round cells, there was a size range (0.25 to 1.0 micrometer). Comparing them to each other, he noticed that the ovals in Experiment V were larger than the tiny spheroids in the pyemic rabbit of Experiment IV. Koch then turned more exclusively to physiology (their biological properties), specifically to metabolism and virulence. Two examples stand out: unlike the tiny hemolytic micrococci, the large ovals neither enclosed nor coagulated red blood cells ("Investigations": 55–56); however, both the small (Experiment IV) and the large micrococci (Experiment V), when inoculated into a series of animals, displayed a constant level of virulence (56).

Precisely in the manner that Cohn had suggested in 1875, Koch analyzed the results from multiple perspectives. Fortunately, for Koch, improved optics made the study of morphology less problematic. Once the organisms could be seen and distinguished from anatomical structures and cellular debris, the investigator could begin to compile physiological data about the life-cycle of the bacteria (its growth stages and colonies), its effect on the host, and the host's reaction to the pathogen (e.g., virulence did not increase through serial passage in both pyemic [Experiment IV] and septicemic [Experiment V] micrococci) ("Investigations": 71).

Koch learned that knowledge of morphology and of physiology, though important, was not entirely reliable. Micro-organisms might be similar in form, size, and conditions of growth, but they might differ from each other in terms of the diseases with which they were associated ("Investigations": 64). Pathophysiology, a more revealing determinant of species, considered functional changes associated with or caused by a disease. Koch observed, for example, that a bacillus introduced into a lab animal's ear entered the bloodstream while a chain-like micrococcus only destroyed local tissue; and, in contact with blood, septicemic and pyemic micrococci in the rabbit affected the tissues differently (65–66). Koch considered the fundamental question: what accounted for these differences in effect?

Understanding the pathophysiology of an organism through serial passage from one animal to another had important medical implications about disease recognition and treatment (Koch, "Investigations": 65). Further proof of a suspected pathogen-pathology connection came to light through dissection and histology; for example, when a disease was induced in a healthy animal through the inoculation of a putrid substance, later tests showed that *only* the organism initially found to be specific for that pathology was present within the affected tissues (italics added; 65). To distinguish one bacterium from another, a scientist could therefore advantageously combine three modalities: morphology, physiology, and pathophysiology (65–66). With these criteria in hand, Koch hypothesized that a particular species of bacteria could be responsible for a specific kind of wound infection (67). His experiments had furnished important new information about Cohn's nebulous 1875 taxon: **Sphaerobacteria/** Genus 1. Micrococcus (Cohn, "Studies": 214).

Koch's investigations contributed to the development of laboratory methodology. Isolating pure strains of a micro-organism was indispensable to experimentation. If a species could be accurately identified and cultured, contaminants could be extracted from the series without having to restart the process ("Investigations": 68). The ability to make adjustments in the course of an experiment, as Koch learned, was another astounding achievement. Staining techniques increased classificatory precision. In 1878, techniques were based on the carmine preparations that H. R. Goeppert and Cohn originally had used in 1849. New dyes, such as Weigert's Bismarck brown and aniline violet, were being developed in Koch's time. Ehrlich's methylene blue (1882), for example, would be instrumental in Koch's 1884 discovery of the tubercle bacillus ("Investigations": 68; Bulloch, *History*: 215–216, 320).

In order to ensure the purity of a culture if adequate staining was lacking, and when working with minute and difficult to identify micro-organisms, one had to experiment *in vivo*, that is, within the living body of an animal (Koch, "Investigations": 69). It was, nonetheless, a demanding task since micro-organisms that grow in a living body were limited in number; and, both distributively and pathophysiologically, each microbe was unique. Koch learned as much from the septicemic and gangrenous processes that he had induced in mice. The contaminated blood injected into lab mice contained two microbes: one causing septicemia; and the other, localized gangrene. These two pathogens were the only ones isolated while all others died. Under the microscope, Koch identified one as a small rod-shaped bacillus; and the other, as a chain-forming micrococcus. Each could be transferred from one mouse to another, the form and physiology of each germ remaining constant and predictable through the process.

Koch devised ways of segregating one microbial species from another. He found that a blood sample containing only bacilli, upon transmission, reproduced itself in another animal, and through transference became pure. If the bacilli and the chain micrococci were inoculated into a field-mouse, something unusual happened: the bacilli disappeared while the nonspecific micrococcus remained, ready to be purely cultured. In this way he determined that the field mouse could resist the bacilli but not the micrococci. This also raised the possibility that not every host was susceptible to the same parasitic organism: the field-mouse, for example, naturally withstood the bacilli, whereas the domestic mouse used in earlier experiments did not survive the challenge.

The bacteriologist, Koch revealed, could control the disposition of these organisms. He learned to cultivate varieties of bacteria together, to separate them from each other, and then to recombine them ("Investigations": 69). He confidently stated that, through *in vivo* animal experiments using the proper method and technology, only one specific form of bacterium was present in each distinct case of artificially-induced infective disease (70).

Koch finally identified three bacterial morphologies in his experiments: granulated chains, rods, and oscillating threads. If more than one of these forms were present, the investigator had before him either more than one infective disease, or his observations were wrong ("Investigations": 70). When multiple organisms were seen in a morbid process, one had then to determine if one or more of the agents were responsible for disease, and if different kinds of morbidity were involved. A pre-requisite to this complex activity was having an absolutely pure cultivation (70). Conclusive results depended on it.

Koch raised the caveat that none of his experiments supplied evidence of value either for or against "the *parasitic* nature of infective diseases" (italics added; "Investigations": 70–71). As I mentioned in the Introduction when discussing Cheyne's 1897 periodization, in the mid-nineteenth century, doctors who were investigating "bacteria-host interactions" had begun gradually to associate parasites with specific diseases (Worboys, "History of Bacteriology": 3). As W. D. Foster has pointed out, advancements in parasitology, by the middle of the century, had been widening the scope of clinical pathology; for example, in 1860, F. A. Zenker reported finding the microparasite of trichinosis in the body of a woman who had died of a typhoid-like condition, the disorder having struck survivors who had eaten the same pork. Eventually, a clinical description was established, other outbreaks in Germany were recorded, and diagnoses were proven through muscle biopsy (Foster, *Medical Bacteriology*: 14; "The Early History of Clinical Pathology": 178).

Convergent Technologies: Abbé, Zeiss, Weigert and Ewart

Along with Lister's pure cultivation of *Bacterium lactis* and Joseph Schroeter's similar achievement with pigment-forming bacteria, a number of other breakthroughs impelled late-nineteenth century bacteriology (Lister, "An Address on the Treatment of Wounds" [August 1881], *CP*. **II**: 278; N. A. Logan: 2). One of the earliest and most momentous was in microscopy. Koch, among others, benefited from the optical systems developed by Ernst Abbé (1840–1905) and from the latest microscopes, such as those manufactured by the Carl Zeiss Company. These instruments were enhanced by the use of advanced oil-immersion lenses. Its technology traceable to the optician and astronomer, Giovanni Battista Amici (1786–1863) in 1840, the oil-immersion lenses antedating Abbé has a complex history (J. R. Leeson: 93; "G. B. Amici"; J. Solliday). A mathematician and optician, Abbé had become associated with Carl Zeiss (1816–1888), manufacturer of optical instruments in Jena. In 1873 Abbé's paper explained the principles of microscopic resolution; and, in 1876, he traveled to London to present his theory of image-formation and advanced instrumentation (gratings, apertures, and the microscope) to the Royal Micro-scopical Society ("Theorie des Mikroskops"; B. R. Masters, "Ernst Abbé"). Of inestimable benefit to microscopy was the introduction of staining methods by the German pathologist/histologist Carl Weigert (1845–1904), and this was a second notable achievement in this area (Bulloch, *History*: 214, 403). The first to use aniline dyes, such as methyl violet, fuchsine, or aniline brown, to identify bacteria in animal tissues, he communicated his method to Koch who further developed staining methods; the latter's preparations, especially of anthrax and of tuberculosis bacilli, were reproduced photographically (214–215).

These technological advances permitted Koch to acquire clear pictures of bacteria that had been otherwise indistinct or invisible (Bulloch, *History*: 147). Most bacteria absorbing these dyes became sharply outlined; however, without the Abbé condenser smaller bacteria, previously visible only as dark points, could not be differentiated from anatomical structures and from cellular debris. Abbé's ingenious device solved the problem. It consisted of a series of diaphragms with widening apertures. According to B. R. Masters, this optical apparatus caused the dark outlines of cells, cell nuclei, the sharp lines of elastic fibers, and blood-vessel walls to become pale and poorly defined. As shadows of bodies in the visual plane disappeared so, too, would granules hitherto mistaken for bacteria. As obscure structures faded into the background, small bacteria, previously seen as indistinguishable black spots, appeared in color and with distinct nuclei; in fact, the entire field of vision brightened as the stained bacteria became highly resolved. As Koch reported, when the last diaphragm limiting the lens aperture was removed, it then was possible, even where micro-organisms were in close proximity, to distinguish one from another, and bacteria from both anatomical structures and stained, nonbacterial cells (B. R. Masters, "Ernst Abbé"; Koch, "Investigations": 23–32; Bulloch, *History*: 148).

The new technologies permitted Koch to improve the micrococcal taxonomy inherited from Cohn. Koch subdivided into four species Cohn's broad taxonomic heading, *Tribe I. **Sphaerobacteria**/Genus 1. Micrococcus*, defining each in terms of morphology, physiology, and pathophysiology:

- *Species 1*/Experiment II (gangrene in mice: chain-forming and thick masses, causing gangrene).
- *Species 2*/Experiment III (abscess in rabbits: small micrococci, colonizing as zoöglœa and forming spores).
- *Species 3*/Experiment IV (pyemia in rabbits: small micrococci, intermediate in size between species *1* and *2*; 0.25 micrometers, and hemolytic).
- *Species 4*/Experiment V (septicemia in rabbits: large ovals, 0.8 to 1.0 micrometer).

A third bacteriological system relevant to Cheyne's and to Alexander Ogston's understandings of micrococcus discussed in chapter 4 was that of the Scottish zoologist James Cossar Ewart (1851–1933). Ewart's purpose, in an 1878 Royal Society paper (published in the same year as Koch's "Investigations"), was to determine if this indistinct microbe was a specific organism or merely a stage in the life-history of a common bacterium (476). Koch emphasized that individual and aggregated forms could characterize a single species. Since cocci were small and indistinct, investigators such as Koch had difficulty tracing the stages of cellular development; and, to a degree, this affected their understanding of host pathology. After much labor, Ewart was able to clone micrococcus into a pure colony. Taking a small pus specimen from a newly-opened abscess, he deposited it in a nutritive drop of bovine aqueous humor (476). On the fourth day of the experiment, he saw moving particles: round, oval, or dumbbell shapes, congregating in pairs and in tetrads (476). Using a high-quality Number XII Hartnack immersion microscope (unequipped with an Abbé condenser), he determined that the different forms were all *phases of the same microbe* (italics added). Thus, ovoids became dumbbell-shaped and then underwent binary fission (a process whereby bacterial DNA duplicated, the parent cell bifurcated, and two genetically-identical daughter cells emerged [Miller and Levine: 476]). Ewart noted that the ovoid daughter cells also became dumbbell-shaped; and they, too, divided horizontally into new cells. Because the medium of bovine aqueous humor was so nutritious, the cells increased in number, rapidly forming a large milky colony. Observing the colony regularly for three weeks, Ewart was certain that the spheres never turned into rods.

In the fourth week of the experiment, Ewart transferred low concentrations of the micrococcus from the original specimen into fresh preparations of aqueous-humor and of sterile turnip-infused flasks. His object was to study growth and reproductive cycles further. Once again, he was certain that none of the cells in the second series were rod-like. After replicating the process, Ewart inferred that micrococcus was a distinct micro-organism belonging to a single taxonomic group, its members having common features; and he outlined the respective phases through which bacillus, bacterium (neither rod-shaped nor ovoid), and micrococcus developed. The spore-like micrococcus, it became clear to him, developed from ovoid to dumbbell shape and then divided into two. The daughter cells propagated asexually through binary fission, and this reproductive process went on indefinitely, as long as the artificial environment supported growth.

In three months, Ewart was able to enumerate four characteristics of the micrococci: unlike the Genus **Bacterium**, micrococci did not become rod-like in shape; they had both static and motile stages; some formed chains; and others amassed in zoöglœa (480). Ewart's study established the distinctiveness of the micrococci and characterized this organism both morphologically (i.e., ovoid, spore-like, dumbbell-shaped) and physiologically (i.e., mitotic reproduction, motility, either chaplet or zoöglœal colonization).

Cheyne's 1879 Micrococci

In his earliest professional contribution to bacteriology, the 1879 paper "On the Relation of Organisms to Antiseptic Dressings" Cheyne researched the theoretical background of micrococcus taxonomy and its relation to aseptic surgery. He knew the German and French scholarship, 1874 to 1878, and, as we have seen, defended the antiseptic system against those who doubted its efficacy (558–560).

Cheyne's inaugural, May 1879 paper was influenced by Lister's bacteriological thought of the period, 1875 to 1879. Lister's work with *Bacterium lactis* prior to the summer of 1877 suggested that bacterial infection was a plausible analogue to bacterial fermentation (R. Richardson; Coutts, *Microbes*: 148). From the latter's 1875 "The Germ Theory of Putrefaction," Cheyne adapted and reproduced Woodcut #18 images. These illustrations reflected the opinion that pathogenic bacteria were cultivable and susceptible to carbolic acid. Cheyne's paper contributed to an international enquiry, involving Pasteur, Lister, Koch, John Tyndall, William Roberts (1830–1899), and others, on the role of micro-organisms in fermentation, putrefaction, and infectious disease

(R. Richardson). It developed from bacteriological research Cheyne had begun in January 1877, while at the Edinburgh Royal Infirmary; and this work, continuing into the spring of 1878 while under Lister's surgical direction at King's College Hospital, culminated in the publication of the 1879 paper.

In 1877, using Pasteur's solution, Lister had obtained a pure culture of *Bacterium lactis*, the organism that caused milk to sour (N. A. Logan: 2; Santer: 59–65; R. Richardson; Coutts, "Illustrating Microorganisms": 2–3). He described the experiment in a 1 October 1877 "Address," "On the Nature of Fermentation," at the Opening Session at King's College (published in April 1878). Lister believed that *Bacterium lactis* was responsible for lactic-acid fermentation, a discovery leading to further enquiries into the fermentative or pathogenic effects of bacteria (*CP.* I: 351–352). He pursued this line of research in the papers "On the Lactic Fermentation and Its Bearings on Pathology" (1878) and "On the Relations of Micro-Organisms to Disease" (12 August 1881; *CP.* I: 387–398). In the latter paper, Pasteur's and Koch's researches on bacterial pathogenesis had converged in support of the antiseptic method.

The Woodcut #18 drawings (Figure 1) indicate that, between 1875 and 1879, Lister and Cheyne were aware that pathogenic bacteria (Groups 4 [right margin] and 5 [left margin]) had unique and variant forms. They could not, at this time, determine much in regard to bacterial physiology, for the cucumber infusion they used was not the ideal medium for viewing growth phases and colonial morphology.

Even though microbiological research in this period was laborious, imperfect, and error-prone, Lister did his best in the laboratory and lecture hall, in order to demonstrate the morphology, color, scale, and behavior of microorganisms as clearly as possible (R. Richardson). Group 1, across the top margin, displays four stages (a. to d.) of colonial growth. He carefully outlined a variety of forms: single spheres, pairs, a fissionable tetrad, chains, and zoöglœa. Because these heterogeneous organisms occupied the same visual space, the growth and reproductive stages of any single organism were difficult to place in sequence. As Lister had learned in 1875, since the chains and group-forms suspended in liquid medium were irregularly shaped, in Brownian motion, and seeming to divide both transversely and longitudinally, gathering physiological information was challenging. To illustrate the difficulty, he thought he had seen two simultaneous modes of fission in the same cell (possibly accounting for the image of quadratic fission in Group 1.d, upper-right-hand corner). Lister's fine microscope, manufactured by Edward Hartnack and Company between 1860 and 1870, likely had two objective lenses and two eyepieces, and its 920x (inscribed on the lower-left corner of the image) had the capacity to reveal cell division and growth. In 1867, Hartnack (1826–1891) had produced

WOODCUT 18.

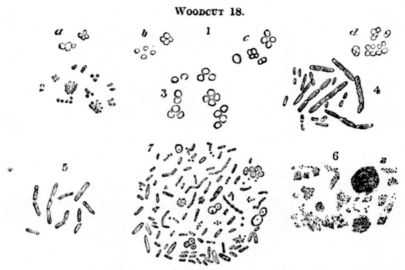

Fig. 1 is copied from Mr. Lister's paper on "The Germ Theory of Putrefaction and other Fermentative Changes," in the 'Transactions of the Royal Society of Edinburgh,' vol. xxvii, 1875, and represents the growth of micrococci as seen under the microscope.

Fig. 2. Micrococci from a wound treated antiseptically, growing in cucumber infusion.

Fig. 3. Micrococci growing in cucumber infusion containing carbolic acid.

Figs. 4 and 5. Forms of bacteria injected into animals, causing death.

Fig. 6. Drawing of the discharge from a wound treated antiseptically.

Fig. 7. Ordinary forms of organisms which may grow in cucumber infusion inoculated from wounds not treated antiseptically.

(Figs. 2—7 are represented as seen with Hartnack's powers magnifying 920 times linear.)

Figure 1. Cheyne had copied these drawings from Lister's 1875 paper "The Germ Theory of Putrefaction." This facsimile is taken from Cheyne's paper ("On the Relation of Organisms to Antiseptic Dressings," *Transactions of the Pathological Society of London*, 1879; 30: 563, https://www.google.archive.org).

a water-immersion objective, the high refractive index of which made for better resolution (Solliday). But even with a Hartnack No. XII and water-immersion optics, Lister's view was limited. Without solid media and advanced staining techniques, he could only speculate about consecutive growth and reproductive stages.

Lister's 1875 drawings, seven figures in all, adumbrated Ewart's 1878 series of Plates showing the life-cycles of *Bacterium termo* and of micrococcus (Ewart:

480–485). In Cheyne's paper, the Woodcut images of Group 1 (a.–d.) comprise a mixture of cells in a variety of reproductive, growth, and colonial formations (from left to right): (a.) a triad and a tetrad of small spherical cells; (b.) enlarged tetrads and separate spheres; (c.) four cellular configurations: a single ovoid, a tetrad, an adsorbent pair, and adherent and dispersed tetrads; and (d.) four additional configurations: two parallel, quadripartite chains, an adherent pair, a T-shaped tetrad, and a single cell undergoing quaternary rather than binary fission. Because of the liquid medium, successive reproductive phases are not easily inferable. Groups 2 (upper left margin) and 3 (right center) also reveal a mixture of physiological stages. Group 2 shows micrococci from an antiseptically-treated wound growing in the infusion. Ten groups of spheroidal microbes in varying configurations are depicted: paired cells, a disjointed triad, a segmented, quaternary chain, and zoögloeal masses. Group 3 (at the center) reveals enlarged microbes in the customary pairs and quarters.

Neither Lister (1873) nor Cheyne (1879) conjectured about the physiology of the bacteria that they had seen. In 1875–1879, they knew that understanding bacterial metabolism, reproduction, and colonization required experimental evidence, the acquirement of which, in turn, depended on technological upgrades in solid media, microscopy, image enhancement, and staining. Despite the technological deficit, Cheyne began with Cohn's limited form-based description of micrococcus as small round cells (less than 1.0 micrometer), either motile or immotile, that developed from pairs into short chains and clusters ("On the Relation of Organisms": 565). Although Cohn was able to trace growth patterns (i.e., from pairs or large cells to colonies) and to differentiate micrococcus from non-spheroidal bacteria, he failed to improve descriptive nomenclatures (*Bacteria* [1872]: 16). Standardized codes and criteria, such as those utilized in successive editions of *Bergey's Manual of Determinative Bacteriology* (1923–2015), however, would eventually bring greater consistency to the nomenclatural system (N. A. Logan: 8).

In sum, Cheyne realized, by 1879, that there were different kinds of micrococci and that fungi could also be spheroidal. Since January 1877, he had been assiduously researching the topic, consulting the writings of Cohn, Koch, Ewart and others; he cited precedents from 1866 to 1878 and knew about the surge in technological development ("On the Relation of Organisms": 565). He referred, for example, to Cohn's Tribe I: **Sphaerobacteria**/Genus 1. Micrococcus; to Koch's observation that micrococcus belong to a distinct Genus having multiple species; and to Ewart's reproductive model for micrococcus that Lister's 1873–1875 work on antiseptic wound management had influenced. In 1878–1879, Cheyne was also familiar with Koch's method and would eventually translate the "Investigations" into English for The New Sydenham Society;

and it would be published in 1880. Koch had described bacterial pathology induced in animal models; and he had outlined methods of preparing specimens, the value of Ernst Abbé's condenser, the optical advantages of the Zeiss objective systems, and the visual acuity afforded by Weigert's staining methods ("Investigations": 22–33). These convergent factors and discoveries, Cheyne acknowledged, in 1878, had made the elusive micrococcus more visible ("On the Relations of Organisms": 566).

Knowing the medical literature implicating micrococci in septic disease, Cheyne and his assistant, Gerald Yeo, while at King's College Hospital, began to conduct pathophysiological tests, using rabbit models, as had Koch ("On the Relations of Organisms": 568). They injected seven cubic centimeters of micrococcus in cucumber solution into a rabbit's jugular vein, while another animal intravenously received six cubic centimeters of a non-spheroidal bacterium. The experiment, repeated six times, had unvarying results: animals receiving micrococci were unaffected, whereas those injected with the bacteria (i.e., non-spheroidal organisms) did not survive (568–569).

In the course of the experimentation, Cheyne had encountered two anomalies that he hastily tried to explain. A small percentage of pus specimens drawn from human abscesses, previously treated by antiseptics, revealed the presence of *live* micrococci (seven of 32 or 22 percent were positive). This was unexpected. A second anomaly was that this strain of live micrococci, upon inoculation, had not sickened the rabbits. It therefore seemed logical for Cheyne to assume (1) either that the surviving "ordinary forms" of micrococci were, for some reason, resistant to carbolic acid; (2) or that the resistant and ubiquitous micrococci were nothing more than harmless concomitants in the abscess.

In view of Koch's finding that multiple forms of micrococci exist, Cheyne was well aware of the risks entailed in making "too absolute a statement," based on conflicting or incomplete information ("On the Relations of Organisms": 569). He did not heed his own advice, however, failing to retract or to exhaustively test assumptions (1) and (2) above. This failure, perhaps a lapse in judgment or in editing, undermined the value of the paper. Assumption (1), for example, contradicted current research. Cheyne's suggestion that the "ordinary forms" of micrococci were benign and, in some measure, antiseptic-resistant, was incongruent with the evidence-based consensus of German researchers who, since 1872, had been documenting the pathophysiology of micrococcal species, and who agreed that many were pathogenic. On the basis of limited testing on a few rabbits, Cheyne had precipitously advanced the opposite view on micrococci, as "simply accidental and not essential to the inflammatory process." Even though micrococcus was a variegated Genus, from 1879 to the early 1880s, he continued to favor the idea that these germs were simply benign concomitants of the disorder under treatment ("On the Relations of Organisms": 574).

One can only speculate as to why Cheyne's 1879 assessment of micrococcus pathogenicity diverged from that of Koch and contemporaries. He was definitely familiar with Koch's 1878 paper, having referred to it twice in the 1879 text ("On the Relations of Organisms": 574–575). Recognizing that micrococci could cause serious disease, he even congratulated Koch in the 1879 paper for having put "forward strong evidence ... that pyæmia in the rabbit is due to the development of micrococci in the blood" (569). Cheyne was unlikely to have ignored Koch's description of spreading abscesses in rabbits, especially its anatomical details, which had a direct bearing on medical practice. In the experiment, described in Part III of the "Investigations," for example, Koch had injected septicemic blood into rabbits, induced abscesses, and the animals died ("Investigations": 45). He had meticulously outlined the ensuing pathology. Citing the bacterial and pathological work of Klebs, Koch had reasonably connected lupine pathology to what occurred in human infectious disease ("Investigations": 45–46). This information should have alerted Cheyne to the possibility of micrococcal virulence in abscess pathology.

Micrococci survived in antiseptically-treated abscess pus, not because they were resistant or benign concomitants, but because weaker antiseptic solutions (<1–20), as the March-April 1878 experiments at King's College had shown, sterilized only 25 of the 32 (78 percent) specimens obtained from London patients. Cheyne should be credited for having conceived of multiple-working hypotheses; however, because the alternative resistance and benign-concomitance hypotheses were neither discounted in, nor expurgated from, the final draft of the 1879 paper, the acid-strength hypothesis, which was the correct conclusion, was overshadowed. Cheyne's competitors and critics would exploit this mistake ("On the Relation of Organisms": 579–581).

The acid-strength experiment deserves close attention. Testing the acid's effects on micrococci with progressively stronger concentrations, Cheyne discovered that some microbes growing in meat infusions, had survived dilute concentrations between 1-to–500 to 1–to–250 parts carbolic acid-to-water ("On the Relations of Organisms": 580–581). His experiments had revealed that, in the 1–250 to <1–20 range, germicidal activity was incomplete. To exceed the 1–20 level, however, was risky. Cheyne understood that a balance had to be reached: the acid had to be strengthened sufficiently to kill bacteria without irritating healthy tissues, interfering with natural healing, or becoming toxic. Developing bacterial cultures in milk, blood serum, egg white, and artificial medium, Cheyne ingeniously determined that asepticity could be approached as carbolic-acid strength was increased to the maximum level of toleration:1–20 ("On the Relation of Organisms": 578–579, 580–582). The acid-strength experiment proved that the micrococci he had originally detected were *not* inherently resistant to the acid (568–569); on the contrary, many but not all had been killed in the weaker medium at levels less than 1–20 strength. Whether "ordinary" forms of the germ were harmless, was a question Cheyne had yet to answer satisfactorily. German bacteriology and the work of Ogston would soon provide the answer.

Cheyne's inaugural paper surveyed bacteriological research commencing in January 1877 at Edinburgh and resuming at King's College, autumn 1877 to spring 1879. In several ways, the paper contributed to the development of systematic bacteriology. First, it reinforced the assertions of Koch and others that micrococci "form a group of organisms quite distinct from bacteria [i.e., all non-spheroidal microbes] as shown by their growth, their reactions to reagents, and the effects of carbolic acid in varying strengths upon them, [and by] their effects on fluids and on the living body" ("On the Relation of Organisms": 582). Second, after 18 months of observation at King's College Hospital (i.e., December 1877 to May 1879), under diverse conditions and using different kinds of fluid media, Cheyne determined that micrococci did not change shape in the course of a life-cycle (582). The 1879 paper would have made a greater impact had the speculative material from the Edinburgh period of experimentation been redacted or disclaimed. The final draft of 1879, with the acid-strength findings as its focus, could then have been retitled "Carbolic-Acid Concentrations, Surgical Dressings, and Post-Operative Effects on Micrococci."

Cheyne carried on with his bacteriological research. In the spring and summer of 1882, when reconsidering the question of micrococcal pathogenicity, he was able to cultivate pure strains with greater facility (*Antiseptic Surgery*: 233). From early March to late June 1882, he employed Koch's techniques in the laboratory (233). With the ability to see specimens more distinctly, he learned that micrococci were not just solitary spheres but heterogeneous, colonial organisms, and, depending on species, were capable of forming triads, quartets, chaplets, and multiple congregations (231). Whereas in 1879 Cheyne suggested that all micrococci were odorless concomitants, in 1882 he recalled that, when present in unchanged dressings, at least one form was responsible for a sour smell, an odor he attributed to a chemical change; the wound in which they were detected healed, as long as antiseptic was employed. In this period, with respect to micrococci, he was routinely making differential assessments.

Cheyne saw the benign-concomitant theory in a new light when the researches of Ogston and of William H. Welch (1850–1934), American bacteriologist/pathologist, confirmed that this Genus was comprised of a variety of species exhibiting different levels of virulence, as Koch had determined in 1878. In the 4 October 1884 essay "Micrococci in Relation to Wounds, Abscesses, and Septic Processes," written after reviewing Ogston's experiments on abscess pathology, Cheyne stated unequivocally that *various kinds* of micrococci could be found in wounds treated aseptically and that each was pathogenically unique (italics added; 648). These organisms grew best en masse, in the presence of oxygen, at body temperature, and only with difficulty without oxygen; and each affected host species differently (648). Incapable of growing in blood, they could cause serious disease by growing in executory canals. In acute abscesses, micrococci were always present and probably caused them. Additionally, in pyemic abscesses, they were sometimes the primary cause of inflammation and suppuration.

4

1878–1925
Alexander Ogston—Landmark III

In the periodization of 1889, Cheyne stressed the importance of Alexander Ogston's discoveries. Ogston who had examined pus from a number of acute abscesses confirmed the hypothesis of Koch and his German colleagues of the 1870s that micro-organisms were abundantly present in these lesions. As Cheyne observed, Ogston was then able to isolate "two different kinds of cocci under the terms [S]taphylococci and [S]treptococci, according to the grouping of the organisms" (*Suppuration*: ix). Indebted in part to Ogston, Cheyne recognized that microbes were responsible for abscesses and for other disorders, known at that time as erysipelas, pyemia, and septicemia. Along with a number of British and European contemporaries, notably the German surgeon, Friedrich Fehleisen (1854–1924), Cheyne would corroborate and elaborate on Ogston's findings (*Suppuration*, ix-x).

Commentary on the history of micrococcus and on Lister's experiences with its forms, in the 1870s and early 1880s, will head this discussion of the scientific relationship between Ogston and Cheyne. Lister's 1870s experiences influenced Cheyne's thinking, and, as late as 1881, both had disagreed with Ogston on the cause of acute abscesses, the etiology of which would be experimentally determined in the 1880s-1890s. With advanced instrumentation and laboratory technology, Listerian bacteriologists contributed significantly to the early development of the field.

1875–1881: Lister's Granuligera

The term *micrococcus*, as noted in chapter 3, had been used generally to signify any minute, spheroidal organism. In 1875 and in the winter of 1881,

Lister had described spheroidal organisms that were unlike any other bacteria he had ever seen ("A Contribution to the Germ Theory," Part I. *CP*. I: 275–313, esp. 281–282). His 1875 drawings depicted these particular cells in masses rather than as autonomous organisms or in serpentine chains. Their mode of growth, as granular configurations, distinguished them from both rod-like and helical shapes that ordinarily grouped in threes and fours (282). For a nomenclature, he created the descriptive term, *Granuligera,* a transliteration of the Greek *micrococci,* meaning "small granule." Believing that he had come across a new germ, Lister's epithet was either of Spanish (*grano* [grain] + *ligero* [light, of little weight]: *Granoligero*) or of French derivation (grain [grain] + léger [light]: *Grainléger*) (*Diccionario*: 106, 127; *Larousse*: 289, 378). He recorded two physiological observations: when cultivated, these immotile germs putrefied organic matter in fluid media.

Unlike Cheyne's 1879 micrococcus, *Granuligera* were unquestionably virulent. An epidemic of these organisms was reported to have broken out in Lister's laboratory, contaminating cultures. Conditions had become so bad that, to solve the problem, he moved his laboratory to the unoccupied upper floor of the building (*CP*. I: 282). Lister had observed that the *Granuligera* were indeed cocci and reproduced by binary fission. In 2009, investigators conjectured that the *Granuligera* infesting Lister's lab were likely to have been *Staphylococcus pyogenes aureus* (Carling and Moore: 82–84).

Lister's second experience with *Granuligera* was documented in the article "On the Catgut Ligature" (1881). Hempen thread suturing a thyroid operation, by the ninth post-operative day, had become thickly pustulant. In a month's time, when the ligatures disintegrated, they were re-examined, found to be acidulous, and loaded with *Granuligera*-like germs, similar in form to those that he had seen in 1875. Since the organisms in both instances clustered in twos, threes, and fours, they seemed unlike the chain-formers. Lister suspected that a morphological affinity existed between *Granuligera* and Cheyne's 1879 abscess organism (*CP*. II: 103). Somewhat like Cheyne's 1879 germ, Lister's had been obtained from infected ankle wounds, and it, too, contaminated culture media. Although Lister isolated *Granuligera* in 1875 and in 1881, the two specimens, though morphologically similar, differed from each other in terms of virulence.

As Lister explored abscess etiology, he had to figure out whether pus was the infectious cause, simply a waste product, or served some other purpose. His doctrine of suppuration, which developed from 1867 to the 1890s, initially held that pus was solely the residue of decomposing blood and tissue in the presence of oxygen ("On the Antiseptic Principle in the Practice of Surgery," *CP*. II: 37). His surgical approach, based on this premise, wrongly assumed that an unopened abscess was not caused by external germs that had invaded a wound; on the basis of this assumption, he saw no need for antisepsis (*CP.* **II**: 48). In August 1881, Cheyne and Koch independently proved, "that in the healthy state of the animal body there are no micro-organisms present among the tissues." This finding was then mistakenly used to support the premise that abscesses arose from physiological disturbances alone and were therefore pre-operatively aseptic ("An Address on the Relations of Minute Organisms to Inflammation," *CP.* **I**: 408). Thus, in 1867, Lister posited that abscess pus nei-ther was produced as an immune response nor was comprised of WBCs (white blood cells); rather, pus collected in an "ordinary abscess" because carbolic acid had "excited" the nerves; and with few exceptions, abscesses contained "no septic organisms" (*CP.* **II**: 40, 42).

Lister had yet to account for the presence of live micrococci in Cheyne's 1877–1878 experiment. In August 1881, he still subscribed to the latter's benign-concomitant hypothesis, "that the micrococci are … a mere accident of these acute abscesses" ("An Address on the Relations of Minute Organisms," *CP.* **I**: 408). The experiments held at King's College Hospital, therefore, had led Lister and Cheyne to formulate an incorrect explanation on the basis of a false prem-ise. Employing "Koch's method," they examined specimens that Lister had taken from an acute abscess. Once its "thick original contents" had been evac-uated, it oozed serum for three days, and healing began naturally (*CP.* **I**: 408). In light of this, Lister thought he had found a thread of consistency between the suppuration doctrines of 1867 and of 1881. The neurological explanation for abscess discomfort seemed perfectly reasonable. Because encapsulated fluid compressed inflamed nerve-endings, opening and draining the boil surgically relieved the pain, hastened healing, and precluded the need for antiseptics. After incision and drainage, the concomitant micrococci, inhabitants of the skin, then migrated into wounds post-operatively and persisted there for a time, since "the system" was temporarily "disordered." Their ubiquity, in Lister's view, neither disproved the neurological theory nor impeded healing (*CP.* **I**: 408).

Lister and Cheyne, in 1881, subscribed to a binary distinction between path-ogenic bacteria, on the one hand, and relatively harmless "ordinary micrococci," on the other (*CP.* **I**: 408). Harmless micrococci, envisaged as nothing more than scavengers, were reduced to secondary invaders of aseptic abscesses: "Mr. Cheyne's idea, therefore, is that when an inflammatory attack is sufficiently severe to produce serious febrile disturbance, these micrococci get in, and, finding in the pus of an abscess a congenial soil, develop in it in abundance" (*CP.* **I**: 408–409). Lister could not have known that an abscess was a site of infection to which the immune system responded: leukocytes, antibodies, and other blood products migrated via bodily fluids to destroy encapsulated bacteria (S. Doerr).

In 1891, in the light of the discoveries of Koch, Ogston, and others, Lister radically altered his earlier doctrine. He finally acknowledged that a strictly neurological interpretation of the acute-abscess formation was wrong, since Ogston had proven that acute boils always contain pus-inducing micrococci, and since Koch, in 1882, had discovered abundant tubercle bacilli in the pyogenic membranes of tuberculosis patients ("On the Principles of Antiseptic Surgery," *CP*. **II**: 347).

1880–1882: Ogston's Methods and Microbes

Alexander Ogston's morphological study of abscess pus revealed the presence of micrococci (granular particles) and of bacilli (rod-shaped particles). He additionally described two micrococcal subtypes. One was *Staphylococcus,* an organism that congregated in clusters and was found in closed abscesses. The epithet combined the Greek words *staphyle* (grape) and *kokkos* (grain, seed]) (G. Licitra, *Staphylococcus: Etymologia; OED.* **II**: 3019). The other was *Streptococcus,* an organism that congregated in helical chains. This epithet, invented by the surgeon and pathologist Theodor C. A. Billroth (1829–1894), united the Greek prefix *streptos* (bent or twisted chain) to the suffix *kokkos* (grain, seed) (*OED.* **II**: 3090; Bulloch, *History*: 237, 353). Many of the micrococci that Ogston encountered were pathogenic.

A major contributor to clinical bacteriology, Ogston had received his medical degree in 1865, studied in Vienna, Prague, and Berlin, and was active in surgical societies such as the German Medical Congress (G. Smith: 9).[1] In 1870, at the age of 26, having been appointed a surgeon in the Aberdeen Royal Infirmary, he learned of Lister's antiseptic system and witnessed its impressive effects. While visiting Lister in Edinburgh, he became versatile in its use and brought what he had learned to Aberdeen, Scotland (G. Smith: 10). A full surgeon in the Royal Infirmary in June 1874 and then a senior surgeon in 1880, Ogston began to investigate the questions of whether micro-organisms caused human disease and whether specific pathogens could be connected to particular disorders (G. Smith: 10–12). He set up his own laboratory behind his Aberdeen home and, in 1879, received a £50 grant from the British Medical Association to study bacterial pathogenesis (I. A. Porter: 355–356; Coutts, "Illustrating Micro-Organisms": 3). Ogston proved to Lister and Cheyne that the Genus **Micrococcus** comprised numerous species, several of which were implicated in abscess development. By 2013, this Genus would be found to contain 16 species (J. P. Euzeby).

The dialogue between Lister, Cheyne, and Ogston, extending from 1870 to the 1880s, weaves through more than one dozen texts. Ogston's experiments during this period are recorded in a number of papers, four of which are of particular interest: "On Abscesses" (1880), "Report upon Micro-Organisms in Surgical Diseases" (12 March 1881), and Parts I and IV of the 1882 series "Micrococcus Poisoning." Ogston benefited enormously from the work of his peers. In his role as a systematist, he adopted the Cohn-Ewart system. Thus, in "On Abscesses," he outlined a taxonomy based on form, comprising **Bacilli** or little rods; **Bacteria** or short bacilli with rounded ends; **Spirilla** or spiral forms; and **Micrococci** or spherical bacteria (122). In his culturing of abscess microbes, Ogston used Koch's method. Pus specimens, dyed purple with methyl-aniline, could be identified with certainty in tissues and fluids. Additionally, Ernst Abbé's lighting and oil-immersion lenses, along with a Zeiss microscope, were essential to the process and gave Ogston an investigative advantage ("On Abscesses": 122).

Acknowledging his debt to Koch, in 1881, Ogston outlined an experimental modality, beginning with form and extending to function. Since the surest way to date of demonstrating the presence of micrococci was to employ Kochian microscopy and tissue-staining techniques, he recommended slides fitted with a 1/18th-inch Zeiss oil-immersion system and accompanying eyepieces; a 155-millimeter long tube; the Abbé sub-stage condenser (without diaphragm); and gaslight illumination, concentrated by a blue-tinted, bull's-eye condenser on the flat side of the mirror. Even with the latest laboratory technology and specifications, however, Ogston knew that one could still overlook or misidentify micrococci if the staining were too faint, or if organisms became decolorized. Nevertheless, it was advanced technology that permitted Koch and Ogston to reveal the fact that abscess pus consistently caused blood poisoning in mice and guinea pigs ("On Abscesses": 125; "Report upon Micro-Organisms": 371).

In "Micrococcus Poisoning," Part IV, Ogston reasserted the importance to bacteriology of Koch's method, of Abbé's optics, and of effective staining techniques, such as Weigert's gentian-violet stain and alcohol preparations (Part IV: 549). So committed was Ogston to the new methods and instrumentation that he considered as unreliable data based on magnifications lower than those provided by the 1/12th-inch oil-immersion objectives and on procedures not following Koch's directions (Part IV: 550). Inclined to accept Koch's idea that micrococci and acute suppuration were connected to each other, Ogston introduced supportive evidence from the examination of infected hospital wounds containing micrococci ("On Abscesses": 126).

In 1882, supplied with experimental proof, Ogston rejected three hypotheses to which Lister and Cheyne had subscribed. As noted above, the first was that fluid tension and inflamed tissue, rather than infectious organisms, stimulated the production of pus ("Micrococcus Poisoning," Part IV: 529, 532–534). The second was the idea that inflammation was exclusively a neurological phenomenon, the result of irritation, burns, or exposure to cold (Part IV: 535, 543, 536–538); and the third, that the reputed absence of micrococci in an acute boil proved that germs of this type were not to blame for infections. Ogston's counterargument to the third point was that micrococci were indeed present in acute abscesses, that abscesses encapsulated pathogens in granular tissue, and that the extracted organisms grew *in vitro* (Part IV: 539–540).

In "On Abscesses," Ogston targeted Cheyne's 1879 conclusion that antiseptic-resistant micrococci were insignificant. The latter's misjudgment, in Ogston's view, was the direct but unavoidable consequence of not having employed Koch's methods ("On Abscesses": 126). Although Ogston agreed with Cheyne that where treatment was careless the micrococci could infiltrate wounds, he criticized Cheyne's hypothesis that live micrococci were impervious to carbolic acid. The truth of the matter, according to Ogston, was that microbes of this kind could, to varying degrees, withstand weak carbolic solutions. This is a curious comment on Ogston's part, coming as it did in 1880. Apparently, Ogston either had not read Cheyne's paper carefully or had ignored the latter's corrective experiment revealing that weak carbolic-acid solutions (initially used for his experiments), in the range of 1–500 to 1–250, were unable to destroy micrococci completely but that the 1–20 solution could sterilize a colony. Cheyne and Ogston agreed that acid strength was the determining factor in the survival of micrococcus.

Overall, on a number interrelated points, Ogston questioned the findings of Lister and Cheyne. Contrary to their conclusions, Ogston affirmed that species of micrococci were actually a frequent cause of acute abscesses; that pus, rather than being solely a waste product, was "very closely connected" with these lesions; that the responsible organism could produce septicemia. All would ultimately agree that the constitution of the laboratory animal or of the patient determined the extent and outcome of a micrococcal infection and that the 1–20 carbolic-to-water ratio was the most effective, tolerable strength.

In the autumn of 1876, while engaged in early investigations at the Edinburgh Royal Infirmary, Cheyne had become aware of German bacteriology's great strides and of how essential adequate facilities, supplies, technology, and methods were to British medical science. In the late 1870s, bacteriological work in British hospitals was not routine, and those who were professionally qualified worked in public health (Coutts, "Illustrating Micro-organisms": 3). Without sufficient funding, facilities, and equipment, Cheyne found it difficult to corroborate German findings and to demonstrate satisfactorily the interdependency of microbiology and medicine ("On the Relation of Organisms": 558–560). That he had read Koch's 1878 "Investigations" at a time when he was technically unable to employ the latter's methodology fully must have been frustrating. No matter how he had labored to link clinical presentations to laboratory practice, while working in the Infirmary's laboratory from January to autumn 1877, he was not always able to differentiate between micro-organisms and cellular debris. Laboratory facilities were better at King's College Hospital, when Cheyne assumed the post of House Surgeon in the autumn of 1877. By April 1882, Ogston who had been making the most of the advanced technology at his disposal extolled the value of the triadic system in microscopy; most importantly, the new technology allowed the investigator to distinguish between bacteria and organic granules. Without the advanced instrumentation and media, he unequivocally stated, an investigator was unable to make a genuine contribution to the field ("Micrococcus Poisoning," Part IV: 550–551).

1879–1925: The Cheyne-Ogston Dialogue

From c. 1881 to 1882, Cheyne and Ogston, both of whom were proponents of Listerism, sparred with each other rhetorically. In "On Abscesses," Ogston challenged Cheyne for having "advanced the thesis (based on observations of wounds) that micrococci are innocuous organisms" (126). Ogston, in 1881, went so far as to imply that the Lister-Cheyne view of micrococcal pathogenicity was out of touch with contemporary research on the micrococcus. Whereas eminent researchers and physicians, such as Billroth, James Israel (1848–1926), Emil Theodor Kocher (1841–1917), and others, had demonstrated that this organism causes blood-poisoning, Cheyne asserted that it was "an innocent organism" (369). In certain respects, Ogston's criticism was justified. If we reconsider Cheyne's tone and rhetoric in 1879, however, one might say that he was ambiguous, perhaps diffident, but certainly not assertive. Despite Cheyne's early disadvantages in technique and instrumentation, he was able to conduct an acid-dilution experiment at King's College Hospital (March-April 1878), proving that, for asepsis to be achieved, the 1–20 ratio was required. The editorial lapse in the 1879 paper, as I pointed out, was twofold: first, Cheyne had failed to emphasize the acid-concentration discovery; and, second, he had failed to expurgate the unlikely hypotheses (i.e., resistance and benign-concomitancy, respectively). Ogston exploited perceived errors in language, logic, and method: that is, Cheyne was self-assured ("We *must* ... conclude that the ordinary forms of micrococci are harmless"), had jumped to a conclusion ("the *ordinary* forms of micrococci [wherever] derived are harmless"), and was indecisive ("On the Relations of Organisms": 582). It is important to recognize that Ogston's criticism was not even-handed. In "On Abscesses," for example, he failed to acknowledge Cheyne's acid-strength discovery as a genuine discovery of great value ("On the Relations of Organisms" 569"; "Report": 372).

Cheyne responded to Ogston's commentary, intermittently, in a series of texts: the 1882 *Antiseptic Surgery*, the 20 September 1884 "Report on Micrococci in Relation to Wounds, Abscesses, and Septic Processes," the 1885 *Manual of the Antiseptic Treatment of Wounds*, and the 1925 biography of Lister. In these writings, he gradually acknowledged the rectitude of Ogston's micrococcal deductions, tried to account for his 1877–1879 misjudgments, and made counterclaims where appropriate.

From 1881 to 1882, even though Cheyne had Hartnack microscopes at his disposal, he lagged behind Ogston whose laboratory had the advantage of the latest upgrades in German optical, staining, and media technology. For this reason, Cheyne still entertained obsolete notions; for example, he thought that all micrococci metabolized decaying organic material, thrived when removed from oxygen, and were unrelated to inflammation and pus. Ogston, on the other hand, modified and disproved these ideas ("Micro-Organisms in Surgical Diseases": 375).

Cheyne gradually altered his views in the light of the latest discoveries, many of which were Ogston's. Thus, he accepted four Ogstonian claims: that micrococci caused acute abscesses, that pus was "closely connected" to these organisms, that toxins caused septicemia, and that physical constitution determined susceptibility to toxins ("On Abscesses": 128). In 1882, Ogston enunciated additional bacteriological principles, derived in part from German antecedents. He restated that common micrococci are the same as the virulent cocci responsible for inflammation (thus, rejecting Lister's neurological idea of abscess causation; and Cheyne's, of micrococci as benign concomitants) ("Micrococcus Poisoning," Part IV: 541). On the basis of micrococcal heterogeneity, Ogston also supposed that some, presumably benign species, did not cause inflammation; that the virulent forms became attenuated (540–541); and that blood-poisoning was the secondary effect of disease in tissues (561–562, 564).

In the 1882 monograph *Antiseptic Surgery*, Cheyne described new experiments of the spring and summer of 1880, undertaken concurrently with those of Ogston (223). Unlike the Edinburgh research of 1876, in the King's College research of 1877–1880, and especially that of March-June 1880, Cheyne began to apply Koch's staining methods to the analysis of surgical discharges. Although live micrococci again persisted in aseptic wounds, with upgrades in instrumentation, culture media, and staining, he was able to see these micro-organisms with greater acuity and to distinguish them from anatomical structures and cellular debris (233). He became skilled in the use of stains such as aniline dyes (especially methyl violet, fuchsine, and aniline brown), at mixing dyes with distilled water until the desired color was attained, at using cucumber infusions, and at preparing slides with Canada balsam, an oleoresin of the balsam fir ("Canada balsam," *Webster's*: 121; *Antiseptic Surgery*: 233–234).

The preparation of specimens required meticulous care at the bedside and at the bench. How a wound was bandaged and routinely disinfected, for example, was essential to the 1880 investigation. Aseptic treatment included careful preparation and changing of dressings; and, if this were done properly, a good outcome was predicted. Putrid discharges ordinarily teemed with bacteria, but, with drainage improvements, the concentration of micro-organisms diminished (*Antiseptic Surgery*: 236, 242). Once an original dressing had become saturated, new ones had to be applied regularly over a six- to eight-day period. Although tests in this period showed, once again, that micrococci persisted in great numbers, Cheyne found no pus, and bacterial colonies did not appear to interfere with healing (239, 243). It is not clear, in this period of time, whether he had been routinely employing higher concentrations of carbolic acid or combining antiseptics to test their interactive effects.

In 1883, Ogston assessed the most recent microscopic evidence. Retaining Cohn's 1875 classifications system as a guide, he placed *all* cocci in a single Genus, his long-range intention being to sort out variants morphologically. With the benefit of up-to-date equipment and knowledge drawn from experiments, he was able to identify a chain-forming organism and a new form called *Staphylococcus*, both of which were micrococcal subtypes—not rod-like but, literally, small spheroidal organisms (Buchanan: 374). By the mid–1960s, through DNA base-composition studies, the differences between *Staphylococci* and other micrococcal forms would be established (Götz, et al.: 5); for example, the Genus **Staphylococcus**, up until the early 1970s, was thought to have comprised only three species: *Staphylococcus aureus*, *Staphylococcus epidermidis*, and *Staphylococcus saprophyticus*. Through chemical and genotypical analyses, 36 species (according to one source) were added to this Genus, bringing its current membership to 49 (Götz, et al.: 5). In addition, the Genus **Streptococcus**, from 1985 to 2002, has increased to more than 50 species and subspecies (R. Facklam).

Like Ogston, Cheyne also hoped to improve microbial classification. His physiological approach, from 1880 to 1882, supplied complementary information. In *Antiseptic Surgery* (1882), on the basis of six criteria, Cheyne physiologically differentiated *Micrococcus* from rod-like bacteria: (1) in strongly alkaline fluids, made so by the addition of carbonate of soda, bacteria (i.e., rod-like germs) died faster than did micrococci, indicating a physio-chemical difference between the two forms (244); (2) as Koch had observed, hematoxylin (a crystalline phenolic compound in logwood) stained micrococci but not bacteria ("hematoxylin," *Random House*: 616); (3) unlike granular albumen which dissolved, and unlike bacteria that either agglutinated or disappeared when exposed at body temperature to strong peptic solutions, micrococci were unaffected; (4) micrococci, unlike other forms of bacteria, divided longitudinally; (5) non-spheroidal bacteria were more sensitive to carbolic acid and other chemicals than were micrococci; and (6) the latter did not transmute into other bacterial forms (244). By 1882, in light of Ogston's research, Cheyne had characterized micrococcus generically: these spheroidal microbes belong to "one well-defined class"; a variety of micro-organisms could co-inhabit a septic wound (245); and the population of a micrococcal colony diminished proportionally as the carbolic-to-water concentration approached the 1–20 level.

In regard to micrococcal metabolism and its effects on living tissues, Cheyne had even more to say. These microbes might not putrefy or decompose tissues, but (in this particular case) they were a hindrance. At this juncture, he realized the need to revise his opinion of the so-called "ordinary" micrococci. Surgical experience had taught him that one should not be indifferent to these germs. If they got into wounds, pus formed, and wounds healed slowly. Unknown to him in 1879, they could also invade bone canals, producing sour secretions and irritation. Cheyne alluded to Lister's troublesome *Granuligera* that had infiltrated and destroyed hempen ligatures after a thyroid operation (*Antiseptic Surgery*: 247). Evidently, Lister and Cheyne had come across at least two distinct species. On further examination, Lister's *Granuligera* was identified as *Staphylococcus pyogenes aureus*, a harmful agent. These facts convinced Cheyne of the need to exclude these organisms during and after surgery (*Antiseptic Surgery*: 248).

By 1882, Lister and Cheyne had arrived at an important milestone in laboratory experimentation. In *Antiseptic Surgery* (1882), with access to the latest technology in microscopy, Cheyne was able to illuminate the physiological nature of micrococci. Three camera-lucida drawings, reproduced from this text and renumbered for our discussion as Figures 2.1, 2.2 and 2.3, feature micro-organisms taken from surgical discharges. Lister and Cheyne, at the time, had a functional, low-tech system for enlarging cultural images and no doubt benefited from the late Joseph Jackson Lister's (1786–1869) improvements in achromatic and compound microscopy ("On the Interdependence," *CP*. **II**: 502; "The Third Huxley Lecture," *CP*. **II**: 515; "Obituary Notice," *CP*. **II**: 543). The instrument they used in these investigations, *camera lucida* (Latin: "light chamber"), consisted of a prism that projected the slide-image through the microscope onto a light-sensitive drawing surface, upon which the viewer traced images in detail ("camera lucida," *Webster's*: 120). But they now had high-powered microscopy (1030x), and Zeiss' water- or oil-immersion lenses. Consecutive stages of micrococcal growth are visible below in selected, juxtaposed plates: Figure 2.1 (Plate I. *fig.* 1) at 1450x; and Figures 2.2 (Plate IV. *fig.* 25) and 2.3 (Plate IV. *fig.* 31) at 1030x.

Figure 2.1 (Plate I. *fig.* 1) exhibits micrococci in various numbers and configurations: doublets or dumb-bell shapes, triads, quartets, tetrads, and a zoöglœal mass with four parallel rows numbering 16 cells. With improved optics, Cheyne was able to display, in Figure 2.2 (Plate IV, *fig.* 25), the organisms' developmental stages from conjugation through successive points in the growth cycle: **cluster a.** (upper left), viewed at 8:55 a.m., reveals four daughter cells, bound in cloverleaf formation (in what modern cell biologists call *Telophase*: the stage of chromosome exchange and the appearance of adherent cells [Miller and Levine: 244–249]);

Figure 2.1 (top, Plate I. *fig.* 1), Figure 2.2. (middle, Plate IV. *fig.* 25), Figure 2.3. (bottom, Plate IV. *fig.* 31). These three images, selected from a set of five plates consisting of 40 images, are reproduced from Cheyne's monograph (*Antiseptic Surgery: Its Principles, Practice, History, and Results*, London: Smith, Elder & Company, 1882, p. 599, https://www.archive.org).

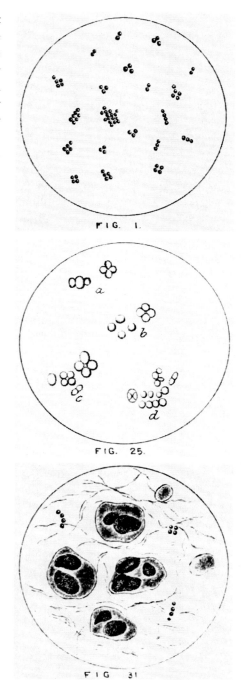

cluster b. (near center), viewed at 9:04 a.m., depicts the individuated daughter cells (referred to as *cytokinesis*: the daughter cells, each with a complete cell membrane, have disengaged from one another); **cluster c.** (lower left), viewed at 9:30 a.m., shows cells developing from a single bacterium into a quaternary group (prior to *Telophase*), and the quaternary cells maturing (during *Telophase*) prior to fission; and **cluster d.** (lower right), viewed at 10:36 a.m., portrays the linkage of daughter cells in chains, a bacterium replicating into four (at *Telophase*), along with confluent cells. Figure 2.3 (Plate IV. *fig.* 31) features pus taken from an acute finger abscess containing micrococci. Although Cheyne did not speculate here on cellular activity or on the relation of pus to the micrococci, the colonies in this image appear to have been destroyed.

Cheyne and Ogston, from 1878 to 1881, still differed on whether pathogenic organisms caused acute abscesses. As reported in the 1879 paper, of the 32 acute and chronic abscesses from which Cheyne had collected discharge specimens, 25 were aseptic; the remaining seven contained live but supposedly-innocuous micrococci. For Ogston, who, in 1880, had the latest preparations, the results were quite different. Pus discharges from acute abscesses, using Koch's solid media, staining methods, and the latest microscopy, *always* revealed live organisms; chronic abscesses, he mistakenly surmised, were devoid of them. By 1882, once Cheyne had been able to stain pus discharges, he too saw micrococci. In view of the evidence, he accepted Ogston's tentative hypothesis that these particular germs "are always present in acute abscesses" (*Antiseptic Surgery*: 254). Both of them still thought that "organisms are … absent from chronic abscesses"; but further research would show that the chronic form also harbored germs (254).

Lister retrospectively credited Ogston for his breakthroughs. In the 1896 Presidential Address, "On the Interdependence of Science and the Healing Art," he acknowledged that acute abscesses always contained micrococci of some kind. Using colonial formations as the main criterion, he was able to differentiate *Streptococci* from *Staphylococci* (*CP.* **II**: 501). Cheyne acknowledged his error about abscess infection earlier in three *seriatim* Reports describing the 1880 experiments, published on 20 and 27 September and on 4 October 1884. As noted, without advanced laboratory technology, prior to 1878, he had been unable to stain pus using the latest Kochian methods. Once this deficit was removed, he acknowledged having missed what Ogston would later discover: that micrococci are in acute abscesses. Nor did he hesitate to credit Ogston for his contributions to micrococcal taxonomy, to the pathogen-pathology linkage, and to the understanding of natural immunity ("Report" [20 September 1884]: 553).

Of great significance to the development of Cheyne's bacteriological thought was the revelation that the Genus **Micrococcus** had many species. Various kinds could live in septic wounds, and these varieties differed from each other in terms of associated pathologies ("Report" [4 October 1884]: 647). The German hygienist Carl Flügge (1847–1923), in *Micro-organisms, with Special Reference to the Etiology of the Infective Diseases* (a text Cheyne would translate in 1886 for the New Sydenham Society), subdivided the Genus into three domains. Interested in the relation between microbes and disease, Flügge chose pathophysiology, rather than morphology in mammalian vertebrates, as the first principle of organization; hence, he grouped organisms under the rubrics *human pathogens* (183–204) and *animal pathogens* (204–212); a third classification was designated for *saprophytic micrococci*, those that live on decomposing matter (212–231). Of the 48 species he had identified, 29 belonged to the Genus **Micrococcus**. Flügge has also been recognized for the discovery of the saprophyte, *Micrococcus roseus*, but he attributed this discovery to the gynecologist Ernst von Bumm (1858–1925), who had accidentally found the germ in atmospheric dust on the surface of nutrient medium ("**Micrococcus**, Genus"; Bulloch, *History*: 237). Bacterial discoveries increased as lines of communication with the Continent were opened up, as British bacteriology acquired modern instrumentation, and as professional meetings were devoted to the study of bacteriology in its medical contexts.

Ogston also emphasized that bacteriology, pathology, and surgery were interdependent disciplines, a regnant idea in the writings of Pasteur, Lister, and Cheyne. Ogston's 1880 experiments, in particular, supported the hypothesis that specific organisms were associated with specific diseases; for example, in cases of erysipelas, colonies grow in the lymphatic system; in those of pyemia, they grow in the blood, forming emboli; and in those of septicemia, they tend only to grow locally and, once infiltrating the blood, release toxins in blood and tissues ("Report" [4 October 1884]: 648).

The 1885 Manual: A Primer in Medical Bacteriology

Cheyne's 1885 *Manual for the Antiseptic Treatment of Wounds for Students and Practitioners* includes discussions on a range of contemporary topics, such as the relation of bacteriology to surgery, systematics, toxicity, suppuration, pathology, and histology. This text showed how Cheyne had assimilated contemporary discoveries and, through experimentation and the interchange of ideas, had revised misconceptions of the 1870s.

In the January 1885 Preface to the *Manual*, Cheyne connected pathogenic bacteria directly to surgical practice. As a surgeon *and* a bacteriologist, he was doubly qualified to explore how the two disciplines complemented each other. Intended to foster "a thorough knowledge of the practical details of … treating wounds," the *Manual* is generically akin to the 1884 "Biological Laboratory," a text he had contributed to *Public Health Laboratory Work*. Cheyne begins by introducing "the scientific basis of wound treatment" (p. v). Nine chapters are allotted to the biology of wounds, two of the seven treating pathogenic disease and disinfection, and the text instructs readers on aseptic surgery (materials employed, modifications) and on the antiseptic system (chemical agents, various methods), discussed further in the next chapter (pp. ix-x). Chapter II, "Bacteria and Disease," to which we shall turn below, is an index to Cheyne's bacteriology in 1885.

Following a brief description of the nature and life history of bacteria, Cheyne systematized these organisms (*Manual*: 11). Not abandoning differential morphology, he constructed a fourfold system, consistent with those of Cohn, Koch, and Ewart: (1) **Micrococci,** round bodies; (2) **Bacteria,** small oval rod-shaped bodies, about twice as long as broad; (3) **Bacilli,** rods of various lengths; and (4) **Spirochaetae,** or spiral filaments. Some of these organisms were motile by virtue of flagella (motion seeming greatest in the presence of oxygen), while others were immotile. When stationary, some tended to grow as zoöglœa (*Manual*: 11–12).

Cheyne investigated bacterial toxicity. Laboratory animals injected with putrid fluid developed fever, became incontinent, and collapsed. A Danish pathologist, Peter Ludwig Panum (1820–1885), with whose work Cheyne was familiar, had tried to analyze the composition of putrid or septic poison (Bulloch, *History*: 131). A number of other researchers had, by this time, discovered that putrid specimens such as these contained "congeries of toxic bodies" (Bulloch, *History*: 134). The surgeon, Ernst von Bergmann (1836–1907), for example, in 1872 had isolated the poison responsible for bacterial toxicity; he called this crystalline substance *sepsin* (Bulloch, *History*: 133; Zimmermann: 552–560). Putrid fluids, Cheyne knew, were produced by bacteria and poisoned an animal's circulation (*Manual*: 15). Pathogenic prokaryotes, biologists point out, produce exotoxins (secreted poisons) and endotoxins (released from disintegrating cell walls) (Campbell, Reece, et al.: 571–572; Peterson, "Bacterial Pathogenesis"). Cheyne was certain that antiseptic treatment minimized the risk of poisoning; and, in septic wounds, "traumatic fever" erupted if toxins entered the blood (*Manual*: 15).

Returning to Lister's idea that trauma rather than infection induced pustulation, Cheyne moderately revised his mentor's doctrine, as had Ogston earlier. In 1885, Cheyne included bacteria in the process, not causally but coextensively: thus, the "effect of tension is … greatly increased if, at the same time, micro-organisms develop in the retained fluids" (*Manual*: 17). But, through experimentation, the sequence of events leading to abscess formation was becoming clearer. He recalled Koch's account of a micrococci injection in rabbits producing "masses of micrococci" that, in every instance, preceded inflammation and the production of pus (*Manual*: 18–19). Discoveries in immunology would gradually reveal that inflammation and swelling at an infection site removed microbes, toxins, and foreign material, preventing their spread, and initiating the repair of tissue (Tortora and Derrickson: 433, 439, 562). Cheyne's perspective in 1885, in the early days of immunology, was that wound inflammation and pus had four causes: (1) decomposition of the discharge from the wound; (2) the application of irritating antiseptics to it; (3) nerve compression from the accumulating discharge; and (4) "the growth of micro-organisms in the tissues at the surface of the wound" (*Manual*: 17). Contemporary research on bacterial pathogenesis led Cheyne, over time, to abandon causes (1) to (3) in favor of (4). Much was being learned about pathogenic bacteria in laboratory and clinic. The interest in *Streptococcus* and *Staphylococcus*, how they were related to each other, grew in the living tissues, and induced suppuration, made Cheyne a serious student and practitioner of medical bacteriology (*Manual*: 17).

To determine if related organisms caused similar tissue damage, he reviewed the latest literature (*Manual*: 18). Aligned with Koch, whose "Investigations" Cheyne had translated in 1880, Fedor Krause (1857–1937), in 1883, had demonstrated that acute necrosis in osteomyelitis was linked to orange-yellow colonies of micrococcus that grew on gelatinized meat infusion or on potatoes (*Manual*: 20; Van Arsdale: 221; Krause: 152–155). In cases of erysipelas, the destructive potential of some micrococcal species was evident. An acute febrile disorder, characterized by localized swelling and inflammation, erysipelas would be directly ascribed to hemolytic *Streptococci* (20, 22). The German surgeon, Friedrich Fehleisen, using Koch's "Investigation" to explore the etiology of human erysipelas, was another major contributor to the bacterial-pathology discussion at Landmark III. His "Die Ætiology des Erysipels [The Etiology of Erysipelas]" (1883) was the first detailed account of the *Streptococcus* strain responsible for this inflammatory skin disease (Bulloch, *History*: 365; Foster, *Medical Bacteriology*: 65). Cheyne recounted Fehleisen's successful cultivation of the micro-organism on gelatinized meat infusions and on solidified blood serum, how he biopsied tissue from the red margins of the lesion, embedded little bits in cultivating material, grew colonies for six days, reinoculated them into fresh tubes, cloned a "pure cultivation," and injected the material into rabbit ears, inciting an infection that, in 36 to 48 hours, spread throughout the animal's body (20–21). The most important histological finding was that sections of the spreading margin from the rabbit "showed the same appearances as the sections of the skin in man," micro-organisms having infiltrated the lymphatic systems in both cases (21). To continue the process, Fehleisen is said to have inoculated a human being with the cultivated micrococci, producing erysipelas (21). Along with having demonstrated the transmissibility of the disease through inoculation across species barriers, he also observed "that a patient was protected for a short time … from a fresh attack of erysipelas" (21). "The Etiology of Erysipelas," Cheyne realized, was a crucial paper that had to be made widely available to the English-speaking world. Leslie Ogilvie translated it from German into English for inclusion in Cheyne's collection, *Recent Researches* (1886). The translation and re-publication of Fehleisen's paper was another instance of how, as editor, translator, and liaison between Britain and the continent, Cheyne exercised multiple roles in the profession.

Cheyne also discussed in detail the importance of Koch's 1878 induction of pyemia in rabbits. The appearance of purulent blood-poisoning showed that bacteriology and pathology were intimately connected and that knowledge of both was essential to the study of disease ("Investigations": 47–53; *Manual*: 23). Koch had understood how experimentation linked micro-organisms to disease. The cocci that he had introduced into rabbits, for example, were minute, grew in colonies, and adhered to red blood corpuscles ("Investigations": 52). Cheyne confirmed these phenomena experimentally, implicating micro-organisms in the formation of clots and emboli (*Manual*: 23). Although, in 1885, the etiology of pyemia in humans was not fully understood, Cheyne whose bacteriological acumen had sharpened considerably since 1879 was able to detect types of micrococci in human blood, in secondary abscesses, and in internal organs (23). Micrococci of some kind were also blamed for septicemia, a condition in which virulent micro-organisms from a localized infection invade and introduce "ptomaïnes" into the bloodstream (23).

In 1925, Cheyne reprised the 1877–1879 laboratory analyses of micrococci. He claimed, once again, not to have seen them because of substandard equipment and inferior culturing methods, and that was one reason why he had hastily pronounced them harmless (*Lister*: 81). To emend this "wrong deduction," by 1925, he was able to link these micro-organisms to their effects; thus, he could state unequivocally that *Streptococci* and *Staphylococci* were responsible for septic diseases; that bacilli caused gangrene and tetanus; and that saprophytes, if introduced into a wound, could set up a destructive process. Rather than being harmless concomitants, scavenging cocci could migrate into a wound through a drainage tube or from an old dressing and impede healing (80). To preclude this complication, operators were obliged to disinfect the skin before and after an operation and at regular dressing changes (82).

In 1891 William H. Welch identified a relatively innocuous micrococcus, *Staphylococcus epidermidis albus*, in what was thought to have been an aseptic wound. After learning of Welch's discovery, Cheyne raised the possibility that, in 1879, the surviving micrococci he had found might have been members of this species; thus, by implication, the benign-concomitant hypothesis might have had some validity after all (*Lister*: 81–82n.). In 1891 Welch identified *epidermidis albus* as a skin scavenger responsible for stitch abscesses. Lying "both superficially and also deeper than can be reached by present methods of disinfection of the skin," *epidermidis albus* was hard to eradicate. That, along with acid concentration, could account for its survivability in aseptic wounds. For Welch, *epidermidis albus* was a ubiquitous pest, inhibiting wound healing, especially when drainage tubes were inserted; and it caused stitch abscesses, even in wounds that had been treated antiseptically or rendered aseptic ("The Infection of Wounds": 415). Cheyne, therefore, had offered two possibilities to account for the 22 percent remnant: if these were indeed a strain of *epidermidis*, perhaps the carbolic acid could not penetrate the wound deeply enough to reach them? The more likely possibility, demonstrated experimentally in 1879, was that the acid was too weak.

5

1884–1889
A. J. F. Rosenbach—Landmark IV

From 1884 to 1889, Cheyne focused on the relation between bacteriology and surgery. His bacteriological activities, by this time, had greatly diversified. Under the auspices of The New Sydenham Society, he translated three major German-language works on microbes and infectious diseases. We have already looked at his fluent rendering of Koch's 1880 "Investigation into the Etiology of Traumatic Infective Diseases," which directly benefited fellow researchers, notably Alexander Ogston. Equally influential in the history of bacteriology were Cheyne's New Sydenham translations of Carl Flügge's encyclopedic 1883 *Micro-Organisms: With Special Reference to the Etiology of Infective Diseases* [English translation: 1886]), and of Anton Julius Friedrich Rosenbach's 1884 seminal paper "Micro-Organisms in Human Traumatic Infective Diseases." In 1889, Cheyne rightly ascribed historical importance to Rosenbach whose work is representative of Landmark IV.

Cheyne's editorial efforts made the clinical and experimental bacteriology of Koch, Fehleisen, Passet, Flügge, Rosenbach, and others, accessible to the English-speaking world. His interest as translator and editor, was to widen "the range of practical works that investigators could use … in teaching and demonstration" (Coutts, "Illustrating Microorganisms": 6). Along with translated monographs, Cheyne edited an essay collection, including Rosenbach's important study, to provide the international medical community with a textual resource through which developments in medical bacteriology, from 1876 to 1886, could be studied. Evidence in the literature had confirmed Cheyne's belief that certain bacteria were "causally related" to specific diseases (Editor's Preface, *Recent Researches*: vii).

While Assistant Surgeon at King's College Hospital, Cheyne translated, from German to English, Rosenbach's 1884 "Micro-Organisms in Human Traumatic

Infective Diseases," and he included it in the 1886 collection, *Recent Researches on Micro-Organisms in Relation to Suppuration and Septic Diseases.* This invaluable collection, which he edited, features a team of 11 German- and French-to-English translators (including Cheyne's Rosenbach translation). The book's 13 sections cover the works of Koch on traumatic infectious diseases, tuberculosis, and cholera; of Gustav Wölffhugel (1854–1899), Georg Gaffky (1850–1918), and Friedrich Löeffler on disinfection; of Gaffky on enteric fever; of Fehleisen on the etiology of erysipelas; of Albert Neisser (1855–1916), André-Victor Cornil (1837–1908), and E. Suchard on leprosy; of Carl Friedländer (1847–1887) on micrococci of pneumonia; of Löeffler and Johann Wilhelm Schütz (1839–1920) on glanders; of Rosenbach and Karl Garré (1857–1928) on traumatic infective diseases and on acute purulent inflammations; of Löeffler on diphtheria; of James Israel (1848–1926) on actinomycosis; of Bernhard Fischer (1852–1915) and Bernhard Proskauer (1851–1915) on disinfection; and of Pasteur and others on the attenuation of viruses and on vaccination (*Recent Researches*: xv-xvi).

The focus of this chapter, therefore, will be the bacteriology of Landmark IV: how Rosenbach, Flügge, and Cheyne, and others characterized four micrococcal species: *Staphylococcus pyogenes aureus, Staphylococcus pyogenes albus, Micrococcus pyogenes tenuis, Streptococcus pyogenes,* along with the saprophytes (organisms that live on decaying organic matter) (Miller and Levine: 537, 1088).

1880: Rosenbach's Microbes

Alexander Ogston had narrowed his attention to micrococci as a primary cause of suppuration and, in "On Abscesses" (1880), made a lasting impression on medical science when he described "some spherical bacteria in groups," the masses of which "looked like bunches of grapes" (123). The image has endured. From the conjunction of the Greek words *staphylē* (bunch of grapes) and *kókkos* (berry) emerged the neologism, *Staphylococcus.* In 1884, on the basis of pigmentation, Rosenbach subdivided these bacteria into two species: *Staphylococcus aureus* (from the Latin *aurum,* gold) and *Staphylococcus albus* (from the Latin *albus,* white) (later renamed *Staphylococcus epidermidis*) (G. Licitra). J. Passet, in 1885, further subdivided Ogston's *Staphylococcus* into three species (Bulloch, *History*: 131): *Staphylococcus pyogenes aureus, Staphylococcus pyogenes citreus,* and *Staphylococcus cereus albus,* respectively (Cheyne, *Recent Researches*: 403n.). The hygienist Carl Flügge had actually begun systematizing the Genus in 1883. The German language and first edition of his book had appeared in that year, antedating the papers of Rosenbach (1884) and of Passet (1885).

Anton J. F. Rosenbach, a German surgeon and bacteriologist whom Cheyne considered representative of this period of discovery, was renowned for his work on tetanus. He had something in common with Cheyne: early in his career, he too had lacked a proper laboratory and performed experiments under makeshift conditions, not in the backroom of an infirmary, but in his own home (Bulloch, *History*: 393). In his early days at Göttingen, Rosenbach might have had access to sophisticated microscopy and staining techniques.

Before exploring bacterial physiology and pathogenesis, Rosenbach had to establish the morphology of *Staphylococcus pyogenes aureus*, a spherical, orange-yellow, pyogenic cell that, in the aggregate, resembled bunches of berries. Sowing pus cells on solid blood-serum, at a temperature of 30–37 degrees Centigrade for 24 hours, revealed to Rosenbach an opaque line of deepening, orange-yellow oil that widened into round masses, three to four millimeters in diameter (403–404). The colony dissolved gelatinized meat, darkened, and then sank to the bottom of the dish. Exposed to the air, it dried up, lost color, and was difficult to use in experiments, although not completely devitalized. In the absence of air, it lived for some time. Rosenbach observed that the younger cocci were smaller than the older ones (404).

Although fragile *in vitro*, these organisms when injected into rabbits and dogs proved very harmful (404). A mixture of five grams of liquefied culture destroyed rabbits, and dogs suffered with abscesses. These virulent organisms, as their metabolism suggested, were not saprophytes. Rosenbach cultivated them successfully at body temperature, using egg albumen and boiled beef and in vacuum flasks, along with organisms that seemed to have anaërobic properties; their partial hydrolysis of proteins left large quantities of peptone (404). From this increasing amount of physiological data, the investigation could move towards etiology.

The second *Staphylococcus* species that Rosenbach, Passet, and Flügge described was *pyogenes albus*. Rosenbach stated that in every morphological respect, except for color (Latin: *albus*/white), it was indistinguishable from *Staphylococcus pyogenes aureus* (405). The pathophysiology of *albus* was also similar to that which was associated with *aureus*. Rosenbach extracted pus from a human knee-joint infected with *albus* and made a pure culture from it, using blood-serum medium. Mixing the colony with water, he transferred it *in vitro* to a fourth generation. He then injected the solution into the knee-joints of two rabbits, causing swelling in thighs and knees, emaciation, and death. Rosenbach also injected the micro-organism into a calf in order to learn about the structure and content of abscesses (405). The strategy was to find a way to establish causality between *Staphylococcus pyogenes albus* and a particular disease.

Evidence defining the organism and its etiology continued to accumulate. When a rabbit inoculated with the organism developed an infection, Rosenbach noted that the exudate smelled like osteomyelitic pus. On the basis of the odor, he suspected that the pyogenic organisms had caused the soft-tissue disease in the knee, as well as bone infection (406). Cultivations were prepared from other lesions in the rabbit, and the organism was detected at various anatomical locations. The results were significant: *Staphylococcus pyogenes albus* had infiltrated the animal's body. Rosenbach had obtained it directly from ventricular blood, the abdomen, and the knee. The pathological changes he subsequently observed in animal models supported the hypothesis that *Staphylococcus pyogenes albus* caused pyogenic disease. The abdominal specimens showed, in addition, that peripheral infections could spread inwardly, *albus* having the ability to enter and form abscesses along the lymphatic system. Blood cultivations, therefore, indicated that the organism had spread (406). Because *Staphylococcus pyogenes albus* was biologically akin to *Staphylococcus pyogenes aureus*, Rosenbach suspected that many other *Staphylococcus* species were also pathogenic.

The third organism that Rosenbach investigated was the elusive *Micrococcus pyogenes tenuis* (406–407). He invented this Greco-Latin trinomial to convey its distinguishing features, as a spherical (*Micrococcus*), pus-producing (*pyogenes*), and fragile (*tenuis*) organism. Solitary and elusive, *Micrococcus pyogenes tenuis* formed barely-visible colonies and differed in certain other respects from both *Staphylococcus aureus* and *Staphylococcus albus*. *Micrococcus tenuis* massed in small, transparent groups when sown on an agar line, and it grew into transparent layers (406–407). Under the microscope, *Micrococcus tenuis* was somewhat larger than *Staphylococcus* species and had darkened ends with lighter material between the poles; but no animal experimentation had yet been conducted using this species (407). Flügge, in the later edition of his book, included Rosenbach's findings and augmented the latter's description slightly: in only 10 percent of cases was Rosenbach successful in detecting *Micrococcus pyogenes tenuis* in the pus of unopened abscesses (189). In his 1886 Classification System, Flügge would most likely have assigned *Micrococcus pyogenes tenuis* to the heterogeneous division, **a. Cells spherical**, and to its subclass of isolated or of chain-forming cells living in "amorphous slimy families" (Buchanan: 33).

Billroth, in 1868, had invented the term *Streptococcus* to characterize the chain-forming micro-organisms that he had discovered (Gaw: 129; L. G. Wilson: 403). By 1884, three species would be identified as being active in traumatic infective diseases, and these constituted the Genus. Koch, in 1878, had shown that one form caused progressive necrosis in murine tissues. As we have seen in chapter 1, Pasteur's May 1879 findings on chain-forming micrococci implicated them in the formation of abscesses (Cheyne's first paper was published on 6 May) ("On the Extension of the Germ Theory": 118–130; Wilson: 410–411). In chapter 4, I surveyed Ogston's most important observations on both *Streptococcus* and *Staphylococcus*, from 1880 to 1882, with respect to acute abscesses. And Friedrich Fehleisen, in 1882, while in a surgical clinic at Würtzburg, also studied *Streptococci*. He published his findings in the 1883 essay "The Etiology of Erysipelas" which, as noted, had been translated by Leslie Ogilvie, and published in Cheyne's 1886 *Recent Researches* (L. G. Wilson: 411–412). In this text, which Cheyne knew well, Fehleisen reported on the examinations of 13 hospital cases, two of which had ended fatally ("The Etiology": 261–286). Skin biopsies from these patients showed that lymph-vessels, subcutaneous cellular tissue, and superficial layers of the corium (mucous membrane corresponding to the dermis) were filled with chain-forming micrococci that had not yet entered the bloodstream, but the biopsies confirmed that *Streptococci* could spread (267). After testing the effects of pure *Streptococci* on rabbits, Fehleisen then took the intuitive leap of experimenting on patients with intractable cancer and other diseases ("The Etiology": 272–281). Physicians had been using inoculum from erysipelas against these diseases, its use as a therapeutic modality dating back to the seventeenth century (272). In some cases, as the erysipelas spread, tumor mass increased in size, as well; but, after the patient recovered from the induced infection, the tumor degenerated. Because the *Streptococcus* infection was a danger to the patient, secondary inoculations were not done (L. G. Wilson: 412–413). In 1884, Cheyne referred to Fehleisen's pioneering work in "Biological Laboratory" (18–19). The American surgeon, William B. Coley (1862–1936), in 1893, published his experimental results on the use of erysipelatous inoculations in cancer treatment ("Treatment of Malignant Tumours": 487–541; E. F. McCarthy, "The Toxins").

Rosenbach also worked to differentiate between coccal species so as to link them to diseases, and to classify them with consistent terminology. He knew of two Streptococcal species: one of them was Fehleisen's *Streptococcus erysipelatous*; and the other was his own discovery, *Streptococcus pyogenes aureus* (407). Nomenclatures such as these encoded the bacterium's morphology, physiology, pigmentation, mode of colonization, and pathological effects on human beings. One example of this sort of codification would be Fehleisen's *Streptococcus erysipelatous*: in the nomenclature, four Greek morphemes convey the images of a twisted chain (*Strepto-*), of spheroidal cells (*-coccus/cocci*) and of reddened (*erythos/erythri-*) skin (*pelas*) ("*Streptococcus*," *Webster's*: 159, 283,

854). In like fashion, the species name, *Staphylococcus pyogenes aureus,* was a Greco-Latin construct encoding form, function, and host pathology: the colony formed a cluster of tiny spheroidal organisms (*Staphyle-* + *coccus*), was golden (Latin: *aureus*), and induced pus (Greek: *pyo/pyon-* + *-genic* [Latin: *generatus*]) ("*Staphylococcus,*" *Webster's*: 854, 696, 348, respectively).

As these examples suggest, bacteriologists such a Fehleisen and Rosenbach constructed species epithets representing the sum of microbial characteristics. Rosenbach's organisms, once in growth media and thanks to new technology, were easily identifiable: *Streptococcus pyogenes'* cells were translucent white dots, no bigger than grains of sand; they scarcely grew on solidified blood serum; at a temperature of 35–37 degrees Centigrade (normal human body temperature), however, they grew rapidly; and sowing them in lines on nutrient media produced an unbroken streak of growth, with a tendency to form small centers. In time, the organisms turned brown (407); and the borders of the colony were spotted with heaps of cocci. Overall, the colony grew over a two-to-three-week period, reaching a maximum breadth of two to three millimeters (408). In addition to etymological constructs, Rosenbach used botanical imagery to distinguish one entity from another: thus, the *Streptococcus erysipelatous* culture had a dendritic, fern-leaf appearance, while the *Streptococcus pyogenes* culture resembled an acacia leaf (408).

In regard to pathology, Rosenbach made a number of perceptive claims. Pus from chronic abscesses, for example, furnished him with evidence to overturn Ogston's claim that micro-organisms, though abundantly present in acute abscesses, were not found in the chronic form. On the contrary, from evidence gathered as early as 1878, he intimated that chronic conditions such as inflammation of the bone were the result of infection traceable to the tubercle bacillus. To prove that pathogens were present in a chronic condition, investigators repeatedly inoculated animals with pus and consistently induced acute tuberculosis (Rosenbach: 410–411).

Rosenbach also discovered that both *Streptococcus erysipelas* and *Streptococcus pyogenes aureus* invaded animal tissue. Fehleisen, however, thought that neither was pyogenic. Rosenbach corrected him after acquiring firsthand evidence that *Streptococcus erysipelatous* insidiously grew in and subsisted on living tissues, was pyogenic, and broke down tissue. Furthermore, both species of *Streptococcus* acted in similar fashion (415). In terms of the different ways they affected healthy tissue, Rosenbach agreed with Ogston, suspecting that *Streptococcus pyogenes aureus* was associated with erysipelas, promoted pus accumulation beneath the skin, and even led to constitutional symptoms (415). Rosenbach observed subtler differences in the ways pathogens behaved. One was that *Staphylococcus* induced pus rapidly but did not penetrate tissues to the extent that *Streptococcus* did (416). He also tentatively connected *Staphylococcus pyogenes aureus* to osteomyelitis (416). Sepsis experiments showed that micro-organisms were also the source of "different kinds of noxious agents," affecting tissues even after the organisms were no longer visible, and constitutional symptoms were attributed to the absorption of toxins (419). Septic symptoms, Rosenbach concluded, were usually the consequence of invasive cocci, of bacilli, and of other organisms (420).

1886: Flügge's Inventory

Carl Flügge, in 1883, founded the first German hygienic institute in Göttingen, and was appointed professor at Breslau in 1885 and in 1887. With Koch, he edited a hygiene periodical, and, in 1886, authored a digest for the study of bacteriological systematics (Bulloch, *History*: 366). Flügge understood that classifying bacteria according to distinctive characters was not easy. One chief reason for this difficulty was their minuteness, the highest magnification available, even with oil-immersion systems, being incapable of guaranteeing a clear view of cell structures (169–170). Another shortcoming was that the successive stages of reproduction were difficult to map, since cells in various developmental stages were simultaneously and actively present in the visual field. To have a better understanding of known species and of unclassified ones, and to locate new organisms systematically, a more efficient method of identification was required. Further improvements in instrumentation and the advent of microphotography progressively met these challenges (Flügge:170). The combination of upgraded instrumentation, the differential methodology deriving from Cohn, and the assessment of bacteria in terms of morphology and physiology, furnished Flügge and his colleagues with an efficient approach to species identification (170).

Following Koch's model (with the exception of Division 4 below), Flügge preferred a conventional form-function assessment of bacteria (171). He proposed a fourfold design: **(1) Micrococcus; (2) Bacilli; (3) Spirilla;** and **(4) Fission Fungi with Variable Vegetative Forms**. Micrococcus, the focus of our attention here, was assigned to the category reserved for spherical or egg-shaped cells. These cells multiplied by fission, replicated without changing shape, and neither moved spontaneously nor formed endogenous spores; and he detected variations in cell diameters (171). The spherical form, he observed, was not distinctly marked during binary fission; two cells, in several instances, appeared to adhere to one another. During division, according to Flügge, round cells elongated into rods, constricting at the center; importantly, the coccal rods, while dividing, could be distinguished from bacillar rods. After cells divided, the resulting micrococci either remained isolated or aggregated (172). Flügge was able to distinguish between two kinds of zoöglœal formations. In one called, *Ascococcus*, cartilaginous masses were observed enclosing the colony; in a second, *Clathrocystis*, the gelatinous mass dissolved to a thin external layer surrounding a fluid-filled cavity (172). Forty-eight species of minute, round bacteria which Flügge assigned to three etiological subdivisions populated his 1883 taxonomy of the Genus **Micrococcus**: *Division I. Micrococci*: (A.) Micrococci Pathogenic in Man (14 species); (B.) Micrococci Pathogenic in lower animals (seven species); and (C.) saprophytic Micrococci (27 species).

Flügge's observations on *Staphylococcus pyogenes aureus* supplement those of Rosenbach. The former reproduced Passet's diametrical calculation for the organism as 0.87 micrometer. The cells grouped in twos, in fours, and in three-to-four-cell chains, as well as in irregular masses. Most importantly, they were Gram-positive when stained with iodine, with iodide of potassium solution, or with alcohol (Flügge: 183). He cultured these organisms on several kinds of media, seeking to identify "the products of their growth," and to ascertain if toxins were released from cells (184–185). Ludwig Brieger (1849–1919), whom Flügge cited, had already isolated an organic base from *Staphylococcus* that he had grown in meat infusion. A German physician and chemist, Brieger was an expert on ptomaïnes and on proteinaceous toxins (Bulloch, *History*: 355). The particular substance Brieger had isolated appeared to have been a product of the organism's growth but seemed to have no effect on animals (Flügge: 185). Flügge reported that subcutaneous injections of the bacteria found they were indeed pyogenic, seemed only to kill mice, and caused abscesses in other animals (185). Intraperitoneal and intravenous injections, on the other hand, killed laboratory animals of all species. Furthermore, the infection travelled, damaging the kidneys, joints, muscles, and bones (186).

Staphylococcus, according to the experiments of Rosenbach and of Passet, occurred frequently in human abscesses. Cheyne who had come a long way from the pathogenesis of 1879 noted that, in 1885, Passet had located four organisms in abscess pus cultures: *Staphylococcus pyogenes aureus*, *Staphylococcus pyogenes citreus*, *Staphylococcus cereus albus*, and *Staphylococcus cereus flavus*. An important discovery in the realm of public health was Passet's detection of *Staphylococcus pyogenes aureus* in household water and of *Staphylococcus pyogenes albus* in raw beef (Rosenbach: 403 and note). Investigators learned, furthermore, that suppuration seemed to be connected in some way to these organisms. *Staphylococcus pyogenes aureus*, for example, rapidly destroyed tissues, as pus built up and spread to the lymphatic system (Flügge: 187). This bacterial process made for a strong case that *Staphylococcus aureus* caused acute abscesses, empyema (pus in body cavities), acute osteomyelitis, and other serious conditions. Karl Garré (1857–1928), Swiss surgeon and bacteriologist, had recently discovered that *Staphylococcus*, cultivated from osteomyelitis pus, could also cause skin infection (Flügge:187; Bulloch, *History*: 368).

Flügge began his section on *Staphylococcus pyogenes albus* with a direct reference to Rosenbach's discovery of both *albus* and *aureus* in pus; and he reiterated the similarity between the two organisms, with pigment as the distinguishing trait (Flügge:187). Borrowing from Passet and Rosenbach, Flügge believed that in human beings *Staphylococcus albus* occurred more frequently than *aureus*, whereas the reverse was true for animals (188).

Overall, Flügge's *Streptococcal* inventory included five species, and his determinations were based on authoritative literature, including the papers of Ogston, of Rosenbach, of Krause, and of Passet. Flügge concentrated largely on physiological characteristics, particularly on the tendency of cocci to divide transversely, and to make looping chains and heaps of more than ten cells (a textual illustration, at 800x, depicts colonial variants within macrophages). These micro-organisms displayed color variations (Flügge: 189–190). When inoculated into mice and rabbits, *Streptococci* were moderately virulent; however, if the hosts were compromised by artificially-introduced toxins, the organisms proliferated and fatalities occurred rapidly (190). In human beings, *Streptococcus pyogenes aureus* was frequently found in pus (191). Fehleisen had discovered, through cultivations and human inoculations, that *Streptococcus erysipelas* (as he called it) was the causal agent of the disease (a finding Flügge commended). Fehleisen observed, moreover, that these colonies were morphologically similar to one another (Flügge: 191); their effects on animals varied from species to species; and, as previously noted, *Streptococcus pyogenes* shrank inoperable tumors (192–193).

Flügge helped to isolate *Streptococcus pyogenes malignus*, obtained from a leukemic spleen. Once cultured, these micro-organisms resembled other cocci and in animal experiments proved virulent. Mice inoculated subcutaneously with small amounts of the germ invariably succumbed in three to five days. The organism could be transmitted to other mice through small tissue samples or blood (193). Flügge was led to suspect that Krause's *Streptococcus pyogenes aureus* was identical to *malignus*; in this case, persistent laboratory work helped to dispel nomenclatural ambiguity (194). The fourth species, *Streptococcus articulorum*, was assigned to Löeffler who had detected it in diphtheria cases. When cultivated, large cocci in chains turned up that were virulent in animals (194–195). The fifth micro-organism, *Streptococcus septicus*, also proved to be harmful to experimental animals (195–196).

Although the five *Streptococcal* species mentioned above were microscopically and culturally similar to one another, each was associated with unique symptoms and pathology. Furthermore, animals that had developed immunity to one strain had not to others. From these data, Flügge deduced that "[*Streptococci*] of very different virulence occur in the various infective diseases of wounds in man." Moreover, *Streptococcal* symptoms varied according to the specific way the tissues were invaded, the physiological functions of the infected organs, and the characteristics of the organism (Flügge: 196–197). Scientists and physicians in this circle influenced the progress of bacterial systematics, and Cheyne understood how indispensable their work was to medical bacteriology.

1884–1889: Cheyne's Work at Landmark IV

The texts of 1884–1889, constituting Cheyne's contribution to Landmark IV, demonstrate his talent as an instructor in the history and practice of medical bacteriology. This section looks at the objectives and contents of the literature: "The Biological Laboratory," Part I of *Public Health Laboratory Work* (1884); the 1885 *Manual* of professional instruction on the antiseptic treatment of wounds; translations, editorial work and bacteriological studies of 1886; and the 1889 republication of his lectures on suppuration and septic disease. These texts were intended for a wide readership, for general readers, students, and professionals, the progress of medical bacteriology.

1884: "The Biological Laboratory"

As publicity agent and instructor, Cheyne supplemented his writings with demonstrations and exhibitions. His work in connection with the 1884 International Health Exhibition in London dealt exclusively with general bacteriology rather than with public health, which was primarily concerned with sanitation, environmental pollution, and the quality-control of consumer goods. In the domain of public health, bacteriology was narrowly focused on diseases arising from insanitation and contamination (e.g., typhoid, cholera, plague). In public health manuals, the study of micro-organisms was subordinate to environmental factors.

Cheyne had the opportunity to contribute to *Public Health Laboratory Work*, which was to function as a standard text for Medical Officers of Health (Worboys, *Spreading Germs*: 177n.). Rather than being a supplement to the public health digest and guide, Cheyne's assignment, "Biological Laboratory" (Part I of *Public Health*), belongs generically and substantively to the canon of bacteriology. In fact, it appears to have been a forerunner of the bacteriological textbook or manual, a type of writing that flourished in the late 1880s and 1890s. As the allied fields of public health and bacteriology developed, they diverged from one another as specializations: the public health books having a chemical emphasis; the bacteriological, a biological one. In fact, by the early twentieth century, as bacteriology matured as a discipline, manuals on the subject could no longer be used as addenda in public health books. Although bacteriological references were still included in the hygiene texts of the 1880s, with the rapidly increasing amount of information these writings pointed to the need for an independent literary genre: the bacteriology textbook.

Cheyne's contribution to *Public Health* and that of his co-authors, W. H. Corfield (1843–1903), Medical Officer of Health, and Charles E. Cassal (1858–1903), Hygiene Demonstrator and Public Analyst, manifested differing perspectives on microbiology and public health ("Cassal"). Corfield and Cassal, on the one hand, emphasized food and water contamination, polluted air, and consumer health over microbiology. A prolific author on sanitarian issues, publishing 11 books between 1871 and 1902, Corfield worked at the intersection of sanitary science and bacteriology. *The Etiology of Typhoid Fever* (1902), for example, is an exhaustive compilation of material on the disease. Corfield's object was to elucidate sanitary measures that would prevent epidemics, such as the one that ravaged British troops during the Boer War (1899–1902), accounting for 13,000 of the 21,000 British fatalities (Kohn: 45).

Public Health, a primer on the subject for visitors to the 1884 International Health Exhibition, became a prototype for later health texts, such as Henry R. Kenwood's (b. 1862) *Public Health Laboratory Work* (1893) that would run through many editions (Worboys, *Spreading Germs*: 177–178n.). More specifically, Corfield and Cassal concentrated on hygienic laboratory work, and they examined the chemical constituents of natural and of artificial materials (93). In their domain were environmental conditions (air quality, water, soil), and consumer goods. Specimens of air, water, and soil were routinely tested to find toxic substances, as were processed food, drinking water, and dairy products. Bacteriological water-analysis had not met up to expectations (Hamlin: 293). References to bacterial pathogens in the Corfield-Cassal part of the book were minimal. The implications of microbial contamination for the food chain and for water resources, notably with respect to cholera, typhoid, and tuberculosis, are outlined concisely, but the emphasis was on environmental factors.

That Corfield's and Cassal's *Public Health* (less Cheyne's Part I) served as a model for later manuals used by Medical Officers of Health, was corroborated in Henry R. Kenwood's Preface to the 1893 first edition of his *Public Health Laboratory Work*. A standard text, it would go through a number of early twentieth-century editions. Kenwood noted that, in content and design, the book was modeled on Corfield's syllabus for the Practical Hygiene course that he taught at University College, London. Kenwood's text, a concise resource, provided the Public Health examiner with practical knowledge.

Like Corfield and Cassal, Kenwood was preoccupied with environmental hazards. He therefore discussed the quality of drinking water, organic toxins, chlorine, toxic metals, phosphates, sulfates, solid residue, chemical sediment, and gases. Drinking water conditions were rated for purity. The type, quality, and constituents of the soil were evaluated, along with atmospheric concentrations of carbonic acid, toxic gases, and particulates. Coal gas and air quality were routinely monitored in industrial centers. Food was also chemically and microscopically tested for parasites, spoilage, and contamination of every sort.

In his public health editions, Kenwood included a variety of bacteriological sections, and it appears that, as editor, he had selected bacteriological information that he thought compatible with, or useful to, the public health professional. In the 1893 edition, he incorporated, as Part VI, Rubert Boyce's (1864–1911) "Methods Employed in Bacteriological Research." Boyce's "Method," a laboratory guidebook for academic use, is procedural. The author was concerned with procedures: Chapters I to IV (of Part VI) cover sterilization, cultivation, and, with a microbial rather than chemical emphasis, the assessment of air, water, and soil samples; and Chapters V to VI, probably of lesser interest to the sanitarian than to the medical microbiologist, defined general morphology and physiology, and catalogued infectious diseases of importance to public health.

It is interesting to follow Kenwood's editorial decisions and how he tried to maintain a balance between the two allied fields. In the second edition of *Public Health* (1896), he substituted for Boyce's how-to manual Christopher Childs' substantial essay interconnecting medical bacteriology and public health. Childs was an expert in both fields, and he might have been concerned that bacteriology was being marginalized. As a medical and public health doctor, he dealt with standard topics—water, air, soil, milk and food—but did so from the perspective of infectious disease. He began, like Boyce, with an academic exercise, guiding students through a six-step laboratory procedure to isolate microbes in air, water, soil, milk, and food. Childs instructed students in the cultivation of specimens, in how to record their findings over time, and in how to identify micro-organisms, guided by the authoritative writings of bacteriologists such as Crookshank (1858–1928), Sternberg, Eisenberg, Frankland and German Sims Woodhead (Kenwood [1896]: 426). As a physician specializing in public health, Childs looked closely at environmental pathogens; and his readers were especially trained in locating them. The micro-organisms responsible for typhoid and cholera, *E. coli* infection, and the tubercle bacilli in milk were each sought (424–444). Childs, it appears, sought to balance the sanitarian and microbiological interests.

In the fifth edition of Kenwood's *Public Health* (1911), the author took a further step in the direction of a book fully integrating public health and bacteriology. W. G. Savage's "Bacteriological Examinations" was included as Part VII. This 60-page study, though informative and thorough, proved extraneous to the needs of the public health professional whose overriding concern was with chemical analyses of environmental phenomena. Like Childs, Savage stressed the importance to public health of etiology: the causes and effects of bacterial pathology, involving cholera, typhoid, and tuberculosis contamination of milk. But, as bacteriology developed into an independent discipline, its informational content gradually exceeded the boundaries of the public health text; the latter had been concurrently expanding to accommodate new information in that discipline. Microbiology was outgrowing the textual parameters of sanitary science.

In the Preface to the 6th edition of *Public Health Laboratory Work* (1914), Kenwood restructured Childs' format to exclude all but occasional references to micro-organisms, and the bacteriological addendum was gone for good (Preface, *Public Health* [1914]: v). Since he hoped to produce "a handy laboratory guide to the *chemical* branch of public health laboratory work," including a section on laboratory bacteriology was deemed unnecessary: "the needs of the public health student in bacteriology are *now fully provided for by several excellent laboratory manuals*" (italics added); doubtless, he had the works of Crookshank, Sternberg, and others in mind. Unsurprisingly, the 7th edition of 1920 was narrowed parenthetically as: *Public Health Laboratory Work (Chemistry)*. In certain areas, sanitation and bacteriology could not be detached from one another: deadly pathogens that caused botulism, food poisoning, tuberculosis, and other maladies were therefore treated concisely.

Now that we have seen how the Corfield-Cassal portion of the 1884 *Public Health* was an antecedent of Kenwood's books, we can explore the idea that, in content and importance, Cheyne's "Biological Laboratory" was a forerunner of the modern bacteriology textbook. Even though Cheyne described equipment and technology in 1884, his emphasis was on biology and medicine, especially on the relation between pathogenic bacteria and living hosts. In that light, "Biological Laboratory," Part I of *Public Health*, paralleled Rubert Boyce's (1864–1911) "Methods Employed in Bacteriological Research," Part VI of the 1893 Kenwood text. Cheyne's "Biological Laboratory" and Boyce's "Methods," though they differed from each other in purpose, were primarily concerned with bacteriology, not with public health.

Cheyne's 1884 essay invites comparison to related material in the Kenwood editions. The former surveys morphology, physiology, ecology, technology, internationally-standardized methods of research, and instrumentation. Whereas W. G. Savage (1881–1917), the County of Somerset Health Officer, studied milk and food contamination and wrote an addendum to a later Kenwood edition, Cheyne was interested in conveying to a general audience the principles and practices of a new discipline (Savage, *Milk and Public Health* [1912]). He did this in writing and through an exhibit of his own making, the details of which were reported in *The British Medical Journal* of 5 July 1884. In the interest of public education, Cheyne had set up a replica of the typical biological laboratory, complete with apparatuses and culture samples. According to a reporter, the aim was to show the public how growth media are made and studied, that bacterial colonies could be identified, that not all bacteria were harmful, and that these cells could be seen microscopically. Cheyne's exhibit was full of hardware: incubators, a disinfecting-chamber, and all sorts of laboratory appliances. The exhibit invited the public into an unseen world, "one of the most interesting portions of the present Exhibition" ("The Biological Laboratory": 27). His project and instruction were considered successful (Coutts, "Illustrating Microorganisms": 5).

With respect to the intended audience, Cheyne's "Biological Laboratory" chapter differs from Boyce's manual. The former was designed to inform a wide audience of specialists and laypersons and to situate laboratory method within the historical context of scientific discovery, while Boyce's "Method" of 1893 was exclusively intended for academic use as a laboratory guidebook. In regard to purpose and audience, Cheyne's mode of discourse contrasts with that of his 1893 counterpart. Basic information for a general audience, in Cheyne's text, was conveyed in a clear prose style. Boyce, on the other hand, pragmatically instructed students and a specialized audience of workers in the technical aspects of microbiology. Each mode of discourse served its respective purpose: Cheyne's, to introduce readers to a new branch of science and to minimize the use of technical language for general readers; Boyce's, to guide advanced students in the use of a new technology. Nevertheless, these texts, as educational resources, complemented each other.

As a survey or comprehensive introduction, Cheyne's "Biological Laboratory" was a lineal precursor, not of public health manuals, but of later works, such as George Miller Sternberg's 877-page *A Manual of Bacteriology* (1893), Arthur Bower Griffiths' *A Manual of Bacteriology* (1893), and Edgar March Crookshank's *Text-Book of Bacteriology* (1896), an expanded version of his 1886/1890 *Manual of Bacteriology*.

Cheyne's *Manual of the Antiseptic Treatment of Wounds for Students and*

Practitioners, which I touched upon in the previous chapter in terms of systematics, pathology, and toxicity, represents his contribution to the new genre. A precursor of larger compendia, the *Manual* consolidates knowledge relevant to all branches of medical bacteriology. Designed for general readers, students, and practitioners, the book covers an array of topics related to medical bacteriology, notably taxonomy (11–14), methods of cultivation (13), pathogenesis (14–24), aseptic versus antiseptic wound treatment (25, 44, 119–120, 125–128), and innate immunity (141–144).

1885: The *Manual*

In the *Manual*, Cheyne was especially interested in presenting the rationale, methods, and modifications of aseptic and of antiseptic surgery. He maintained that any treatment claiming to prevent "the causes of wound diseases" was by definition antiseptic. The complete exclusion of organisms from wounds was the aim of the aseptic method (33). The antiseptic methods denoted "the addition of various antiseptics to the discharge, either in the wound, or after it flows out," rendering it "an unfit soil for the development of organisms" (*Manual*: 33–34, 35). For greater clarity, Cheyne redefined the concepts simply in 1925: antiseptics that kill micro-organisms, preventing bacteria and their exotoxins from poisoning the bloodstream, were the surest means to the aseptic end (Lister: 113–114). To achieve a state where bacteria cannot damage tissue or poison the bloodstream, one turned to chemical regimens: carbolic acid, boracic acid and ointment, salicylic acid and cream, chloride of zinc, corrosive sublimate, sulfuric acid, iodoform, and variants of bichoride of mercury (*Manual*: 25–29).

Cheyne even outlined an aseptic operation through its various phases. To dispel misconceptions about antiseptic surgery, he conducts the reader through important preparatory stages, with disinfection a priority throughout. He describes disinfection of the skin using 1–20 carbolic and of the surgeon's and dressers' hands using 1–40 carbolic (in 1889, he would recommend 1–2000 sublimate solution to follow, with a nail brush [*Suppuration*: 92]); the antiseptic preparation of instruments (*Manual*: 49) and of ligatures (49, 52); sterile drainage methods (56–63); ways of fixing and changing dressings (70–73); and how to prepare and apply sterilized gauze (43, 47, 69) and bandages treated with carbolic acid (43, 67, 70–71). To stress the importance of controlling microbial contamination, he cautions the readers that, if an operator during the course of a procedure was so much as to reach for an instrument out of the carbolic mist continually being sprayed to keep the space germ free, he had to repurify his hands before touching the wound (50). These sprayers, either manually operated or steam-powered, were eventually discontinued as cumbersome, noxious, and unnecessary since the skin of the patient, not the air, was determined to be the source of harmful bacteria (53). Lister's excessive caution about maintaining the aseptic state of a wound and of using antiseptics judiciously if infection occurred stemmed from his experiences with hospital infections which, time and again, afflicted patients whose surgeries had otherwise gone well.

1886: The "Report" on Infection

The 31 July 1886 "Report," an important contribution to Landmark IV, was a preliminary account of recent experiments in pathogenesis, performed while Cheyne served as Assistant Surgeon at King's College Hospital and as Research Scholar of the British Medical Association. In this paper, he described extensive research, the design, scope, and perspicuity of which resembled aspects of Koch's 1878 "Investigations." Studying the occurrence of microorganisms in human disease and their relation to infection, Koch had asserted that human infections were parasitic in nature ("Investigations": 64); and that pathogenic bacteria differed from each other in form and in physiological characteristics and effects on hosts (65, 67).

Cheyne also formulated a law based on a concept derived from previous investigators: that blood infected with pathogenic bacteria increased in virulence when transmitted through successive animals (Koch, "Investigations": 70–71). His "Report" on infection presented a series of experiments of September–October 1885 to 16 April 1886. Over this seven-month period, he had tried to unravel the complexities of infection, introducing specific microbes through serial-inoculations of varying strengths into rabbits, guinea-pigs, and mice. In the first series of experiments (eight in total), September-October 1885 to 29 January 1886, he injected 28 rabbits with varying concentrations of the bacterium *Proteus Vulgaris*. From this experiment, he learned about parasite-host interactions. Rabbits, it turned out, were not very susceptible to *P. Vulgaris*. When Cheyne injected this bacterium along with its products, presumably its toxins as well, the cellular resistance of the rabbit was impaired, and the bacteria gained a foothold. In contrast, if a high concentration of the bacteria were introduced in certain anatomical locations, they grew for a time but were then attacked by the rabbit's immune system; however, traces of poisonous material were left ("Report on a Study of Certain of the Conditions of Infection" [31 July 1886]: 200). When the concentration of bacteria was high, the poisonous by-products destroyed tissue, enabling the micro-organisms to spread over a large area until layers of granulating cells gathered around the colony; the resulting abscesses barred further infiltration. The extent to which the bacteria spread appeared to depend upon three factors: (1) the number of cells; (2) the quantity of toxic material initially injected; and (3) the health of the animal. It was distinctly possible, Cheyne thought, that the large infusion of bacteria and their metabolic by-products were sufficient to kill the rabbit in a few hours. He concluded that an increased quantity of the toxic products allowed the bacteria to proliferate over a larger area and to produce a larger measure of toxin (200).

At this point in the research schedule, however, Cheyne had to take a two-month hiatus since he found himself without laboratory accommodations. He then accepted an invitation from Koch to work in his Berlin laboratory. The ensuing phase of the investigation was to ascertain how various pathogenic organisms behaved. Cheyne replicated Koch's 1878 experiments on traumatic infective diseases in a series of nine Berlin experiments: the first, from 25 to 29 March 1886, tested anthrax on guinea-pigs ("Report on a Study of Certain of the Conditions of Infection" [31 July 1886]: 200–201); the second, from 31 March to 2 April, Koch's bacillus of mouse septicemia; the third, from 1 to 20 April, the bacillus of chicken-cholera; the fourth, from 7 to 18 April, the bacillus of rabbit septicemia; the fifth, from 9 April to 17 June, *Staphylococcus pyogenes aureus* on rabbits and mice (204); the sixth, from 18–20 April, *Staphylococcus pyogenes albus* on rabbits (204); the seventh, from 12 to 26 April, *Sternberg's Micrococcus* on rabbits (204–205); the eighth, from 17 to 22 April, the tetanus bacillus on rabbits (205); and the ninth, from 7 to 28 April, *Micrococcus tetragenus* on guinea-pigs (205–206).

Cheyne drew five conclusions from the dosage investigations, three of which he elevated into laws. The first was that, the greater the predisposition of the host, the lower the quantity needed to infect it; and conversely, infection occurred if a large enough dosage overcame immunity ("Report on a Study of Certain of the Conditions of Infection" [31 July 1886]: 206). In animals moderately susceptible to a particular pathogen, the severity of an infection varied in direct proportion to the amount of bacteria inoculated into the animal model. This law, in turn, had three corollaries: small doses of bacteria were unproductive; at an intermediate stage, a local infection arose, its extent contingent on the bacterial dosage; and, at the third stage, a large dose caused death. A third law claimed that, at a certain point, the length of the incubation-period varied inversely to the amount of injected bacteria; a fourth point (not a law), exhibited in several experiments, was that a small dosage provided some immunity to the fatal effect of a larger inoculum. A fifth point, deduced from Cheyne's investigations, was that the effect of the bacteria depended, in some cases, on the anatomical site of injection (206).

1889: *Suppuration and Septic Diseases*

The synoptic commentary on *Streptococcus pyogenes aureus* in Cheyne's 1889 *Suppuration* monograph was factually indebted to Rosenbach, Passet, and Flügge. These cocci congregate in bunches (*Suppuration and Septic Diseases*: 4). A comprehensive physiological description of this bacterium, in Cheyne's view, had to take into account optimal temperature for growth, color spectrum as index to growth stages (white, to light yellow, to bright orange-yellow), effects on media (gelatin, agar, and potato), metabolic by-products (peptonized albumen and identifying odors), and duration of vitality (4–5). The investigation of pathogenesis was also extensive. The process of cultivation led to serial transference. *In vivo* injections varied with experimental conditions, with dosage, and with the species of laboratory animal (5). Suppuration appeared rapidly in inoculated rabbits, as did organ abscesses when the microbe was injected intravenously. If the bacteria were introduced into injured bone, suppuration also resulted. Even if moderate amounts of inoculum were injected into knee-joints of rabbits, inflammation, pustulation, and death occurred (5). Major sources indicated that dogs survived the challenge despite suffering from abscesses and bone infections.

Cheyne included in the *Suppuration* monograph valuable data on morphology, physiology, and etiology, precisely what the medical readership would need to know when confronted with the pathogen clinically. *Streptococcus pyogenes aureus* had fortunately become easier to identify through advancements in laboratory technology. Surgeons now had an idea about its associated pathology and could test a variety of antiseptics on it. Cheyne's synopsis consolidated the most important bacteriological data for the practical use of medical students and surgeons.

Although Cheyne contributed fewer than 100 words to the early assessment of *Staphylococcus pyogenes albus*, he did more than merely translate *verbatim*; for example, he explained why *Staphylococcus pyogenes albus* was white rather than transparent or yellow. Whereas Rosenbach and Flügge thought that, for animals, *Staphylococcus pyogenes albus* was not as virulent as *aureus*, Cheyne disagreed: in his experience, *albus* was more virulent than *aureus* (*Suppuration*: 5). Further on in *Suppuration*, Cheyne acknowledged that differentiating one strain from another was labor intensive. Two microbes might look alike under the microscope, might grow in a similar manner in various culture media, and might affect certain vertebrate species in the same way, yet they might be completely different from each other (13). The only way to move the classificatory process forward would be to inoculate these organisms into human volunteers, an ethically proscribed practice (13). Learning how these organisms behaved remained a medical priority (14–15).

Cheyne linked bacteria to acute suppuration on the basis of several well-known factors. Pus in acute abscesses teemed with micro-organisms, and certain microbial species were associated with certain kinds of tissue inflammation. The coextension of *Streptococci* with widespread suppuration, and of *Staphylococci* with circumscribed abscesses, for example, warranted further study (*Suppuration*: 17). Human experimentation, he observed, could provide absolute proof of the causal connection between virulent micro-organisms and pyogenic diseases. This statement was referring to actual human induction experiments using material drawn from abscesses and impetigo (18). Pasteur, for instance, had conducted furuncle experiments in May 1879 ("On the Extension of the Germ Theory": 118–130). Investigators, such as Karl Garré, had massaged diluted mixtures of the pyogenic bacteria into human skin, and infections resulted (*Suppuration*: 20–21). Abscesses formed after the gynecologist Ernst von Bumm (1858–1925), Director of the *Frauenklinik* in the Charité Hospital in Berlin, auto-injected and then inoculated two volunteers subcutaneously with *Staphylococcus pyogenes aureus* (21–22). The evidence was consistent and compelling. Cheyne remarked that an anti-causation position was becoming untenable: even if organisms were not found in some cases, one could not be absolutely certain that pathogenic microbes were absent (24–25).

A related point of discussion was how inflamed aseptic tissue related to suppuration. Cheyne no longer accepted the Listerian belief that mechanical injuries devoid of open wounds produced pus (*Suppuration*: 25). Furthermore, experiments by independent observers had shown that in animal models, "if organisms are absent, suppuration does not follow" (26–27). Rather than to accept the claim that suppuration could occur in inflamed but aseptic lesion, Cheyne blamed human error; most likely, organisms were actually present in the pus but either were missed in an imperfect examination or had died out before the abscess had been incised (27). Aligned with Ogston, Cheyne stated: "I have never yet seen ... pus coming from the surface of a wound without finding micro-organisms in it" (29).

Cheyne characterized, and contributed to, the period he called Landmark IV, which was an abundantly inventive moment in the history of bacteriology. He educated the public with lectures and displays in 1884 at the International Health Exhibition; wrote a synoptic guide to the discipline in 1885; surveyed, catalogued, translated, and commented on the latest information in the field in 1886; and conducted original experiments, with upgraded methods and instrumentation at King's College, along the lines of Koch's 1878 investigations of traumatic infective disease.

6

1879–1895
Élie Metchnikoff—Landmark V

In the 1897 Address "On the Progress and Results of Pathological Work," Cheyne assigned Landmark V to immunology research and referred to concepts, on which Élie Metchnikoff and early immunologists had been working (587). In several contexts, Cheyne expressed his views on a theoretical debate between opposing interpretations of innate and adaptive immunity: that of the cellularists who maintained that phagocytic blood cells, digesters of bacteria and of other invasive organisms, constituted the body's primary defense; and that of the humoralists who contended "that only the soluble substances of the blood and other body fluids could immobilize and destroy invading pathogens" (Silverstein: 25). Rather than being mutually exclusive, these theories would prove to be reinforcing armaments in a biochemical system of natural defense. From 1879 to 1897, Cheyne subscribed to the binary hypothesis but thought that, in importance, humoral exceeded cellular (phagocytic) immunity. By the mid–1890s, humoral substances were finally recognized as being "cellular products," resolving the opposition between the two schools of thought (Bulloch, *History*: 272). A succession of discoveries revealed that the two theories described branches of a common defensive system involving white blood cells (WBCs) and antimicrobial substances, such as proteins, enzymes, and oxidants.

1862–1891: The Line of Research

Metchnikoff received Landmark status in Cheyne's historiography, chiefly on the basis of his 1884 paper "A Disease of Daphnia Caused by a Yeast: A Contribution to the Theory of Phagocytes." Neither the concept of phagocytosis nor the Greek neologism, it is interesting to observe, originated with Metchnikoff. Carl Claus, of Vienna (1835–1899) created the term by joining the Greek prefix øαγο to the suffix κγτος ["phagocytosis," *OED*. **II**: 2150]); the German cognate, *fresszellen*, means "cell-devourer" (Tauber: 138; *fresszellen*, *New Cassel's*: 165–166, 581–582). In 1882, while in Vienna, Metchnikoff had presented his ideas to Claus who then encouraged him to publish the 1883 paper "Investigation into the Intracellular Digestion of Invertebrate Animals" (J. M. Cavaillon). Metchnikoff's investigation began in Messina, Italy, in 1882; and the paper "Disease of Daphnia," was published in 1884.

The concept of phagocytosis belongs to a line of inquiry, begun in 1862 when the marine biologist Ernst Haeckel (1834–1919) had described the way molluscan leukocytes absorbed India-ink particles (Stossel: 13). T. Langhans (1870), B. V. Birch-Hirschfeld (1872), Sir William Osler (1875), and Robert Koch (1876) had also noticed this function in leukocytes and lymph sacs (Bulloch, *History*: 259; Ambrose, "The Osler Slide"). The pathologist Julius Cohnheim (1839–1884), in the 1877–1880 "Lectures on General Pathology," recorded that pus-corpuscles appearing in exudations grew, "by taking up whatever substances they find ready at hand"; and "the enlargement of individual corpuscles by the consumption of others, is but a special application of this principle" (357–358). Metchnikoff was, therefore, at the forefront of an exciting new field.

While in Messina, Italy, from 1882 to 1883, Metchnikoff had learned that certain mesodermic cells in starfish had digestive properties, and he wondered if these properties existed on the cellular level as an immunological reaction to pathogens (Bulloch, *History*: 259). His acclaimed research began with the water flea, *Daphnia magna*, to which he fed spores of the yeast, *Monospora bicuspidate*. As expected, ingested spores penetrated the water-flea's intestinal wall, budded from the membrane, and filled the body cavity with yeast cells. Reacting to "the intruder," blood corpuscles attached themselves to the parasite ("A Disease of Daphnia": 133). The amœboid blood corpuscles, Metchnikoff found, spread freely in the flea which lacks a circulatory system. As spores accumulated in the body cavity, innumerable blood corpuscles migrated to the area and an inflammatory reaction ensued (133). When Metchnikoff witnessed the cellular defenders ingest the fungal spores, he described the digestive process. Once spores were absorbed, they thickened, yellowed, and their outer membranes became jagged; finally, the spores swelled into spheres, turned brownish yellow, and eventually became granular. As the budding spores disintegrated, the blood corpuscles united into "a fine-grained, pale plasmodium," retaining their amœboid motility (135).

Metchnikoff deduced from these observations that, since secretions from the blood corpuscles had destroyed the spores, the corpuscles were actually a function of the cellular-defense mechanism (136). The transparent *Daphnia* model encouraged him to test the hypothesis on frogs and rabbits in order to learn more about "the battle of phagocytes against bacteria" (136). If this were a universal property, he suspected that this knowledge could illuminate the pathological processes of mammals (136, 137).

The mechanism of cellular immunity, he soon learned, was complex and the parasites not defenseless. In the *Daphnia* experiment, for example, blood corpuscles with ingested spores sometimes burst open, freeing absorbed parasites. This unexpected result, Metchnikoff ascribed to the spore's own chemical defense, dramatizing the Darwinian idea that infection and illness involved "a battle between two living beings—the fungus and the phagocytes" (136–137).

Further studies on phagocytosis in cases of anthrax and of other bacterial diseases were to follow. In the 1891 "Lecture on Phagocytosis and Immunity," read at the Institut Pasteur, Metchnikoff enumerated three subtypes of the leukocyte and stated that most macrophages originated from endothelial tissues (213). This was an influential discovery. Biologists now know that as microbes invade the body three kinds of WBCs respond through phagocytosis and blood products. The most numerous cellular defenders are components of the WBCs: the *neutrophils* (60–70 percent of all WBCS) destroy bacteria using enzymes and oxidants; in the second category are *lymphocytes*: T cells, B cells, and natural killer cells (20–25 percent of all WBCs) mediate antigen-antibody reactions, secrete proteins that neutralize toxins, agglutinate or adhere to bacteria, and precipitate antigens. The monocytes which comprise 3–8 percent of all WBCs were Metchnikoff's chief interest. After transforming into fixed or wandering macrophages, they ingest invaders. Neutrophils and lymphocytes, 80–90 percent of all WBCs, have the greatest bactericidal power, whereas monocytes that engulf invaders account for 10–20 percent of the immunological activity (Tortora and Derrickson: 51–52, 359, 363, and 366). Metchnikoff's cellular theory was therefore concerned exclusively with the study of the *Daphnia*'s amœboid equivalent to the human monocyte. Metchnikoff had recognized that phagocytes, the focus of his research, were one element of a unified system.

The crux of the debate was the prevailing misconception that wandering macrophages were the exclusive defenders and that Metchnikoff subscribed to this interpretation. On the contrary, both Metchnikoff and Cheyne recognized that phagocytes were part of a larger defensive system. The humoralist concept of immunity held that bodily defense relied predominantly on cells that produce antimicrobial proteins, enzymes, and oxidants, which are present in blood and lymph, and only secondarily on microbe-hunting macrophages. This allocation of activity was fundamentally correct. By the turn of the century, however, the blood chemistry of immunity would be sharply redefined: antimicrobial substances were secreted by WBCs, while phagocytes, either mobile or fixed, joined the fight as discrete cells. Metchnikoff, in the mid–1880s, reflected that his work had only begun to illuminate the complexities of natural immunity, and that the ingestion of invasive organisms was one aspect of bodily defense. *Bacillus anthracis*, for example, became attenuated in the blood of immune sheep without intracellular digestion. Thus, Metchnikoff had a sense of how bodily defenses other than that waged by phagocytes combated bacterial infection, and these other properties included blood-borne cells, inflammation, and fever.

1879–1895: Conceptual Unity

From 1879 to 1895, Cheyne whose commentary was concurrent with Metchnikoff's correctly surmised that the humoralist theory was largely responsible for bodily defense, and he made an attempt to reassess the system in terms of its binary components. As early as 1879, he had had clinical experience with the humoral phenomenon (Bulloch, *History*: 256–257). While experimenting on abscesses, for example, he recognized that in a healthy state, the blood and tissues had the inherent ability to destroy invasive organisms ("On the Relation of Organisms to Antiseptic Dressings": 573). Another point contemporary researchers had made was that phagocytosis, a factor in defense, was not constant in all vertebrates. Carl Christmas-Dirckinck-Holmfeld (1860–1915), whose research Cheyne consulted, had discovered that the inoculation of virulent organisms in mice and rabbits did not always stimulate a cellular reaction; neither pus nor phagocytosis, for example, was detectable in the rabbit-mice inoculations; however, if rats were used (juveniles rather than adults), purulence ensued. To explain the inconsistency, Holmfeld who was a humoralist was inclined to believe that "a chemical biological action," rather than cellular digestion, interrupted bacterial growth in some but not in all rodent species. He declared that micro-organisms, in these cases, succumbed to "some chemical material in the pus," rather than to phagocytosis (Cheyne, *Suppuration*: 38). Other humoralists whom Cheyne cited in 1879 were Andre-Victor Cornil (1837–1908), V. Wyssokowitsch (d. 1912), V. Baumgarten (1848–1928), and Hugo Ribbert (1855–1920) (36–39). These workers elucidated the interrelationship between the two systems. They demonstrated that WBCs did not ingest every pathogen. When invasive organisms were ingested, humoral proteins preceding the phagocytes had already attenuated the invaders; generally, phago-cytes played a role secondary to antimicrobial compounds (Silverstein: 33).

From 1879 to 1895, Cheyne acknowledged the importance of Metch-nikoff's discovery but considered phagocytosis to be subsidiary to the blood-and-tissue neutralization of pathogens. As early as 6 May 1879, Cheyne had made a number of important observations in this regard; one was that micro-organisms did not occur in the blood or tissues of the healthy body and that the blood was innately aseptic ("On the Relation of Organisms to Antiseptic Dressings": 571, 588–569). In *Antiseptic Surgery* (1882), he reiterated that blood and tissues are innately aseptic and capable of maintaining that state against invaders (252). He had little doubt that leukocytes play an important part in destroying micro-organisms and in limiting inflammation; nevertheless, he valued Metchnikoff's theory as only one aspect of a larger system; phago-cytosis was a real phenomenon (he writes in 1888) but "not always the whole process" (*Suppuration*: 36). Metchnikoff realized this, too, when he had shown that *Bacillus anthracis* became attenuated in the blood of sheep. The animals were rendered immune to the disease through vaccination: ovine resistance having occurred *without* phagocytic activity. According to Cheyne, Metchnikoff had already suspected that immunity was a multifarious reaction, involving both the cellular uptake of invasive organisms and foreign material, along with the biochemical inhibition and destruction of parasites (*Suppuration*: 37). Rec-ognizing the truth about both the humoral and cellular theories, Cheyne fused them conceptually. Experimentation had proven that the humoralist and cel-lularist hypotheses were neither contradictory nor exclusive of one another. Cheyne's immunological thought in the 1890s was in accord with, and adum-brated to a degree, the researches of Bordet (1894), of Wright and Douglas (1903), and of Ehrlich (1908), whose theories are surveyed below. Their efforts made for a coherent understanding of innate immunity.

The Belgian bacteriologist and 1919 Nobel Prize recipient Jules Bordet (1870–1961) had been appointed to the Institut Pasteur, in 1894, as *preparateur* in Metchnikoff's laboratory. The 1895 paper "Leucocytes and the Active Prop-erty of Serum from Vaccinated Animals" might have influenced Cheyne's 1895 observation, in *Treatment of Wounds*, that in some cases phagocytes were sub-ordinate in antimicrobial effect to the products of blood and tissue (144–145). Cholera research had reinforced the credibility of the humoral theory, directly connecting "the appearance of an antiseptic power in the serum [to] the appear-ance of immunity in the animal." In vaccinated animals, this bactericidal effect, called acquired immunity, became accentuated (145). An unvaccinated animal was not entirely defenseless. Its tissues, too, possessed an inherent property that destroyed "actively diverse types of vibrios"; but its action was not as powerful or as specific as that produced by serum vaccination (146). It is now known that Bordet's *complement system* comprised some 30 blood proteins that intensify the inflammatory response, enhanced phagocytosis by attacking harmful organisms, and neutralized extracellular pathogens (Tortora and Derrickson: 438). In Bordet's conceptualization of the immune response, the humoral catalyzed the cellular/phagocytic system.

Acknowledging that Metchnikoff's claims for phagocytosis were "more or less correct," Cheyne believed that innate and natural immunity depended for the most part on humoral activity, but that the cellular-phagocytic and the humoral were complementary, and that both stemmed from WBCs (*Treatment of Wounds*: 19). He correctly viewed the initial immune response to be biochemical: "the first action of the bacteria is a chemical one, and is due to the effect of anti-bacter[ial] substances which are present in the serum of the blood and in the fluids effused through the inflammatory process" (19). Antibacterial compounds of the blood and tissue biochemically attenuated harmful bacteria, thereby enhancing phagocytosis. Cheyne acknowledged, but did not overestimate, the power of non-phagocytic immunity to destroy saprophytic and, to a limited extent, pathogenic organisms (19).

In 1897, Cheyne recounted how the humoral-cellular dialectic had arisen. Both theories described bodily mechanisms of natural defense, aspects of innate and of acquired immunity. The problem was that the proponents of each theory thought that their understanding excluded, rather than complemented, that of the opposition. As further information was revealed, however, general conceptions of how blood components worked against invaders became progressively more specific. Cohnheim's earliest publications on corpuscular immigration convinced many, at the time, that the leukocyte was the exclusive defender of health—that is, until new discoveries revealed "that other cells of different origin must be taken into consideration" (Cheyne, "On the Progress": 588). Metchnikoff's theory inadvertently led many to assume that "all wandering cells were phagocytes" (588). This, too, was incorrect, as continuing research showed that cells differed greatly in their "phagocytic action"; a more appropriate word in this context would have been "anti-bacterial," for both cells and proteins destroy bacteria but in different ways.

The immunity dialectic also had a philosophical dimension. In "On the Progress and Results of Pathological Work" (4 September 1897), Cheyne perceived that the concept of interactive cellular-defense mechanisms could be taken to imply that immunity had a teleological or immanent mechanism, either innate or extra-biological. To describe "the protective arrangements of the body" in teleological terms would, in this view, have been a spurious account of how phagocytes, proteins, and other defensive components functioned.

Such a presupposition assumed that the cells possessed "an acting intelligence," a vitalistic propensity to cooperate in the body's defense. For the surgeon who faced infection on a daily basis, innate biological mechanisms resistant to parasitic invasion were purely natural phenomena (588).

We have already seen how the phagocytic theory emerged and when Metchnikoff became its chief proponent. In the light of scientific discoveries, made from 1890 to 1908, the humoral was gradually subsumed under the cellular (phagocytic) theory. The resulting unified theory held that the source of the immune response against parasites was by definition *cellular*; the defensive activity involved both scavenging cells and chemical proteins, such as globulin, that neutralized invasive micro-organisms and toxins.

From ca. 1890 to 1908, when Cheyne was defining his binary interpretation of immunity, the cellular and humoral theories converged. Hans Buchner, E. H. Hankin, J. Denys, J. Havet, and others had been developing the humoral theory with its emphasis on blood chemistry coincidentally with phagocytic research extending from Haeckel to Cohnheim, Claus, and Metchnikoff. The works of six noteworthy humoralists whose focus was the biochemistry of microbial substances converged in this period. The bacteriologist/immunologist Hans Buchner (1850–1902), of Munich and Liepzig, theorized in 1891 that antibacterial proteins called *alexines* existed in the blood serum (Bulloch, *History*: 271). Similarly, the British bacteriologist Ernest Hanbury Hankin (1865–1939) posited that defensive proteins, "ferment-like, albuminous bodies," occurred naturally or were stimulated through vaccination to neutralize microbes or toxins ("On Immunity": 340). And, in 1894, J. Denys and J. Havet proved that a dog's blood serum had bactericidal properties (Bulloch, *History*: 272; "Sur la part des globules blanc"). In 1903, Almroth Wright and Stewart Rankin Douglas (1871–1936) moved the discussion closer to theoretical coalescence, determining that humoral antibodies interacted with micro-organisms to make them susceptible to macrophages (Silverstein: 37). In the paper "An Experimental Investigation of the Rôle of the Blood Fluids in Connection with Phagocytosis," they conclusively proved that blood fluids modified bacteria, enhancing phagocytic activity. They defined this phenomenon as the "*opsonic* effect" (from the Latin *opsono*: "I cater for; I prepare victuals for"); and from this term derived the plural noun *opsonins*, "to designate the elements in the blood fluids which produce this effect" (366).

The Wright-Douglas experiment proved that cells and non-cellular elements worked in unison to destroy pathogens, their combined activity having "bactericidal" effects. Each antimicrobial activity exerted a digestive or destructive power "on bacteria brought into contact with them" (369). During the opsonic process, through which blood elements and phagocytes work together, complement proteins adhere to microbial surfaces to promote phagocytosis (Tortora and Derrickson: 438). From an immunological viewpoint, then, Wright and Douglas understood "a condition of immunity" to mean the effect produced by three interactive phenomena: "the phagocytic activity of white blood cells, the agglutinating and bactericidal activity of blood fluids, and the opsonic effect" (Wright and Douglas: 369).

Coiner of the term *antibody* (*AntiKörper*), Paul Ehrlich (1854–1915) in the first of his 1908 Harben Lectures described the biological unity of the cellular and the humoral views of immunity. Proteins secreted by plasma cells, called *antibodies*, bound and neutralized bacteria. Some caused bacteria to clump together. Others activated Bordet's *complement*, "a group of at least twenty inactive proteins in plasma" responsible for (1) destroying pathogens, along with (2) causing inflammation, (3) *opsonization* (proteins binding to the microbial surface), and (4) *chemotaxis* (chemically attracting phagocytes to an infection site) (Tortora and Derrickson: 438, 440, 442, and 450).

These discoveries supported Ehrlich's belief that the opposition of the cellular to the humoral theory was an artificial dynamic. Both phagocytosis and the activity of antibodies were cellular in nature and origin. The action of antibodies, as Ehrlich points out, "by no means excludes phagocytosis; destruction of the bacteria outside the cells and their assimilation by the phagocytes are processes which may take place alongside each other, and, by their simultaneous action, increase the protective power" (339). Specific antibody reactions are, in fact, the basis of "phagocytosis," a phenomenon Metchnikoff perceptively studied (339). Opsonins and cytotropic substances make bacteria susceptible to phagocytic attack, exemplifying the interdependence of humoral and cellular processes, and redefining the humoral as a cellular process (339). Like Cheyne, Ehrlich recognized that narrow arguments restricted scientific inquiry: "One cannot, therefore, go to work in a one-sided way when analyzing and judging the various forms of phenomena, but must carefully consider together all the factors in question" (340).

7

1882–1920
Tuberculosis and Anti-Vivisection

From 1879 to 1895, as Cheyne treated tuberculosis cases, he faced two integrated problems: one related to bacteriology; and the other, to political ethics. Animals were essential to his search for the bacteriological cause of the disease; however, the Anti-vivisectionists campaigned, uncompromisingly, to end animal experimentation.[1] Published distortions of Cheyne's work and character did not dissuade him from conducting research on the cause and treatment of the disease. He consistently employed animals in his research, humanely and in accordance with the law, despite those who unjustly questioned his professionalism.

An opponent of Cheyne who placed the rights of animals above, or on par with, those of human beings could find a compromise allowing experimentation on nonhuman vertebrates difficult to accept. The infliction of pain and killing of animals in the interests of legitimate medical research, however, often was impossible to avoid; and the use of animals in research was a complex ethical problem. Lister's 1853–1863 physiology experiments exemplify the dilemma. On the one hand, his experimental use of frogs and mice was permissible. Consequently, he learned much through amphibian and murine studies and produced a number of papers on contractile tissue of the iris, on muscular tissue of the skin, on involuntary muscle fiber, on the cutaneous pigmentary system, on nerve fibers, and on blood coagulation and circulation. Of Lister's 16 physiology papers, 12 were published between 1853 and 1863 (Part I. Physiology; *CP*. I: 1–199; E. Howard). On the other hand, in order to study blood coagulation in higher vertebrates, the physiology of which approximated that of human beings, Lister had to use ruminants, such as sheep, oxen, and horses, the jugular veins of which he incised, catheterized, and bled. Critics would likely have objected to how domestic and farm animals were to be used. In the 16 November 1863 Croonian Lecture, for example, Lister points out that the moment of blood coagulation after death could only be determined if the animal were killed instantly with a pole-axe blow to the skull or neck, and if its blood were then subsequently monitored for clotting ("On the Coagulation of the Blood," *CP.* **I**: 109–134, esp. 122). Only by inducing sudden death by pole-axing a horse or an ox could he confirm the hypothesis that, over a 48-hour post-mortem period, the blood neither putrefied nor clotted. As a result of these experiments, Benjamin Ward Richardson's (1828–1896) contention that ammonia was responsible for coagulation would be overturned; and, in the process, much was learned about vascular disease (B. W. Richardson: 450; "On Spontaneous Grangene," *CP.* **I**: 69–84).

A series of Animal Cruelty Acts—of 1837, 1849, 1850, 1854, and 1876—were intended to resolve the ethical dilemma of animal rights advocates and medical professionals. The 1876 legislation, in particular, had a powerful impact on medical research in many areas and especially on Cheyne's campaign against tuberculosis. Before detailing the main points of the 1876 legislation, I would like to consider a historical event that set the tone for Anti-vivisectionists in the mid–1870s.

1861–1874: The Norwich Incident

The Royal Society for the Protection of Animals had convened a court of inquiry to look into an infamous experiment, performed at the 13 August 1874 Norwich Meeting of the British Medical Association ("Prosecution at Norwich": 751–754). The Prosecution charged Dr. Valentin Magnan (1835–1916), a French physiologist, along with four British colleagues, with animal cruelty. Magnan's purpose had been to test the toxic effects of alcohol and absinthe on two dogs. The demonstrators had strapped the animals to boards and, through the femoral arteries, injected alcohol into the first and absinthe into the second. The dog that received the alcohol became unconscious, but the other died convulsively.

The Prosecution rested its case on the claim that alcohol and absinthe toxicity were already well understood and that, consequently, Magnan's experiment was not breaking new pharmacological ground. All of this was true. In 1861, a French newspaper article had reported that the government, considering absinthe worse than opium, had already prohibited its use in the army and navy ("Absinthe and Its Poisonous Effects"). In 1864, Dr. Marcé reported to the Academy of Sciences that absinthe liqueur contained a toxic ingredient:

essence of wormwood (20 grams of wormwood to 100 liters of alcohol). Petitions of 1867 to the French Senate called for its prohibition. At this time, Magnan and Bouchereau had begun to experiment with the substance to learn about the physiological effects ("Absinthe," *Scientific American*). To distinguish between the adverse effects of pure alcohol and those of absinthe, they put a saucer filled with concentrated essence of wormwood in a glass case and then put a guinea-pig inside. The guinea-pig breathed the fumes, became agitated, lethargic, and died convulsively, foaming at the mouth, and displaying symptoms of epilepsy. The other guinea-pig, housed in a glass case with pure alcohol, had symptoms consistent with alcohol consumption ("Absinthe").

Much of the human evidence proscribing absinthe was clinical and available for review (E. E. Walker). An 1868 article in the *American Journal of Pharmacy*, for example, warned that even moderate consumption of absinthe could lead to trembling, muscle contractions, tingling, numbness, yellowish skin, nightmares, hallucinations, general paralysis, brain lesions, and death ("Absinthe"). Modern biochemical pharmacology and other disciplines have established that wormwood is associated with kidney failure, porphyria, muscle deterioration, optical disorders, mental deterioration, and seizures (Fetrow and Avila: 579–580).

In view of these facts, was Magnan's 1874 canine experiment at Norwich a logical extension of his guinea-pig test in 1864? In the Norwich demonstration, was he moving gradually up the scale of physiological complexity from guinea-pig to dog? Knowing something about his absinthe-related research from 1864 to 1874 will help us to determine what motivated him.

A review of Magnan's research and publications suggests that Norwich was a sideshow and that the absinthe-poisoned dog to which the operator was coldly indifferent was expendable. Magnan was fully aware of what would happen if a high dose of absinthe were introduced directly into a dog's bloodstream. While he had been a student, the effects of alcohol and absinthe had been the focus of intense investigation, beginning with the researches of Meynier (1859) and of Marée (from 1864). Before Norwich, Magnan had published papers and a monograph on the etiology of alcoholism and of absinthisme. His published research antedating Norwich were: "Accidents déterminés par l'abus de la liqueur d'absinthe" (4 and 9 August 1864); "Étude expérimentales et Clinique sur l'absinthisme et l'alcoolism" (1871); and the above-cited monograph, *De L'Alcoolisme; Des Diverses Formes du Délire Alcoolique et de Leur Traitement* (Paris, 1872, 1874). Magnan had already performed canine experiments, in 1872, similar to the Norwich demonstration, and he documented the initial, graphic experiment.

The reason why Magnan had strapped the Norwich dog to a board and bound its jaws shut becomes clear after reading the 1872 account. The Parisian dog which was friendly and quite tame had neither been fettered, nor gagged, nor bound. Fully aware from earlier experiments of the effects of even a small injected dose of absinthe, Magnan intensified the 1872 attack with a lethal 4.0-gram dose. After a short period of giddiness, the dog suddenly fell to the ground, convulsing in the first stage of seizure. Neck muscles contracted violently. Muscles on one side of its body contracted so far as to distort the animal into an unnatural bow-shape. The animal's head turned downward, muzzle to chest; it fell, turned over, and endured a second attack that flexed and extended its head; mercifully, it died after the third attack. The reason why the jaws of the Norwich dog were clamped shut was because tonic convulsions would have caused it to foam, snap wildly, and bite its tongue. Epileptic seizures followed. A dog overdosed with absinthe (5.0 grams abdominally injected) was on record as having hallucinated so badly that it attacked a hallucination (*De L'Alcoolisme*: 22–27; Magnan, *Reviews and Notices*: 346).

In 1874 Magnan knew exactly what to expect and might have sought to impress his colleagues—all at the expense of the dog. Uninformed medical witnesses in Norwich were naturally taken aback by the spectacle of two dogs grotesquely strapped to boards (one is described as cruciform); and they were appalled by the femoral-artery injections of alcohol and of absinthe, respectively, into each animal. Dr. Haughton, of Trinity College Dublin, at the 13 August 1874 Court-Hearing condemned the effects produced in the epileptic dog and thought the experiment unnecessary and cruel ("Prosecution": 752). He had grounds for believing so. John Colam, Chief Prosecutor and Secretary of the Royal Society for the Prevention of Cruelty to Animals, also condemned the demonstration because the dog had suffered, and presumably no new information had been ascertained, since the effects of absinthe on human beings was well known and the use of dogs in this way had no ostensible, veterinary purpose (752). Magnan's supporters, however, testified that the demonstration had an unprecedented outcome: that absinthe caused epileptic seizures. Two possibilities can account for this statement. Either they had not consulted the literature of 1859–1874, or they were feigning ignorance.

A corollary to the protesters' argument was Magnan's choice of a painful intravenous route to intoxication, considered as cruel and needless. The arterial injection of the dog saved time and intensified the adverse effects. Some suggested at the Hearing that the animals could have been force-fed. Others disagreed, saying the dogs would have regurgitated the substance. The dog that was mainlined the alcohol "became dead drunk," slept it off, but woke up the next day, "as well as ever." The absinthe dog, however, suffered grievously, and for that Magnan was thought to be culpable. Some wondered why anesthesia had not been given; the rebuttal was that a sedative or anesthetic would have skewed the physiological results (752).

If the aim had been to educate the medical community of Great Britain about the dangers of alcohol and absinthe consumption, then Magnan's demonstration might actually have had some public health legitimacy. The dangers of absinthe, already public knowledge in France, could now be loudly communicated in Britain, the scandal ironically dramatizing the dangers of the liqueur. All agreed that vivisection was cruel, but many argued, semantically, that the Magnan demonstration was not vivisection in the strictest sense of the word since no live animal was dismembered. The magistrates dismissed the case because proof of a violation was lacking, and Magnan walked away without paying a shilling (754).

1876–1879: Animal Cruelty Legislation

The content of the 1876 Act surprisingly did not play to radical sentiments (*Cruelty to Animals/Anti-Vivisection Act 1876*). The Act's stipulations were extended to include cases in which animals employed in medical, physiological, or scientific research were subjected to pain. It strictly proscribed or limited painful procedures on animals, except in instances when the experimental benefits to humanity clearly outweighed the rights of nonhuman vertebrates. An experimental procedure involving animals, at the outset, had to be justified as a means of prolonging human life and of alleviating human suffering (39 and 40 *Victoriae*. 4, c. 77; 3 [1.]); the investigator who wished to experiment on living vertebrate animals had to apply to the Home Secretary and to receive the endorsement of leading medical bodies (e.g., the Royal Society of London), along with that of a medical professional (3 [2.]; R. D. French: 143). Animals to be used had to be anaesthetized (3 [3.]); if an animal emerged from the experiment in pain or with serious injury, it was to be euthanized (3[4.]); and no experiment could be performed simply as a teaching tool or illustration in medical institutions.

In the interest of scientific progress, the 1876 Act included a sensible proviso (3[5.] and 3[6.]): the regulations could be superseded if experiments on living vertebrates, or if physiological knowledge and instruction, were "absolutely necessary" to the interests of human health. If an animal's sensibility had to be maintained to validate a crucial test, even anesthesia could be withheld (3[5.2]). Even if the test animal emerged from anesthesia in pain or with injury, euthanasia could be delayed, "if killing the animal frustrated the object of the experiment"; once the object of the experiment had been achieved, euthanasia was then to be performed (3[3.]).

In regard to dogs, cats, and horses, experimenters had to abide by special restrictions on pain. If a certified, scientifically-justified experiment could not be undertaken, either because no animal model, other than dog, cat, or horse, was available, or because one of these animals had been established as specific, or as indispensable, to the procedure, then domestic and farm vertebrates in the above group could be used, provided that the foregoing criteria regarding pain and injury were observed (5.). The use of curare as an anesthesia (which asphyxiated an animal), and the public exhibition of painful animal experiments for payment or gratuitous entertainment, were absolutely prohibited ([4.] and [6.]).

Though 39 and 40 *Victoriae* was balanced legislation, its enforcement caused considerable confusion and rancor. One area of concern was that the location of the experimental facility had to be approved, registered, and made subject to inspection ([7.]). On this point, bureaucracy interfered with scientific practice. As final arbiter, the Home Secretary, who required progress reports from the licensees and who delegated on-site inspectors of labs, was given the authority to withhold a license under this Act, to grant it as he saw fit, or even to revoke it ([8.], [9.] and [10–11.]).

The 15 August 1876 Act, though it had been revised to meet the concerns of medical scientists, appeared to have been reasonable, ethical, and humane. It recognized that curing intractable diseases might depend on animal experimentation and on the creation of vaccines and medicines. It acknowledged that inflicting pain on animals, though regrettable, could be the only way to manage an infectious disorder.

The implementation of the 1876 Act (and of its 1879 successor [42 and 43 *Victoriae*]) severely hampered medical research at a crucial moment in the history of bacteriology ("Cruelty to Animals Act" [1879]). The Home Office, seemingly insensitive to the demands of the medical community, delegated to civil servants or politicians the authority to ignore the recommendations of learned medical advisors; and a Certification system in animal research, ranked A to F, was formulated (French: 192; Tansey: S21-S23). Prominent researchers, as a result, were denied licenses to practice; and experiments were curtailed (194–195). The newly-formed Association for the Advancement of Medicine by Research asserted that vivisection and related modes of experimentation were essential to the advancement of science and that unqualified persons should not have been appointed, "to exercise judgment in a matter involving esoteric professional knowledge" (217). The source of the controversy in this period, therefore, lay not in the letter of the law, but in its implementation.

Of the many examples of animal models in research, one stands out. From November 1882 to 1885, the spinal cords of 25 rabbits furnished Louis Pasteur with the means to cultivate the rabies virus that would ultimately be used to save nine-year-old Joseph Meister's life and, for more than a century, the lives of untold numbers of human beings, and of domestic and farm animals ("Prevention of Rabies": 379–387). Pasteur recognized the pathology of rabies in brain tissue, but he could neither interpret what he saw nor isolate the cause. But he was able to contain and grow the mysterious pathogen in rabbit cords, a method that had already been established as a "convenient host for experimental rabies" (Waterson and Wilkinson: 54). If not for the availability of experimental animals (25 rabbits and 50 dogs), Pasteur would not have been able to produce the immunizing vaccines (379–387).

1882–1884: Koch's Investigations

The use of animals in tuberculosis research antedated the work of Koch and Cheyne. Julius Cohnheim (1839–1884), Carl J. Salomonsen (1847–1924), and Paul von Baumgarten (1848–1928) demonstrated that tuberculosis was an infectious disease that could be communicated to lab animals via inoculation into the anterior chamber of the eyes; and Hermann von Tappeiner (1847–1927) produced evidence that the disease could be spread through inhalation (Kaufmann: 107; Koch, "The Etiology of Tuberculosis": 109). In a 24 March 1882 lecture, Koch describes tuberculosis as it naturally occurs in a variety of animals. He artificially infected mice, rats, guinea pigs, rabbits, hedgehogs, hamsters, cats, and dogs with the tubercle bacillus (Kaufmann: 107). His self-experimentation with this material and the ensuing severe reactions convinced him that investigators had to be careful when extrapolating in this way from animals to human beings (107). Nevertheless, animal models, as Koch's experience indicated, were indispensable to the search for the mechanisms underlying natural and acquired resistance to tuberculosis (107). The content of Koch's original papers on the etiology of tuberculosis, surveyed below, influenced Cheyne's understanding of the disease.

The 1882 paper "The Etiology of Tuberculosis" illustrated how indispensable animals were to Koch's work. To demonstrate that a parasitic organism rather than a bodily structure was the cause of the disease, he needed to locate these presumptive organisms in the tissues of an animal that had died of tuberculosis, either naturally or under laboratory conditions; and to accomplish that he needed a suitable staining method for microscopic examination. Using a carefully-prepared solution of methylene blue and other compounds, he was able to differentiate between the brownish tint of animal tissues and the bluish tint of the tubercle bacilli (Koch, "Etiology" [1882]: 110). On the basis of two points, Koch then inferred the high probability that tubercle bacilli caused the disease: these organisms were most frequently found when the disease was developing; and they disappeared when it became quiescent (111). But, to prove that the bacilli penetrated the tissues as would a parasitic disease, they had to be extracted from the body and then purified through the serial elimination of extraneous micro-organisms and of organic by-products. Once this process had been completed successfully, the isolated bacilli had to induce the very same disease when transferred to living animals through the inoculation of pure cultures (111). In Koch's experiments, animals had three purposes: dying naturally or from inoculations, they provided specimens for seed cultures; blood serum from cows or sheep when sterilized and solidified provided an ideal culture medium; and purified cultures could be transferred to healthy animals to reproduce the disease ("Etiology" [1882]: 113).

Koch's experiments produced important results. Tuberculosis pathology in guinea-pig tissue, whether of the rare or laboratory-induced kind, he discovered to be pathologically invariant. Furthermore, suppurating bronchial lymph nodes, along with cheesy decomposing lungs, routinely resembled similar disease in human lungs ("Etiology" [1882]: 113). But, in certain respects, naturally- and artificially-occurring tuberculosis differed from each other. In the artificial induction, the abdominal inoculation-site swelled and, because a large quantity of tubercle bacilli had been introduced at once, the infection moved more rapidly than it would have in the spontaneous form. Since tissue damage was more extensive in the inoculated animals, it was therefore easier to differentiate artificial from spontaneous cases (113).

While uniformly proving that the inoculation of pure tubercle cultures into guinea pigs caused tuberculosis, Koch recorded each phase of the disease: eight days after the injections, the animals deteriorated, and the tissue damage, documented four weeks after they had been euthanized, was predictable. The pathology of the spleen and liver differed from that of lab animals injected with pathogens other than tubercle bacillus ("Etiology" [1882]: 114).

Positive results supporting causation were obtained by Koch even when the micro-organism had been transferred between different species. Thus, pulmonary tuberculosis from apes, upon transference, sickened guinea-pigs, as had brain and lung material from deceased human beings; and nodules from the peritoneum of cows stricken with bovine tuberculosis also infected guinea-pigs. No matter what the species, the disease developed in the same way. Specimens harvested from diverse animals and from human beings were identical and caused similar pathology ("Etiology of Tuberculosis" [1882]:114). To make major strides, Koch had to use animals extensively. Of the 15 pure cultures of tubercle bacilli made, four were from guinea-pigs infected with ape bacilli; four were from guinea-pigs infected with bovine bacilli; and seven from guinea-pigs inoculated with human bacilli. These results reinforced the conclusion that the bacillus, if present in great numbers, was the cause and not a concomitant of tuberculosis (115).

In "The Etiology of Tuberculosis" (1884), Koch described an experimental methodology, the rudiments of which had been adumbrated by Jacob Henle in 1840, and developed in 1883 by Koch's assistant, Friedrich Löeffler (1852–1915) (Henle: 76–79; Alfred Evans, "Henle-Koch Postulates"; and Worboys, *Spreading Germs*: 177).[2] In "On the Anthrax Inoculation" (1882), Koch had acknowledged the great value of Löeffler's work on immunization in mice, rabbits, rats, and guinea pigs. In addition, Löeffler's experiments had shown that immunity to certain bacterial diseases occurred after an earlier attack. But, as his animal models indicated, this was true for some but not for all bacterial pathogens. Since he had not used the ovine model that Pasteur had employed, he could not conjecture beyond that point. Pasteur had maintained that, without exception, inoculations of attenuated anthrax induced ovine immunity (Koch, "On Anthrax Inoculation" [1882]: 102). Without the use of animal models, as these historical instances show, Koch and his contemporaries could never have uncovered the etiology of anthrax (in 1876) and of tuberculosis (in 1882).

What Koch had learned about the pathogenesis of anthrax in 1882 was directly applicable to his 1882–1884 study of tuberculosis. In 1884, he made it clear that if the blood of an animal dying of anthrax were injected into a healthy one, the recipient died of anthrax, and its blood then revealed the presence of the characteristic bacilli ("The Etiology of Tuberculosis" [1884]: 116–117). But this was not definitive proof of causality since the possibility remained that an unidentified substance or germ other than anthrax bacilli had been responsible for the disease. To move in the direction of causality and according to the Postulates, the bacilli had, first, to be isolated from the body and tested, as was done in 1882. The first step in obtaining a pure culture was via transference from generation to generation. By the third or fourth transference, the culture was considered free of original blood substances and of anything that was foreign; and by the fiftieth transference, the culture was deemed pure. Inoculating a lab animal with the end-product induced fatal tuberculosis. Animals inoculated with blood directly from a naturally-infected animal or from purified culture in glass exhibited identical symptoms and died. Blood from these carcasses revealed high concentrations of the anthrax bacillus, further proof that the bacilli caused the disease (117). All that Koch had learned through anthrax experiments was then applied with great success to the understanding of tuberculosis (117). These achievements were made possible through the judicious use of animal models.

Anti-vivisectionists targeted Koch's immunological research. Although the continental biomedical community was outside the jurisdiction of British legislation, nevertheless, Pasteur, Koch, and others were vilified in the activist literature (R. D. French: 260). As W. F. Bynum points out: "Although to many Pasteur and Koch were international heroes, to others they represented all that was cruel and dangerous in a world threatened by materialistic, godless science" (168). Two events in the 1870s had caught the public's attention, galvanizing the Anti-vivisectionist campaign (Bynum: 168–169). The first, in 1873, was the publication of the two-volume, *Handbook for the Physiological Laboratory*, edited by John Scott Burdon Sanderson (1828–1905). This text features substantial essays by Emanuel Edward Klein (on histology), whose cholera theory will be discussed in the next chapter, and by Sanderson, Michael Foster, and T. Lauder Bunton, on physiology. Comparative experiments on animal physiology, described throughout, were intended for use in medical curricula. Experiments were conducted on amphibians and small vertebrates (guinea-pigs and rabbits), very rarely on puppies and cats, but not on domestic and ruminant animals. Rabbits were utilized routinely in studies of bone growth (65), muscle (130), liver function (140), corneal function (173), respiration (250), vagus nerve and cardiac function (318), body heat regulation (342–344), salivary gland function (465), mastication (467), bile pigment (498), and diabetes (514).

The second political catalyst, as we have shown, was the gratuitous 1874 Norwich dog-killing involving alcohol and absinthe. The demonstration, as some argued, had been obviated by Magnan's 1872 canine experiments in Paris, proving that alcohol acted on "the Tenth or pneumogastric nerve" ("Dr. Magnan of Paris": 211; Henry Gray, *Anatomy*: 738–739); concurrently, "the active principle of the absinthe" was also known to affect the same nerve in the stomach. It was the combined stimulation of these nerve-endings, according to Dr. T. P. Lucas, that produced "epileptic symptoms" (211). Presumably, the *a priori* argument that the procedure was part of an epilepsy inquiry had not been made.

The Anti-vivisectionists argued their case relentlessly. In the anthology, *Physiological Fallacies*, an anonymous author who had been assigned to survey recent developments in tuberculosis research dismissed Koch's 1882 discovery outright. Even though the latter's results "accord with sound reasoning and established facts," they were not accepted as presented since Koch's British endorsement had come from John Tyndall whose judgment the activist press questioned ("Tuberculosis," *Physiological Fallacies*: 100). According to the Anti-vivisectionist writer, although Koch's experiments were "scientifically interesting," they were reputedly inconclusive and subject to interpretation (104). The opposition, it appears, had ignored current research and the persuasive evidence that, in inoculated animals, the tubercle bacillus incited tuberculosis. The method, and not its humanitarian aim, was central to the Anti-vivisectionist argument. Koch (in their view) was doing nothing more than spreading "a painful and fatal disease in animals" (105).

The aim of Kochian research was to develop medical, immunological, and surgical interventions that would mitigate human suffering. In the 29 April 1882 editorial, "Recent Researches on Bacteria," *The British Medical Journal* reviews the success of Koch's methods with respect to tubercle bacilli, stressing that "the pathogenic properties of the various forms of organisms which are found may be ascertained by experiments on animals" (624–625). The author of *Physiological Fallacies* saw the matter differently. Physiological experiments on tuberculosis were being conducted illegitimately or even for profit.

1882–1883: Cheyne Visits Toulouse and Berlin

In 1882, the newly-established Association for the Advancement of Medicine by Research (A.A.M.R.) had to solve two problems: (1) some believed that Britain was lagging behind in research, although Lister's linkage of microbial metabolism to lactic-acid fermentation in the 1870s suggested that Britain actually had progressed as far as the Germans in this field; however (2), if a gap persisted, then the Anti-vivisectionists threatened to widen it by interfering with laboratory experimentation (Worboys, *Spreading Germs*: 175–177; Coutts, *Microbes*: 170). On 22 April 1882, John Tyndall informed the medical community that Koch, in "The Etiology of Tuberculosis," had systematically proven through animal models that the tubercle bacillus was the cause of tuberculosis (175). In May 1882, Cheyne and Koch's private assistant displayed the bacillus at King's College ("Bacilli in Tuberculosis": 797; Worboys, *Spreading Germs*: 176). As the evidence continued to accumulate, the A.A.M.R. in the summer of 1882 assigned Cheyne to a fact-finding mission to France and Germany, requiring stays in Toulouse where he was to evaluate Honoré Toussaint's (1847–1890) claim that a microbe was responsible for tuberculosis; and, in Berlin, to avail himself of Koch's methods and to acquire samples from him for testing upon his return to Britain (Worboys, *Spreading Germs*: 177; Coutts, "Illustrating Micro-organisms": 4). Cheyne connected the investigations of Toussaint and of Koch to the researches of J. S. Burdon Sanderson (1828–1905), of Theodor Albrecht Klebs, and of others ("Report to the AMRR"). Returning to England, he presented his findings on 1 February 1883, in the "Report to the Association for the Advancement of Medicine by Research on the Relation of Micro-Organisms to Tuberculosis." In this document, published in the April 1883 edition of *The Practitioner*, he systematically invalidated the claim that tuberculosis had no specific bacterial cause. Cheyne assessed the laboratory work of Toussaint and of Koch and then conducted his own investigation, with the aim of proving that the tubercle bacillus was the inciting organism.

The first round of the investigation in Toulouse, from 28 August to 7 November 1882, consisted of five experiments to see if tuberculosis could be induced by any one of a variety of substances: in Experiment I, thread was implanted in lab animals (rabbits and guinea pigs); in Experiment II, calf lymph was injected; in Experiment III, human lymph; in Experiment IV, dead tubercle, cork, and wool; and in Experiment V, human pus. Upon dissection, no tuberculosis was detected, not even in the dead-tubercle test. Across the board, Toussaint's tests in Experiments I–IV disqualified these materials. But the negative results that Cheyne accrued through the process of elimination proved that neither inorganic nor putrid matter was implicated as a cause.

While Cheyne was carefully eliminating presumptive pathogens, he worked under the shadow of the activists. Vivisectionist-related intrigue might have been afoot during the first round of research. Cheyne repeatedly claimed that the carcasses of lab animals had been "stolen"; an instance of this involved two rabbits used in Experiment III (28 August-7 October). The animals had actually been improving from a prolonged, non-tubercular infection caused by the injection of human lymph serum into their left eyes, a standard procedure ("Report to the AAMR" [1 February 1883]; "Abstract of the Report": 444–445). "Both animals" were reported as stolen "on October 8th." Cheyne reported stolen carcasses occasionally while in Toulouse, and one can only speculate as to why this was happening, and as to why he documented it in an official report. Was evidence being stolen to incriminate Cheyne, or was he disposing of evidence as a precautionary step?

Anti-vivisectionists definitely had an eye on Cheyne's efforts. This is evidenced by their response to Experiment V in which human pus, injected into the left eye of a rabbit, brought on an ocular infection; the animal gradually recovered from it, but then died on 10 January 1883 after living 64 days in traumatic captivity. In the "**Cheyne, Wm. Watson**"-entry of *The Vivisectors' Directory*, a roster of European experimental pathologists (a derisive label in the activist vernacular), Experiment V of 7 November 1882 is summarized. The corporate author redacts, decontextualizes, and highlights its content for emotional appeal: the "Experiment with pus from the wound of a patient suffering from pyemia. The pus was thick and foul smelling. **1.** One minim was injected *into the left eye* of a rabbit, causing widespread inflammation, and the animal was ill for some time. It gradually recovered and, in December, was apparently well. It died on January 10th, 1883. Lived 64 days" (italics added by *Directory* author; *The Vivisectors' Directory* [1 December 1882]: 215–218; and [1 January 1883]: 240–242).

Through a series of experiments performed from 22 July 1882 to 8 February 1883, Toussaint cooperated with Cheyne, as he examined "portions of the organs of various animals ... and also tubes containing cultivations themselves" ("Report to the AAMR"). Toussaint had employed a variety of species to see if the suspected organism developed and if the infection struck different nonhuman vertebrates. On 22 July 1882, Toussaint had squeezed fluid from the muscle of a cow, presumably infected with bovine tuberculosis, and injected it into a pig which subsequently died of extensive tuberculosis; a rabbit, in turn, was injected with porcine tubercle material, and it, too, died of the disease. Cheyne verified Toussaint's results. Histological examinations of the lung, spleen, and lymph nodes revealed to Cheyne that the animal had indeed died of the disease. Toussaint was even able to infect a pig with the organs of a tubercular cow; the pig had been fed infected bovine organs and developed the disease; its blood, in turn, was injected into the peritoneum of a cat which also developed the disease. Dissections revealed in multiple organs and anatomical sites the presence of the tubercle bacillus. Toussaint's experiments disclosed that tuberculosis could affect multiple vertebrates in the same way, although later studies would uncover several exceptions: that from species to species, there were pathological distinctions; and that human beings could get the disease from animal products.

Cheyne turned to Toussaint's questionable hypothesis that micrococci, a spheroid organism distinguished from bacilli, could generate tuberculosis. To eliminate this possibility, Cheyne worked with two tubes of Toussaint's micrococci which he cultured to a state of purity. In Experiment VI (29 August– 2 November 1882), he inoculated the eyes of three rabbits and one guinea-pig; no tubercle was found there or in dissected organs ("Report to the AAMR"). To cover every possibility, Cheyne had cultured Toussaint's micrococci on solid media (blood serum), injected rabbit eyes with the bacterial solution and a guinea-pig abdominally, but nothing tubercular appeared. The animals died from causes other than the injected, non-tubercular bacteria. Once again, there was evidence either of political intrigue or disposal: "On October 9th the three rabbits and one guinea-pig were stolen."

The Anti-vivisectionists had failed to appreciate, or had disregarded, the importance of Cheyne's tests. Once it had been made clear that micrococci were not the cause, support for Koch's tubercle bacillus in scientific circles increased. Anti-vivisectionists should have rejoiced that trial-and-error experimentation on tuberculosis etiology was nearing an end. Cheyne's verifying tests ironically benefited their cause. If the inciting organism had remained a mystery, then experiments on innumerable microbes and with inorganic substances would have been randomly pursued; as a result, the search for the cause of tuberculosis, as a hit-and-miss endeavor, would have caused unnecessary suffering and have wasted lab animals. Cheyne's sound procedures and compliance with the stipulations of the 1876 Act obviated misguided or blind experimentation on animals.

With incriminating intent, the *Directory*'s authors, in the November–December 1882 edition, fixated on graphic procedures in order to evoke a public outcry. Thus, Cheyne was exposed to public derision for having optically-injected micrococci cultures into six of seven rabbits, guinea-pigs, a mouse, and a cat. Since none of the animals developed tuberculosis, the uninformed public was being led to believe that the experiments were pointless. The *negative result*, however, actually fulfilled Cheyne's stated aim: to disprove Toussaint's suspicion that micrococci caused tuberculosis. Once all other possibilities had been exhausted, it was time to use live tubercle bacilli in animal models.

Late in 1882, Cheyne worked with Koch in Berlin, as he had with Toussaint in Toulouse. There, Koch gave Cheyne samples of tubercular tissues and bacilli cultivations, both animal and human, that had been gathered from the great experiments described in the 1882 "Etiology of Tuberculosis." Cheyne took these back to Britain in order to retest the specimens in live vertebrates. From 23 September to 31 December 1882, ocular injections and subsequent post-mortems of rabbits and guinea-pigs showed extensive tuberculosis disease and revealed tubercle bacilli. In Experiment XII, of 28 November 1882, Cheyne established the infectiousness of the bacilli when he transferred tubercular lung tissue from the Experiment X-rabbit into the healthy rabbit and guinea-pig. Both animals developed the disease ("Report to the AAMR": 32–33).

Experiment XIV (2 November 1882) was the most interesting of the series, in terms of both medical bacteriology and political ethics. Cheyne prepared Koch's bacilli with distilled water, readying them for injection into the eyes of rabbits ("Report to the AAMR"). An active ingredient was in each syringe. In the first, he mixed a caustic disinfectant: 1-part bichloride of mercury and 1,000-parts water. Koch had shown that bichloride of mercury or corrosive sublimate had germicidal action. Lister had praised Koch's discovery that a 1–20,000 parts sublimate-to-water solution destroyed highly-resistant anthrax spores ("An Address on Corrosive Sublimate," *CP*. **II**: 293–308, esp. 295–296). Cheyne was now interested in seeing if the sublimate could inhibit or destroy tubercle bacillus and, if either was possible, whether its germicidal property had even wider uses. The rabbits received strong concentrations of the sublimate, perhaps because Cheyne wanted to learn immediately if the chemical had an effect. It did not in the first animal. In the second rabbit, he optically injected a 1 percent solution of resorcinol, a crystalline phenol found in various resins and dyes. Although tubercular inflammations occurred in both eyes, the organism did not appear to proliferate. This was an encouraging sign. In the third rabbit, the bacilli were mixed with a 5.0 percent aqueous solution of carbolic acid. The mixture was injected into the left eye. The right eye actually showed the initial signs of tubercular infection, and it was eventually destroyed by the disease. The left eye also gradually showed signs of infection. When the animal died on 7 January 1883, Cheyne found upon dissection tubercular infiltration in the lungs, kidneys, liver, spleen, and intestine. This was a surprising result: carbolic acid, a mainstay of Listerism, had had no inhibitive effect. Experiment XIV corroborated Koch's hypothesis that the tubercle bacillus was the cause of tuberculosis; but it also showed that antiseptic research had a way to go before a viable regimen could be relied on ("Report to the AAMR"). Resorcinol, however, was promising.

The Anti-vivisectionists were also interested in Experiment XIV but for a different reason. They published a 200-word excerpt from the "Report to the AAMR"—once again, out of context and propagandistically intended. Their aim was to denigrate Cheyne's experimental series as pointless and immethodical: "Experiment XIV., November 2nd, 1882.—The bacilli were rubbed up with boiled distilled water as usual. A little of the pure material was injected *into the right eyes* of three rabbits. Into the *left eyes* … [antimicrobial chemicals] were injected" (italics by *Directory* author[s]; *The Vivisectors' Directory* [*The Zoophilist*; 1 January 1883]; III. Series 14: 240–242). The focus in this passage is on the sensitive injection site; and the emphasis, on morbidity rather than on the dual purpose of the Experiment: to verify Koch's core findings and to test the effects of a germicide. Once again, Cheyne approached his work foresightfully, assessing ways of controlling the micro-organism, and the disease with which it had been associated.

The Society for the Protection of Animal's from Vivisection tried to vivisect Cheyne's career. Referring to his "Abstract of the Report on the Relation of Microorganisms to Tuberculosis," the unidentified author included Cheyne's curriculum vita, his Licensing status, excerpted and abstracted passages that emphasized the number and species of animals employed, and graphic references to the ocular injections of live bacteria the animals had received. But no reference was ever made to the importance of both negative and positive results or to the etiological aim (*Vivisectors' Directory*: 19–20;

"Abstract of the Report on the Relation of Microorganisms to Tuberculosis" [17 March 1883]: 444–445). The negative associations and editorial sleight-of-hand, noted here, were intended to discredit Cheyne: (1) the significance of the project to tuberculosis research was omitted; (2) the bacteriologist's professional status, as physician, surgeon, and laboratory investigator, headed the redacted summary of experiments, ironically portraying the operator as unprincipled or as a well-educated quack; and (3) the calculated effect was to impugn his professional judgment and that of the societies and institutions with which he was associated.

As might be expected, *The Zoophilist*, of 2 April 1883, in an advanced notice of an article on Cheyne's new "Report," refocused on the ocular procedures and not on the germicidal test. They failed to note that the procedure was commonplace in tuberculosis research, as Koch's predecessors had shown: "We only here point out that the inoculations practiced on twenty-five animals (one of which was a cat) were, as Mr. Cheyne himself tells us, 'made into the anterior chamber of the eye, whenever this was practicable.'" They assert that the ocular procedure was more readily tolerated on the continent, and that is why Cheyne was sent abroad (*The Zoophilist* [2 April 1883]; II. [4]: 52). Cheyne had been commissioned to study Toussaint's and Koch's experiments, not to torture animals. In the Preface to the 1884 edition of the *Vivisectors' Directory*, Frances Power Cobbe (1822–1904), the leader of the Victoria Street Society (French: 117), actually explained the anatomical rationale for ocular inoculation: "The eyes are chosen as the special seats for inoculation, because, through the transparent body the processes of disease can be most easily watched" (Preface: v). Despite the intelligent admission, she reverts to the party-line: "Watson Cheyne injects micrococci into the eyes" (Preface: v). No mention was made that Watson Cheyne had been testing Toussaint's micrococcus experiment. Negative results from these inductive tests rendered further experimentation, using various materials and micro-organisms, scientifically and legally unjustified. Cheyne's experiments ironically preserved laboratory animals by pre-emptively identifying dead-end lines of inquiry.

Instead of clarifying the pharmacological purpose of Cheyne's investigation, the editors redacted the 1884 *Directory*-entry to the point that the infliction of pain appeared to be a perverse indulgence or an effect to which the practitioner was indifferent. Cheyne's key discovery, bichloride of mercury as a promising bactericide against tuberculosis, was purposefully ignored or simply not understood. To have acknowledged this finding would have credited Cheyne with success and justified the procedure. So the most important part of the experiment was left out. To evoke an emotional reaction from the reader, the Anti-vivisectionists reproduced, from Clark's 1 January 1883 edition of

The Zoophilist cited above, these italicized statements: "a little of the pure material was injected *into the right eyes* of three rabbits. Into the *left eyes* the following [antimicrobial chemicals] were injected" (*Vivisectors' Directory*: 19–20).

The harassment was unrelenting. The Anti-vivisectionist John H. Clarke, who was a London physician, viewed carbolic acid and the antiseptic system as the "fruits of vivisection." In an ironic move, he selected Cheyne's 1879 benign-concomitant idea as a weapon to discredit both the antiseptic system and medical research requiring animal models (J. H. Clarke, "Antiseptic Surgery": 40). If we recall, in 1879, not being able to account for the survival of 22 percent of the micrococci in carbolated pus, Cheyne posed three possibilities, two of which he proved to be untenable: (1) the remnant survived and even thrived in the acid; (2) the remnant, though immune to the acid, was a harmless concomitant; and (3) the antiseptic was too weak to kill them all. Clarke used points (1) and (2) against Listerism in order to fell with one shot the germ theory of infection, the antiseptic system, and animal experimentation, the means through which the system had been developed. Apparently, Clarke had not read Cheyne's 1879 paper through to the part where the latter had proven point (3), that 1–250 was the acid's lower and 1–20 its higher level of bactericidal activity. Points (1) and (2), as noted earlier, proved to be incorrect. Cheyne's failure to expurgate these hypotheses permitted Clarke to attack the antiseptic system: since Lister's protégé had proven "that germs can flourish in the most 'aseptic' of wounds, and sometimes delight in solutions of carbolic acid," then Lister's theory, in Clarke's judgment, was "all wrong" (40–41). In another article, Clarke highlighted what an earlier commentator had called "the sickening details" of the ocular injections involving 68 animals at Toulouse. The purpose of Cheyne's laboratory work, it seems, was immaterial (J. H. Clarke, "Returns": 217–218).

The attack against Cheyne also questioned the moral integrity of the AAMR, an organization dedicated to the defense of medicine. The *Directory* implied that, if the advancement of science by research meant afflicting animals with painful disease and (allegedly) with no clear purpose, then the operator and his professional affiliations were equally culpable. In the 1884 edition of the *Directory*, mention is made that Cheyne held "a License for Vivisection at King's College, London" and, from 1880 to 1883, consecutive "Certificates with Obligation to Kill." This statement, however, does not agree with Richard D. French's sources which indicate that Cheyne had been denied twice—in June 1881 and in May 1886— but relicensed "considerably later" (184n, 186). It is likely that, at some time, Cheyne had to pursue his research without a license.

1891–1899: Goats and Kids

In the "Notes and Notices" section of the 1 July 1891 edition of *The Zoophilist*, the editors present Sir Andrew Clark's (1823–1896) assessment of contemporary tuberculosis research. According to the President of the Royal College of Surgeons in London, Dr. Clark, a prolific medical writer and specialist on respiratory disorders, proclaimed Koch's tuberculosis cure a failure (Cheyne would find it to have limited dermatological applications). Furthermore, he wrongly concluded that any effort to produce the disease in animals was to no avail since the disease was "a totally different thing than natural phthisis in human beings" (Vol. XI, No. 3: 37). Cheyne's prize-winning "Lectures on the Pathology of Tuberculous Diseases of Bones and Joints," published in April 1891, exposed Clark's error. Yet, even after Clark's opinion was shown to be erroneous, the Anti-vivisectionists continued to deride Cheyne, despite his demonstration that the tubercle bacillus caused the disease and that its multiform pathology stemmed from a single bacillary species.

In Lecture II of "the Pathology of Tuberculosis Diseases," published as two installments on 11 and on 18 April 1891, Cheyne proved through animal experimentation that the same pathogen was responsible for a number of anatomically-distant conditions (e.g., the arcane designations *strumous* disease of the joints and bones and *phthisis* of the lungs). What medical scientists were only beginning to learn was that tuberculosis was not exclusively a pulmonary disease. Rather, it could spread through the body, destroying soft tissue and bones. Thus, if one were to inoculate animals with morbid substances drawn from infected joints, tuberculosis occurred, as it would if infected lung tissue had been transferred ("Lectures on the Pathology of Tuberculosis" [11 April 1891]: 791). Typically, tuberculosis developed in rabbits and guinea-pigs when Cheyne used infected synovial membrane and pus from swollen human joints and injected these cultures under the skin or into the eyes of the test animals. Injected phthisical or consumptive sputum also induced tuberculosis if inoculated into the bones and joints of test animals, as did chronic tubercular abscess-pus and cultured tubercle bacilli; the inocula in every instance had come from human cases. The bacillus, constant in its infectious properties, crossed vertebrate-species barriers. Thus, what had commonly been called strumous disease of the joints indiscriminately affected human beings, rabbits, and goats.

Cheyne tested virtually every theory on tubercular causation that he had come across. An interesting example, especially from the standpoint of the ethical climate he navigated, was his testing of Max Schüller's (1843–1907) claims. Schüller, in 1876, had brought the disease on in animals in an unusual way: he not only injected tubercular material directly into the airways, but he also manually injured their joints, presumably to test the theory that an injury could make an animal (or a person) more susceptible to tuberculosis infection. It would also show that tuberculosis contracted via a respiratory route could travel to distant parts of the body. Cheyne tried but abandoned Schüller's method since no positive results were obtained after injecting a chloroformed guinea-pig subcutaneously with tuberculosis sputum and then contusing its knee joint. His thoroughness in attempting to replicate Schüller's findings, in the short-term, fueled the Anti-vivisectionists' propaganda machine. But the idea behind Schüller's discredited experiment paid off later. In 1899, Cheyne reconsidered Schüller's idea that an injury predisposed man or beast to tubercular infection, recalling instances when an injury to the joint or bone where tuberculosis was quiescent or subsiding compromised the patient's resistance, allowing the disease to recur aggressively. Even spraining or overusing a tuberculosis-compromised joint, in Cheyne's view, could open the way for an otherwise quiescent or chronic state of the disease to become acute ("On the Treatment of Tuberculous Diseases" [1899]: 9).

For Cheyne, pathological, histological, and bacteriological research established the fact that tuberculosis was a virulent, metastatic disease. The use of animal models, surgery, and laboratory work led him to this conclusion. In two experiments on the infectivity of phthisical sputum, he showed that pulmonary secretions could infect bones and joints. Into the right knee-joints of rabbits, he introduced pulmonary sputum that had come from an advanced human case. One experiment, of 8 November 1887, caused swelling and systemic pain for one rabbit, euthanized after 22 days. A second rabbit-experiment, of 8 November 1887, more impressively revealed extensive bone deterioration, bacilli colonies, and abnormal cellular formations.

Cheyne's studies, in 1891, overturned Andrew Clark's thesis that the effects of tuberculosis on animal and on human tissues were intrinsically different from one another. The degenerative processes that the rabbit endured resembled what happened to human beings, afflicted with bone and joint tuberculosis. Cheyne gathered convincing evidence from dissections. The tubercular material in his experiments had infiltrated the rabbit's tibial epiphysis, breaking through the cartilage, and penetrating the ossifying layer. At the end of each long bone are centers of ossification (the natural process of bone growth) or the epiphyses, structures uniting with the shaft as bone grows (Henry Gray, *Anatomy*: 35). These areas, at the extremity of the injected rabbit's tibia, had deteriorated so badly that the bacilli were able to infiltrate the connective tissue ("Lectures on the Pathology of Tuberculous Diseases" [11 April 1891]: 792). This was a particularly important observation. Cheyne and his contemporaries could, for the first time, closely examine the destructive effects of tuberculosis in animal models and compare the results: the bone and soft-tissue damage in the rabbit was strikingly similar to what he had seen in human decedents. Cheyne pointed this out in "Lecture II. Tubercular Changes in Bone" and in "Tubercular Diseases of Bones and Joints" ("Extracts from Three Lectures on Tubercular Diseases of Bones and Joints" [13 December 1890]: 1348–1353).

In "Lecture II," he outlined three subsets of tubercular pathology in bones and joints, several of which involved progressive bone-cartilage disease in human patients. What he had seen in the dissected rabbit limb reminded him of an examination of bone pathology in a child who had died of the disease. The child's femur presented with tubercular infiltration, and the typical cheesy deposit had apparently broken through the cartilage and into the joint. On each side of the bone, he had seen cartilage fragments, but farther away "the articular surface" had lost all healthy cartilage ("Extracts from Three Lectures" [13 December 1890]: 1349). These pathological features comparatively linked the human autopsy to the rabbit dissection, the common infectious agent being the tubercle bacillus.

In a relatively short period of time, Cheyne had compiled impressive data on tuberculosis pathology. In the lecture "Extracts" of 20 December 1890, tracing the course of the disease in lupine knees and joints, he clarified the process; from its early stages, rendering it predictable: first, the cartilage was invaded and destroyed by vascular tissue containing tubercles; second, the invasion extended to superficial fibrous tissue and lamellae that formed a lattice-work on the bone (Henry Gray, *Anatomy*: 24); third, fibrous and sclerotic tissue formed on the trabeculae (bars or fibrous bundles), portions of which had broken away; and, fourth, surface-bone destruction continued ("Extracts from Three Lectures on Tubercular Diseases" [20 December 1890]: 1418–1422; esp. 1419).

Cheyne's tuberculosis inductions confirmed that in nonhuman vertebrates the disease could affect different parts of the body as "local affection[s]" (*Tuberculous Diseases*: 10). From a primary deposit, the infection disseminated widely, appearing in the tissues and on bones as cheesy patches (11). Since animal models provided useful parallels to human tuberculosis, studying bone-joint pathology in nonhuman vertebrates, he realized, could help medical scientists understand the disorder in human beings, and this knowledge would be the foundation upon which surgical and medical interventions could be more effectively developed. Of the rabbit in Experiment IV, Cheyne writes: "Here we have all the appearances of tuberculous osteomyelitis as it occurs in man set up in this case by the injection of phthisical sputum into the bone" ("Lectures on the Pathology of Tuberculous Diseases" [11 April 1891]: 792). Furthermore, the lupine disease corresponded exactly to pathology found in strumous diseases of human bones and joints (794). He made the crucial observation that lupine and human tuberculosis mirrored each other. In both man and beast, the synovial membrane thickened, new tissue underwent caseation, cheesy material filled up the joint, the cartilage was destroyed, and, if the tuberculous growth penetrated into the bone, soft caseous deposits developed and the osseous trabeculae thickened (794).

Cheyne's goat experiments described in the 1891 "Lecture" did not gratuitously replicate the rabbit tests ("Lectures on the Pathology of Tuberculous Diseases" [18 April 1891]: 840–844). Goats were used in the second round of investigation because of their large accessible arteries. In Experiment XXIII, the injection of emulsified tubercle bacilli into a tibial artery brought on pathology that was, macro- and microscopically, consistent with what he had seen in rabbits, guinea-pigs, surgical patients, and post-mortems. The breakdown of cartilage matched what occurred in human patients (841). If tubercle bacilli were injected into the nutrient artery of the bone, the disease was relentless and predictable, causing bone necrosis of a cheesy consistency, thickening of the synovial membrane, and destruction of the articular cartilages as soft caseous deposits in the bone burst into the joints (841). All of the observed changes were characteristic of tubercular disease of bones and joints in human beings which he had seen in practice and in post-mortems (841). Just as in the lupine experiments, the caprine revealed that "all the characteristics of tuberculous diseases of bones and joints in man can be faithfully reproduced by the injection of tubercle bacilli into the nutrient arteries of bones, into the bones themselves, or into the joints" ("Lectures on the Pathology of Tuberculous Diseases" [18 April 1891]: 843). The ensuing morbidity was grossly consistent with what happened to human beings, despite histological distinctions between human and animal pathology.

The Anti-vivisectionists acknowledged none of Cheyne's contributions to the bacteriological, pathological, surgical, and medical comprehension of the disorder. Instead, they incited public outrage by presenting in the starkest terms the role animals played in his work, not in the context of public health and service, but as an exercise in mutilation and in the wanton induction of disease. The anonymous article "Some of Last Year's Vivisections," published in *The Zoophilist* of 2 May 1892, exemplified this tactic. The content of Cheyne's April 1891 lectures "On the Pathology of Tuberculous Diseases of Bones and Joints" implied how, in their opinion, he routinely misused animals. *The Zoophilist* took aim at "Lecture II," which had appeared as two installments in the 4 and 11 April issues of *The British Medical Journal*. One of Cheyne's best papers—an incisive, 11,000-word model of biomedical discourse—"Lecture

II" contains two statistical tables and eight pathology photographs. Intended by the propagandist for public consumption, the *Zoophilist* author(s)' critique argued that Cheyne's endeavor was pointless. The "Lecture," in their corporate perspective, was nothing more than a free-floating list of zoological atrocities, performed by a scientist aimlessly replicating the procedure; the allusion to Magnan is implied. Though the Cheyne experiments conformed to the 1876 Act, out of scientific context the injection of human secretions into the eyes, bones, and joints of rabbits, guinea-pigs, and goats was an unspeakable perversion of medical science. The Schüller method was cited for effect: a chloroformed guinea-pig was given knee-joint contusions with a mallet or hammer.

Cheyne's experimental source of tubercle bacilli was often drawn from human beings who had died from the disease. Statistics that Cheyne garnered showed that the knee-joint was the primary focus of tubercular disease in young people: 39 percent of children under ten who had been struck by the disease suffered crippling and painful synovial damage; 61 percent (ages 10 to 20) of these children and young adults developed primarily osseous disease; between the ages of ten to 20, 49 percent had synovial infection, 51 percent osseous; and those older than 20, 33 percent were synovial sufferers, 65 percent osseous ("Lectures on the Pathology of Tuberculous Diseases" [4 April 1891]: 740). "The frequency with which the disease commences at various periods of time," he points out, "differs greatly in the case of different joints, but as a whole it is most frequent in childhood" ("Lectures on the Pathology of Tuberculous Diseases" [25 April 1891]: 899).

Cheyne archived diseased tissue at King's College Hospital. These tuberculosis specimens were examined and the results carefully recorded in case histories. In children and adolescents, immature connective tissue and undeveloped bone stricken with the tubercle bacillus deteriorated rapidly, crippling survivors (Henry Gray, *Anatomy*: 74). Studying diseased synovial membranes microscopically, Cheyne learned much from a tissue section of the *ligamentum teres*, a triangular band of connective tissue attached to the femur (Henry Gray, *Anatomy*: 242). The specimen came from a five-year-old girl who had died of tuberculosis ("Extracts from Three Lectures on the Tuberculous Diseases" [6 December 1890]: 1283). In a demonstration of tubercular bone disease, he presented a caseous deposit that had been dissected from articular cartilage of the tibia; the disease had spread to the child's spine ("Extracts from Three Lectures" [13 December 1890]: 1348). A specimen removed from a swollen and abscessed knee-joint was photographed and anatomically described. It had belonged to a sixteen-year-old girl (1350–1351). Tuberculosis of the skull bones, though rare, affected young adults primarily; this condition caused brain swelling and abscesses ("Lectures on the Pathology of Tuberculous Diseases"

[4 April 1891]: 741). Disease of the elbow-joint, too, was primarily "an affection of young adults," impairing the bone directly (741). Especially in children, the spine was the most common location of tubercular bone disease (742).

Experimentation on live animals and the pathology specimens collected from King's College Hospital patients provided Cheyne with insight into tuberculosis pathology and helped him to devise surgical procedures. One such procedure was atherectomy: a bladed catheter was inserted into bones and joints to remove tubercular tissue efficiently. He was able, for example, to remove a mass of tubercle bacilli from a six-year-old's humeral condyle. The child's surgical wound was disinfected with carbolic acid, the disease eradicated, and the joint healed. Similar results were obtained for a three-year-old girl, from whose swollen joint he had evacuated a marble-size mass of tubercular tissue, along with part of the synovial membrane. Recovery was complete, as it was for a seven-year-old boy whose three-month struggle with tuberculosis of the hip had crippled him. Cheyne had incised the outer part of the child's trochanter, trephined the bone, removed tubercular deposits near the epiphysis, sponged out the wound with undiluted carbolic acid, and left it open to drain. After six months, the wound finally closed, and the child recovered considerable movement and was disease-free ("Abstract of 'Lectures on the Treatment of Surgical Tuberculous Diseases'" [2 July 1892]: 13). Cheyne's contribution to the understanding and treatment of tuberculosis was considerable, and, as I have tried to show, the use of animal models was essential to a strategy that benefited pediatric and adult medicine.

1890–1891: Koch, Cheyne and the Substance

Cheyne's experience with tuberculosis also had diagnostic and limited therapeutic aspects. Koch announced, in an 1890 keynote "Address" at the Tenth International Congress of Medicine in Berlin (4 August 1890) that he had been developing an anti-tubercular vaccine. Claiming to have derived the agent from tubercle bacilli, he affirmed that it was capable "of arresting bacterial development *in vitro* and in animals" ("Historical Perspectives Centennial"). He did not say it destroyed the bacilli or was a cure. I would like to examine precisely what Koch had said and then consider the results of Cheyne's analysis. Although the use of laboratory animals in tuberculin trials was the foundation of research in this area, therapeutic tests on the substance were undertaken in clinics and hospitals.

At the close of the "Address," Koch reviewed a list of substances that had failed to restrict the growth of tubercle bacilli. The list included animal dyes and chemical compounds, none of which had any effect on tuberculous animals. These failures did not prevent Koch from claiming to have a "substance" that halted the micro-organism's growth in test tubes and in animal bodies ("On Bacteriological Research": 383). He had been working on the agent for the better part of the year but had not yet reached definite conclusions. Despite the tentative nature of the work, he felt confident enough to declare:

> In spite of these failures I have not allowed myself to be discouraged from prosecuting the search for growth-hindering remedies, and I have at last hit upon a substance which has the power of preventing the growth of tubercle bacilli, not only in a test tube, but in the body of an animal.... [M]y researches on this substance ... are not yet completed, and I can only say this much about them, that guinea-pigs which ... are extraordinarily susceptible to tuberculosis if exposed to the influence of this substance, cease to react to the inoculation of tuberculous [bacteria], and in guinea-pigs suffering from general tuberculosis even to a high degree, the morbid process can be brought completely to a standstill, without the body being in any way injuriously affected [383].

Since the guinea-pig results were intriguing, the possibility that tuberculosis could be arrested urgently demanded verification. In the 22 November 1890 "Further Communication on a Remedy for Tuberculosis," Koch reported on similar efforts of scientists in England and Germany, notably of Cheyne's ("Koch's Discovery of the Tubercle Bacillus"; Kaufmann, "Robert Koch's Highs and Lows"). Testing the effect of injections in healthy individuals, even on himself, he estimated that a patient could tolerate as much as 0.01 cubic centimeter of the mysterious tuberculin. Local reactions, swelling, sensitivity, and redness, occurred in those whose bones, joints, and glands were affected (1193–1194). Those who suffered from cutaneous tuberculosis or *lupus vulgaris* seemed to benefit from a series of inoculations, as their lesions tended to crust over and resolve superficially (1194).

Koch could not substantiate the curative value of tuberculin, but he was reasonably certain of its diagnostic value. Because a patient afflicted with tuberculosis reacted severely to the initial dose of vaccine, perhaps as an allergic reaction, the presence of active tuberculosis in the body was thereby confirmed. In cases where lungs or joints had been infected, and when the patient was thought to have been cured, an injection of tuberculin could indicate the presence of residual disease if a febrile and inflammatory reaction followed (1194). As a cure, however, the tuberculin vaccine had not been established. In fact, its active ingredients remained vague and its effects not satisfactorily tested. On the treatment of infected bones and joints, Koch was encouraged by rapid cures in new cases of moderate cutaneous tuberculosis, and he even ventured to say that some improvement was observed in severe cases (1194). Exhaustive investigation was plainly needed.

Anomalies turned up. Those with pulmonary tuberculosis, for example, coughed and expectorated at the first application of tuberculin; then symptoms waned ("Further Communication": 1195). Perhaps unintentionally, Koch had put too much stock in anecdotal and incomplete results. He writes, for example, of four to six weeks of treatment for early-stage pulmonary tuberculosis, rendering patients "free from every symptom of disease." On the basis of this single remission, he overreached by suggesting that they "might be pronounced cured." Before any more could be said, clinical data had to be compiled, correctly interpreted, and anomalies had to be explained. Another example of the unpredictability of the disease was that patients with large lung cavities had improved symptoms (1195). Koch continued to make unsubstantiated statements to which suffering patients clung. One egregious example was that, through this treatment regimen, pulmonary tuberculosis "in the beginning can be cured with certainty by this remedy" (1195). Did he mean that tuberculin had the potential for a cure *only* in the earliest stages of disease? He made it clear, however, that a cure for tuberculosis ultimately depended on a combined medical strategy. For patients in the convalescent or advanced stage of the disease, Koch suggested that they might benefit from a strategy combining tuberculin vaccine, surgery, and "other curative methods" (1195).

In the autumn of 1890, the Council of King's College appointed Isaac Burney Yeo (1835–1914) and Cheyne as visiting researchers to Berlin. Supplying Cheyne with sufficient vaccine, Koch commissioned him to demonstrate its action in London (Koch, "A Further Communication" [22 November 1890]: 1197, 1198). Cheyne's findings were received on 2 March and communicated on 28 April 1891. Based on histological data and clinical experience, Cheyne did not agree that in its present form tuberculin was a curative agent—not in early-stage lung disease (as Koch had suggested), and only rarely in cutaneous disease. The agent, in Cheyne's view, had value as an adjunctive therapy in a broader treatment plan that regulated vaccine doses, controlled secondary infection, employed surgery for tuberculosis of the skin, and recommended convalescence, especially to improve pulmonary function.

The object of Cheyne's tuberculin experiments was to test the veracity of Koch's assertion that, in lab animals, as long as treatment was continued, tuberculin was able to arrest the progress of the disease ("On the Value of Tuberculin": 254). Cheyne corroborated, and elaborated upon, Koch's positive findings in regard to the pulmonary form of the disease in patients whose vaccinations were continued for some time (1.0 to 2.0 decigram/0.10–0.20 grams of tuberculin per week). The aim, in Cheyne's judgment, was to slow down the disease long enough for the body to encapsulate morbid tissues and, over the course of months, to allow antimicrobial products of blood and tissues to neutralize the bacilli (257). Adjusting dosages per patient to prolong the stand-still of internal disease could work for five months or more, giving the immune system time to destroy the bacteria (258–259). If the physician managed the disease satisfactorily, the surgeon could then "expedite the cure" (258–259). Having to continue the tuberculin injections indefinitely, however, was a serious obstacle in hospital practice since patient compliance was irregular and the cost of treatment high (259).

With tuberculin alone, cures for the cutaneous form were rare ("On the Value of Tuberculin": 260). The reason for its limited effectiveness was that cutaneous lesions composed of tubercular nodules were embedded in dense fibrous and scar tissue (261). Ordinary methods for clearing the diseased skin tissue were to scrape or burn tubercles off with acid, but isolated ones tended to recur. Since tuberculin was able to destroy isolated tubercles but was inef-fective against nodules, it made perfect sense to combine these methods. Cheyne contended that once preliminary injections of tuberculin produced a local reaction, nodules were easier to localize as the swelling and inflammation gradually receded; at this point, the physician scraped off tissue that the tuber-culin had softened. Following curettage, Cheyne then applied nitric acid (HNO_3) to the area, increasing the vaccine dosage gradually, and treating for at least six months to destroy rather than to retard the bacilli. In only a few days, wounds tended to heal, at which point Cheyne covered the skin with a salicylic plaster (261).

Through limited clinical experiences, Cheyne arrived at several provi-sional conclusions on the value of Koch's vaccine. Overall, he believed that tuberculin could cure the disease but only in its earliest stage when a few tuber-cles were visible ("On the Value of Tuberculin": 263). For tuberculosis of bones and joints, Cheyne advised that tuberculin be used post-operatively to further wound healing; for example, after Cheyne performed a complete excision, he stitched the wound closed and waited to see if healing commenced. If the wound remained septic, and if tubercles returned (as was often the case), he then used tuberculin to support the healing process. The most important point to keep in mind in these cases was that the regimen had to be scrupulously fol-lowed for at least six months until scarring was complete (265).

Pulmonary tuberculosis in an early stage, according to Cheyne, should be treated with tuberculin without delay. Gradually increasing doses were called for, not to achieve a cure, but to reach a state of standstill; dosage (one deci-gram/0.1 gram) should be gradually diminished so that the patient did not inadvertently acquire tolerance to the substance. For relapsed cases, tuberculin was a last resort ("On the Value of Tuberculin": 267).

Cheyne did not dismiss tuberculin as an abject failure. Rather, he elaborated on its potential and described its limitations. Clinical practice provided an abundance of information about which patients might benefit from tuberculin and about how to use the remedy effectively and with less risk ("On the Value of Tuberculin": 267). Cheyne also reflected on tuberculin's constituents. It certainly contained an active principle of some worth. Was it possible, he thought, that the compound needed refinement—that several chemicals in tuberculin could work synergistically? This question of synergy could only be answered once the active agent had been isolated and tested in animal models. Ending on a constructive note, Cheyne recognized that, in its present form, tuberculin had a place in the treatment plan of tuberculous disease; however, its use was subordinate to surgery (268).

1897–1920: The Dogs' Protection Bill: An Ethical Dilemma

As the Anti-vivisectionist campaign and the struggle against tuberculosis extended into the early decades of the twentieth century, Cheyne again found himself at loggerheads with the activists and with how their political tactics threatened modern medical research. The Dogs' Protection Bill, a postscript to Cheyne's public health campaign, was a late example. The Bill reached its second Parliamentary reading in the summer of 1919. If passed, it would outlaw *all* medical experimentation on dogs. The Bill had medical backers and dissenters, many justified in their indignation. One American critic, Dr. George Hoggan, in a 2 February 1875 letter to *The Morning Post*, was reported as having exposed unjustified and inhumane physiological experiments on dogs and other animals (French, Appendix II: 414–415). Cheyne and eight colleagues, at a Parliamentary Hearing, requested that the current Bill proscribing canine models in the laboratory be revised and made congruent with the letter of the 1876 Act. On 3 May 1919, Cheyne and medical colleagues contended that the 1876 Act and later legislation had already provided safeguards for animals and strict regulations for ostensibly medical purposes ("Dogs' Protection Bill": 552). The tactics and propaganda campaign waged by the Anti-vivisectionists, however, continued to cross the line between civil discourse and ideological propaganda and, in certain instances, approached defamation of character. According to Cheyne and colleagues, the Anti-vivisectionists had promoted public appeal through "deliberate misrepresentations, by partial quotations, by speaking of medical men who support the agitations as 'eminent' when they [had] no title whatever to such an appellation" ("Dogs' Protection Bill": 552). They had been persuading prominent philanthropists that doctors were conducting unnatural procedures, "directly contrary to the truth" (552). Cheyne and colleagues refuted the distortions and underhanded tactics because the prohibition of the canine model in research would set back medical and veterinary care.

To support their counterargument, Cheyne and colleagues distinguished between two classes of canine experimentation: one, neither painful nor lethal, used subcutaneous injections to investigate distemper and jaundice; in this class were feeding experiments, conducted to learn about rickets. A second class, involving incisions, was more controversial. In accordance with the law, these experiments had routinely employed anesthetics. Two results were foreseen: one was that a dog was painlessly killed without recovering consciousness; and in another, the animal was allowed to recover from the anesthetic and kept alive, unless its suffering called for euthanasia, as prescribed by the 1876 Act. The majority of cases fell into the former category. Pain and suffering on the part of the dog would vitiate the experiment and violate the law, so this was scrupulously avoided. Compliant with the law, the doctors in the second group justified vivisection, broadly defined as any painful or life-threatening procedure, and they were on firm medical ground.

Experiments on dogs had been accruing considerable value in physiology, with respect to the cardiovascular system, the processes of digestion, neurological functions, endocrinology, and nutrition. In the conquest of rabies, canine experiments aided immensely in diagnosis and treatment. The physiologist Ivan Petrovic Pavlov (1849–1936), from 1890 to 1900, investigated new treatments for digestive and intestinal disorders, for the management of edema, diabetes, and for organic and functional disorders of the heart, especially in soldiers returning from war ("Pavlov"); in addition, efficient ways to locate and treat brain and intestinal injuries had been the result of canine experimentation ("Dog's Protection Bill": 552–553). Canine experiments improved veterinary medicine. Vaccines were being produced with the use of dogs to protect them from distemper, and drugs had been developed through animal research of this kind to treat canine blood parasites (553).

According to Cheyne and colleagues, at least four reasons supported the employment of canine models for research: the canine vascular system was suited to arterial surgery; canine and human tissue structure and metabolism approximated each other; dogs tolerated anesthesia better than did lower vertebrates; and, like man, they subsist on a mixed diet. Cheyne was not exaggerating when he said that, if canine research were prohibited, "the advance of physiology and pathology and medical treatment" would be seriously impeded and, in some areas, brought to a halt (553). In refuting Sir Frederick Banbury, M.P. (1850–1936), who advocated the Bill's passage in the House of Commons on 19 March 1920, Cheyne reiterated that the House "should refuse to proceed further with a measure which would impose an unnecessary and serious obstacle to medical research" ("The Dogs' Protection Bill": 739).

In the May 1920 edition of *The Starry Cross* (formerly the *Journal of Zoöphily*), Stephen William Buchanan Coleridge (1854–1936), co-founder of the National Society for the Prevention of Cruelty to Children, scoffed at Cheyne's Parliamentary testimony that the absolute proscription of canine models would impede human and veterinary medicine. Coleridge anecdotally paraphrased Cheyne's argument: "out of millions of dogs in this country something like four or five hundred dogs at the outside were dealt with." Coleridge's modest proposal compared Cheyne's canine allusion to child abuse: "out of the millions of children in this country something like four or five hundred at the outside are tortured, maimed, starved, are beaten to death; why fuss about it? There are forty million people in the British Isles: does that justify the ill-treatment of four hundred of them? But anything goes when you are going to talk a bill out" (73).

Cheyne had implied in his 1897 refutation of the Anti-vivisectionists' argument that their philosophy, and not professional medicine, was inhumane: "it is sufficient answer to the antivivisectionists who oppose the use of intelligence and observation and experiment, to point to the saving of human life and the relief of suffering"; moreover, their tactics threatened to restrict bacteriological research: "If these deluded people had their way, the result would be that experiments would be limited to man, and everyone to whom a new idea occurred would apply it without any previous investigation—surely an appalling prospect, whether for physician or patient" ("On the Progress and Results of Pathological Work": 588). As late as 1920, a reasonable compromise between animal-rights advocates and the medical profession had not been brokered.

8

1881–1886
Asiatic Cholera

    The Asiatic cholera pandemic of 1881–1896, the fifth in what would be a series of six such health crises, erupted in India where the disease originated and had become endemic (Kohn: 11–18). The focus of the 1881 outbreak was Lower Bengal. From there, the epidemic spread in 1881–1882 to the Punjab and Lahore regions of northwest India, into Southeast Asia, to the South Pacific, and then to Egypt as Muslim pilgrims carried the disease from Mecca to Damietta, in Egypt's Nile River Delta. Robert Koch headed a German investigative team while Isidore Strauss (1845–1896) and Emile Roux (1853–1933) led the French contingent (Worboys, *Spreading Germs*: 247). The British government, in July 1884, dispatched the histologist Emanuel Edward Klein and colleagues to the region in order to determine the source of the disease. Klein and the pathologist Heneage Gibbes (1837–1912), co-authors of, *An Inquiry into the Etiology of Asiatic Cholera* (1885), delivered their Report to the Royal Society. With tissue samples from cholera decedents, Klein employed the Henle-Koch method to test for disease causation. After conducting experiments and autopsies, he rejected Koch's claim that the so-called comma-bacillus was responsible for the disease.

    The five subsections of this chapter survey the movement from eclectic to bacteriological understandings of cholera; the origin of the Vibrio concept, its form and function, up to Cheyne's time; descriptions of the life-cycle, along with the reproductive and metabolic aspects, of the bacillus; investigations into the regional incidence, distribution, and control of cholera; and the debate between E. E. Klein and Cheyne on its pathogenesis.

1852–1886: Eclecticism versus Bacteriology

From the 1840s to the 1880s, a bacteriological understanding of cholera gradually supplanted eclectic formulations that had tried to explain the origin, nature, and host-effects of the disease. I am using the word *eclectic* in this context to signify a conflation of incongruous elements. Pre-bacteriological theorists, many of whom were sanitarians, created models of this kind. In 1884–1885, when the Koch-Klein debate over cholera etiology and pathology was at its peak, a coherent bacteriological account of the disease began to emerge. The cholera germ would eventually be classified as an obligate parasite, taxonomically in the Class **Gammaproteobacteria**, Order **Vibrionales**, and Genus **Vibrio** (Farmer and Janda: 493–494). No longer existing as an eclectic phenomenon, the cholera pathogen now bears the binomial epithet, *Vibrio cholerae*.

The epidemiologist and medical statistician, William Farr (1807–1883), in the 1840s and 1850s, represented the cholera pathogen eclectically. On the one hand, he depicted it as spontaneously generated, saprophytic, fermentative, and reproductive. On the other hand, he considered it a "nonliving organic poison," a kind of atmospheric pollutant transmissible, not through person-to-person contagion, but on wind currents (Eyler [1973]: 82–84; and [2001]: 228; Worboys, *Spreading Germs*: 114, 129–131). Farr was correct to assume that the agent had toxic properties, an intuition proven to be true one century later. In 1951, biologists genetically identified not one but three such poisons, abbreviated **Ctx**, **Zot**, and **Ace**. As of 2009, a total of 12 choleraic poisons have since been identified (S. N. De et al., "Experimental Study"; J. McDowall; M. Trucksis, et al.; K. Todar; Chauduri and Chatterjee: 6). A soil-bound particle called *cholerine*, Farr's agent had properties more in common with the semimetallic element, arsenic trioxide ($As_4O_6As_4O_3$) than with organic poisons, composed of the proteins, enzymes, and polysaccharides of living organisms. From another perspective, the toxic property of Farr's cholera agent resembles the poisons produced by soil-bound, sporiferous micro-organisms, such as *Bacillus anthracis* (Miller and Levine: 51–52, 1073; Campbell, Reece, et al.: G-15; Chauduri and Chatterjee).

By 1866, as John M. Eyler has shown, Farr's cholera nosology changed radically. His statistical study of municipal public works in London during the 1849 outbreak and the 1854 epidemic, along with his interest in bacteriology, revealed to him that water was the medium of cholera transmission. By the mid–1860s, he would begin to question the sanitarians' atmospheric accounts of how cholera spread ([1973]: 98–99).

Max Josef von Pettenkofer (1818–1901), to whom I alluded earlier, also developed a choleraic etiology combining incongruous elements. He correctly inferred that pure water and sanitation could prevent cholera outbreaks and, in 1854, helped to manage an epidemic in Munich through a new water supply and drainage system (Worboys, *Spreading Germs*: 116–117). As late as 1875, he understood that stagnant water carrying putrefying material into soil was biochemically conducive to the development of cholera: "Nothing which can impregnate the ground with decomposing matter should be left lying for any length of time on the surface" (*Cholera*: 39).

As a sanitarian, Pettenkofer did not consider cholera to be transmittable from one person to another. Investigations in India, he reported in July 1874, found no evidence to substantiate the claim that cholera poison was in human excreta (*Cholera*: 31). Human beings, in his view, could not contract the poison or "generative agent" from parcels or slimy food, transported from affected districts. Nor could sick travelers spread the disease outside the outbreak zone (23, 31. 71). These assertions would be proven to be completely wrong.

Von Pettenkofer's etiology, like Farr's, features an atmospheric component. The surrounding air is "the principle vehicle of the generative agent of Cholera," affecting those in "close vicinity of the focus of emission" (*Cholera*: 72). The generator in this system, as in Farr's, was a pulmonary hazard, having properties analogous (in my view) to asbestos, coal-mine dust, or arsenic trioxide. The three loosely-associated components of von Pettenkofer's cholera etiology, therefore, were stagnant water, noxious gases, and poison dust.

In 1866, James L. Bryden (1833–1880), epidemiologist and Anglo-Indian medical advisor, accepted Farr's theory that cholera was transmitted through polluted drinking water. But Bryden incongruently subscribed to the airborne-transmission theory, as well, likening the cholera germ, not to a noxious gas or inorganic poison, but to a seed, the cycles of reproduction and decay of which depended on environmental conditions, and on air-current transmission during monsoon season (Harrison, *Public Health*: 101). Thus, like von Pettenkofer, Bryden used the seed-soil metaphor to describe how the disease spread (Worboys, *Spreading Germs*: 6–7). Although this metaphor misrepresented cholera's actual etiology, it accommodated the toxic property of the disorder. In keeping with the seed metaphor, Bryden's cholera agent is portrayed as a botanical poison, the analogues of which are the complex nitrogenous substances and organic acids found in seeds and roots. The poisonous alkaloid strychnine, for example, can be derived from the ripe seeds of the Asiatic *nux-vomica* tree, while apple, apricot, peach, and other fruit pits contain amygdalin, a chemical that, when exposed to hydrochloric acid in digestion, releases cyanide (A. C. Dutta: 166–167; C. Claiborne Ray, respectively).

By the mid–1860s, an intermediate phase in the development of cholera etiology from the eclectic to the bacteriological model had been reached. Closer to the natural reality were eclectics who, eschewing figurative representations, confined their etiological analyses to medicine and microbiology. One of them was the botanist Ernst Hallier (1831–1904), of Jena, who initially thought that cholera was a fungus (Isaacs: 283–284). He conjectured, in 1867, that cholera germs originated from ingested fungal spores of the species, *Schizosporangium*, and then developed into micrococci, a term of general use in early bacteriology. Hallier's murky implication that a fungus developed *into* a small spheroidal bacterium revived the obsolete Vibrioid taxonomies of Müller and of Ehrenberg. On the basis of vibratility, if we recall, they had grouped protozoa, algae, and bacteria together generically. Hallier's eclecticism also retained the pleomorphic notion that a fungus could metamorphose into a bacterium (Bulloch, *History*: 178, 202, "pleomorphic," *OED*. II. P-Z: 2208). In an interview with the British investigators, D. D. Cunningham and T. R. Lewis, Hallier would later recant the notion that cholera involved fungal-to-bacterial transformation ("Scientific Investigations," III: 77).

The 1868 research of the tropical disease specialist Edmund Alexander Parkes (1819–1876) was another index to the shift from eclectic to bacteriological understandings of cholera etiology. Although Parkes' views reflected the continuing influence of eclectic systems, he embraced new possibilities, dissatisfied as he was with chemical accounts of zymosis (a term nearly synonymous with fermentation), and with miasmatic and effluvial theories of transmission (Worboys, *Spreading Germs*: 116, 129). Parkes avoided unsubstantiated assertions. As to the origin, propagation, and pathophysiology of cholera, he frankly acknowledged that little was known for sure (*A Manual of Hygiene*, I: 134). He cited D. D. Cunningham's and Timothy Lewis' discrediting of Hallier's fungus theory and thought the windborne-transmission theory, "opposed to all we know of its spread" ("Scientific Investigations, III": 77; Harrison, *Public Health*: 51–54, 99–116; Isaacs: 283–284). Although Parkes rejected the fungus theory, he remained partial to von Pettenkofer's soil-based theory (*A Manual of Hygiene,* I: 133, 136,138).

John Murray (1809–1898), Inspector-General of the Bengal Medical Service, had collaborated with Bryden on a cholera report but disagreed with his transmission theory. In the 1874 *Observations on the Pathology and Treatment of Cholera*, Murray repeatedly stated that a poison was responsible for cholera pathology. His work contributed significantly to the paradigm shift from eclectic to an exclusively microbiological understanding of cholera. He suspected that a particular micro-organism was to blame (15). Convinced that the bacterial agent was spread through human excreta, and not by air currents or poisonous seeds, Murray called for restrictions on large population movements and for the regulation of the water supply (Harrison, *Public Health*: 101, 106).

1773–1885: Morphology

Since the cholera bacillus belongs to the Genus **Vibrio**, I will outline early attempts to understand these unique micro-organisms and how one member of this class of bacteria was associated with *cholera*, a term adapted from the Greek word χολέρα ("bile") ("cholera," *OED*. **I**: 404). Otto Friedrich Müller (1730–1784), equipped with a compound microscope, in 1773–1774, is believed to have been the first to have applied the epithet Vibrio to a bacterial Genus (i.e., *V. bacillus*) (Winslow: 294; Bulloch, *History*: 162, 171). Further investigation revealed that organisms other than bacteria could vibrate; and, in 1786, Müller incorporated the protozoan *V. elongatum* and the algae *V. rugula* into the Genus **Vibrio** (Kent: 199; Buchanan: 16; Guiry and Guiry). Using vibratility as a defining trait, Müller created a heterogeneous taxon, including the bacterium *V. bacillus,* four protozoan species (*V. elongatum, V. lineola, V. undula,* and *V. proteus*) and the algae *V. rugula* (Bulloch, *History:* 171).

In the early nineteenth century, vibratory protozoa continued to be assigned to the Genus **Vibrio**. The French naturalist Jean Baptiste Bory de Saint Vincent (1778–1846) used the new term *Vibrionides,* in 1826, to describe worm-like protozoa, devoid of appendages (Buchanan: 17–19). In 1838, the proto-zoologist Christian G. Ehrenberg (1795–1876) calling quivering organisms *Vibrios,* assigned them to the mixed category, *Vibrionia* (Bulloch, *History*: 54).

In 1849, progress in the characterization of the choleraic germ came when the French physician and naturalist Félix-Archimède Pouchet (1800–1872) found quivering organisms in the rice-water excreta of cholera patients. For the first time, vibratory rods had been associated with cholera. Although Pouchet had correctly associated the disease with this micro-organism, he misidentified the quivering germ as Müller's protozoan, *V. rugula* (Bulloch, *History*: 162). Despite Pouchet's deference to Müller's system, gradually, naturalists learned to distinguish bacteria from protozoa more accurately. They also noticed, not only that several kinds of bacteria were vibratory, but also that each of them was uniquely classifiable. For two reasons, then, Pouchet's discovery was a turning point in the effort to isolate the cause of cholera: for the first time, an unusual vibratory micro-organism and choleraic disease had been coextensively associated, and this particular micro-organism was easily differentiated from larger protozoa or infusorian.

Following Pouchet's discovery, the most important contribution to Vibrio classification and to the characterization of the cholera bacillus was that of the Italian anatomist Filippo Pacini (1812–1883). In 1854, Pacini identified the cholera microbe as a *vibrioni,* confirming Pouchet's discovery of Vibrio in choleraic dejecta (Pacini, "Osservazioni," *Gazetta Medica Italiana*: 401). Pacini associated the micro-organism with the disease on the basis of coextension: "non può essere contagione senza essere parasite [contagion cannot occur without this parasite]" (*Osservazioni Microscopiche*: 28). In addition, he designated this bacillus a distinct species. Its uniqueness would soon be confirmed, and today it bears the nomenclature, *Vibrio cholerae Pacini* (J. P. Euzeby).

Investigators in the later nineteenth century rightly suspected that the importance of vibratility as a defining biological trait had been exaggerated. The great difference in size between protozoa and bacteria indicated that they were taxonomically different forms, even though vibratility was common to species in both biological Kingdoms. In bacterial classification, however, vibratility still remained a useful criterion. For example, H. A. Richardson, in 1870, used movement *and* shape to differentiate one bacterium from another. For Vibrio, Richardson listed three defining characteristics: S-shape, "serpentine movement," and vibratility. An organism displaying all three was assigned exclusively to the Genus **Vibrio** ("Vibrio," *OED*. **II**: 3623). In 1875, Ferdinand Cohn tried but failed to improve the classification of Vibrio. His fourfold Tribal classification for bacteria, as noted earlier, had been based on form: Sphere, Rod, Filament, and Corkscrew. The problem with Cohn's paradigm was that he grouped Vibrio (quiverers) and Bacillus (rods) in the same class ("Studies on Bacteria": 214). By 1875, the relationship between the cholera pathogen and the disease itself was in its earliest phase, but the ability to differentiate between vibratory bacilli and protozoa was a significant advancement.

In 1854 John Snow (1813–1858), an anesthetist, inferred from epidemiological work that a disease-causing entity, either living or chemical, had been present in south London's downstream water during a cholera outbreak (Bynum: 81). In addition, he determined that this material reproduced itself during the crisis; that it could be filtered out or killed by boiling; that sewage-contaminated water in which the agent proliferated had to be kept away from public consumption (Bynum: 81). Theorizing that a poison traceable to the waste products of victims was contaminating the water supply, he presented his findings in *On the Mode of Communication of Cholera* (1855) (Winslow: 271–272). Pacini, also in 1854, had independently identified a unique micro-organism in choleraic excrement (Bynum: 81). The Snow and Pacini investigations provided etiological information that, by the 1880s, might have had a strong influence on the coextensive relation of the micro-organism and the disease. Unfortunately, their work had not convinced their peers (81). Although the chief investigators of the mid–1880s knew of the water-borne theory, they did not allude to Pacini's observation that a bacillus was present in cholera secretions and that the organism and the disease always affected the same anatomical site. Cholera, as Snow had argued, was communicated through polluted water and had toxic properties.

Thirty years later, Robert Koch causatively related the organism to the disease. In the 30 August 1884 Cholera "Address," delivered before the German Board of Health in Berlin, he reported having observed from autopsies "a special kind of bacteria" abundantly present in diagonal sections of intestinal mucosa and in necrotized epithelial tissue (404). These slightly-curved organisms were about one-half to two-thirds bigger and thicker than tubercle bacilli. Microscopic examinations at 600x revealed live organisms on two-day-old damp linen, contaminated with human secretion. To Koch, the putative microbe on the stained linen did not resemble conventional bacilli but looked more like "long spirals"; in length and overall appearance, they could be mistaken for the Spirochete of relapsing fever ("Address": 404). As earlier systematists had observed, morphologically Koch's organisms did not match any established classification. Judging from their rod-like and spiral shapes, he conjectured incorrectly that the organism was a transitional form between bacillus and Spirillum (i.e., curved flagellate bacteria) (404). He even resorted to insect imagery, describing these curved, vibrating rods as "a swarm of dancing midges" (404). In substance, Koch's August 1884 "Address" recapitulated Pacini's 1854 findings.

In the 6 September 1884 Berlin "Address," Koch reiterated that, morphologically, these micro-organisms did not fit any known Genus. His suspicion that they were implicated in cholera, however, had increased. Finding them in the tissues of cholera decedents, as had Pacini, he adopted a temporary nomenclature: the phrase, *comma-bacilli*. The descriptor *comma* (German, *komma*), of course, refers to the curved punctuation mark. On the basis of quivers, curves, and motility, Koch asserted that this germ was visually unique and therefore represented a distinct "species," easily distinguished from other bacteria ("Address" [6 September 1884]: 452). Koch and others continued to suspect that the organism was behind the current pandemic.

Seven months after Koch's 6 September "Address," Cheyne reported on what he had learned about this mysterious microbe. In his 25 April 1885 "Report," he described what he had seen in the secretions of five cholera patients. Unlike Koch, he was very reluctant to classify germs, primarily or exclusively, on the basis of optical evidence. In case #2, for example, though Cheyne recognized Koch's comma-bacilli in specimens of a nearly "pure cultivation," he could not account for the fact that less than half of the bacilli could be absolutely identified as such (823). Another specimen, drawn from a cholera decedent (case #9) was even more perplexing. In it, he saw no organisms fitting Koch's description at all. These inconsistencies revealed to Cheyne that the microscopic identification of bacteria had serious limitations ("Report" [25 April 1885]: 823).

Gradually, it became clear that the comma-bacillus had various subtypes and correlatives, two of which, in particular, had nothing to do with cholera disease. Not every curved rod, as Cheyne learned, was pathogenic. Investigators found that there were curved organisms that were microscopically similar to one another. Two distinct kinds of non-choleraic ones intermingled with each other and with Koch's germ. The two innocuous Vibrios were Finkler's comma-bacilli and Flügge's cheese-bacilli. Complicating the picture even more was the observation that suspected cholera bacilli varied in shape and size, a phenomenon that would later be explained as phases in their life-cycle ("Report" [25 April 1885]: 823). Cheyne reminded readers that, although morphology was a legitimate way to begin an investigation, visual evidence could be deceiving: for example, variations in size and form might be seen in the same organism in different culture media (821). The only way to answer these questions was to follow established procedure. Thus, before animal testing could commence, he had to acquire a pure culture. Upon his return to Paris in the spring of 1885, he sent several tubes of cholera to Koch, requesting that the latter certify both their purity and point of origin as his laboratory. To both requests, Koch responded affirmatively. With authentic seed cultures in hand, Cheyne could then move directly to more revealing tests.

With Koch's pure cultures in hand, Cheyne set about establishing the size and behavior of the comma-bacilli and contrasting these characteristics to those of other Vibrios. To prevent contamination of the culture, he looked at aggregates rather than at isolates. By examining an entire colony, he could compare variants to each other and confirm the presence of "the well-marked curve" that Koch had thought a signature feature. Intraspecific variations required careful scrutiny. Cheyne already knew that comma-bacilli had an immature linear phase, making ocular identification challenging. The mature microbe, in his view, also displayed a variety of forms—both S- and other nonlinear shapes—which recalled H. A. Richardson's idea of formal variants ("Report" [2 May 1885]: 877). Once the limits of morphology had been reached, both Koch and Cheyne independently turned to physiology and then to etiology (478).

1884–1885: Physiology and Etiology

We return, at this point, to Koch's 1884 investigation of the physiological properties of the bacillus. He had reported, in the 30 August 1884 "Address," that comma-bacilli could be cultivated in meat-broth, milk, blood serum, food-gelatin, and on Ceylon moss, mixed with meat-broth and peptone (404–405). The gelatin medium was especially useful when observing colonies. A microscopist could localize colonies better in solid media because it provided a three-dimensional view of cells suspended in the viscous gel (404–405). In a meat-broth culture viewed at 600x the colony grew luxuriantly to the point of resembling a circular droplet, irregularly bordered, hollowed out, and jagged. As it grew, it became granulated, like "a little heap of pieces of glass." The gelatin medium, as a food source, gradually liquefied, at which point the colony expanded; and, sinking into the medium, it formed a funnel, visible from above as a whitish point (405). It became clear that each organism and the colony as a whole had a developmental pattern.

Koch recorded additional physiological data. The bacilli, if nourished, grew abundantly within a temperature range of 30 to 40 degrees Centigrade (86 to 104 degrees Fahrenheit) ("Address" [30 August 1884]: 405, 406). Below 16 to 17 degrees Centigrade, however, the colony shrank and then disintegrated (405). Without air, the bacilli could not live (405). Considering the devastation that the cholera pandemic was causing, it must have been rather ironic to learn that the colonies were not very durable and that unless kept moist would rapidly deteriorate. Koch, in fact, could not preserve the bacilli in a dry state longer than 24 hours (407). Thus, these fragile but deadly germs had to be nurtured if anything was to be learned about metabolism and the life-cycle. The pH levels also had to be carefully monitored; for example, if meat-broth or gelatin were to become acidic, the comma-bacilli became stunted and ceased growing altogether (406). Even with luxuriant growth in a pH-balanced environment, the bacilli lasted only two or three days, and faster-growing species then dominated the artificial environment (405).

Comma bacilli did, indeed, have a life-cycle and competed with other species (405). Koch deduced that, because the cholera bacilli died off so quickly, chemical disinfection was not always necessary. Competitors, such as putrefactive organisms, could eliminate the comma-bacillus in a cesspool. In view of this phenomenon, Koch surmised that the use of chemical disinfectants could actually upset the natural balance in floral ecology, killing off innocuous forms and permitting the unhindered cholera germs to propagate to dangerous levels (406). Germicides were not entirely proscribed. If cholera were found in a sink used for drinking water, for example, then disinfection was needed immediately (406). Nor did Koch obviate the testing and employment of chemical antiseptics for human beings. Among the inhibiting agents tested were aluminum, camphor, carbolic acid, peppermint, sulfate of copper, quinine, and corrosive sublimate; the last, as the most effective antiseptic, could be used internally but only if very diluted (1–100,000) (407).

Koch continued to gather clues, some of which did not make immediate sense. For example, though the bacillus died off rapidly, cholera morbidity persisted. He inferred from this phenomenon that the dying micro-organisms exuded toxins. If this was the implication, then Koch was trying to explain the effects of lipopolysaccharide components of the cellular membrane, released when cells die and their walls break down; scientists have since determined that cholera bacilli release exotoxins: proteins that stimulate intestinal cells and that release chloride ions into the gut, leading to dehydrating diarrhea (Campbell, Reece, et al.: 571–572).

Observing the bacillus in culture had led Koch to conclude that medical intervention against the germ might not be indicated in every instance since it rapidly died in a dry environment ("Address" [30 August 1884]: 407). But the greatest danger lay in the organism's exotoxins (and possibly in its endotoxins): the "complex of symptoms of the attack proper of cholera, which is inspissation of the blood, is ... to be regarded essentially as poisoning" (Koch [6 September 1884]: 456). A physician and Medical Officer, Sidney Vincent Richards, of Goalundo and Kooshtea, Bengal, had already suggested to Koch that cholera bacilli produced toxins in the gastrointestinal tract (*Indian Medical Gazette* [April 1884]; Koch, "Address" [6 September 1884]: 455). Animal testing in S. V. Richards' laboratory, however, was positive only in porcine models (Koch, "Address" [6 September 1884]: 456). The tentative finding was that substances in choleraic secretions, under certain circumstances, could poison pigs (456). The poisons mortified intestinal epithelia and mucosa and, once absorbed, appeared to paralyze the circulatory system (456). The line of inquiry, from Snow and Pacini, to Richards, Koch, and Cheyne, supported the idea of cholera toxins. In 1951, S. N. De definitively established that the culture filtrate of *Vibrio cholerae* contained exotoxins ("An Experimental Study": 707–717; Takeda: 149).

In the spring of 1885, Cheyne's first step in the physiological analysis was to isolate the bacillus. The blood and tissues had to be studied. The microorganisms on various media had then to be carefully observed and their growth patterns recorded. Furthermore, the bacillus had to be sought in "morbid products," other than in dejecta. Without stating so, Cheyne was attempting to answer questions central to Pacini's research: if a species of bacteria were found to be constantly present in cholera or in other diseases, was it the "true cause" of the disease? Moreover, was it present if the disease was quiescent ("Report" [25 April 1885]: 821)? Experimentation on suitable animal models was underway to learn whether the putative organism reproduced the original disease in a healthy animal (821).

Cheyne found meat-infusions, solidified by the addition of gelatin, to be the best culture medium ("Report" [25 April 1885]: 821). Koch's purpose in devising this method, as noted above, had been to facilitate pure cultivation experiments: bacteria planted on solid, rather than in liquid, media were easier to study. In case of contamination from extraneous microbes, for example, an investigator working with solid media could transfer the pure colony to another receptacle without having to restart the experiment (821).

Cheyne adeptly used three culture methods: (1) sterilized test-tubes containing gelatin in which pure cultivations of cholera bacilli were sown with a platinum needle (the organisms growing along the needle tracks); (2) the "slide-cultivation" method in which sterile wire sowed bacteria along tracts on liquefied jelly for viewing under a low-power light microscope; and (3) the "glass plate" cultivation in which different kinds of bacteria were separated from one another, introduced into liquefied jelly in test-tubes, and diffused through the medium. Once the jelly solidified, the bacteria were seeded throughout and, when colonies formed, species could be differentiated from each other under low-power magnification. These methods could be supplemented by cultivations in alternative media and through successive inoculations in animals ("Report" [25 April 1885]: 821–822).

Consistent with the findings of Cohn and others, Cheyne was well aware that, to ascertain the nature of a micro-organism, one had to consider form, mode of growth, and effects on vertebrate hosts ("Report" [25 April 1885]: 822). Reliance on any single determinant was incomplete and error-prone. Cheyne also cited Koch's experience with morphologically-similar bacilli. In contrast, using various growth media and new methods, one could identify comma-bacilli more accurately on the basis of form, of function, and of specific traits exhibited in cultures (822).

Never ignoring precedent, Cheyne read the work of fellow researchers and conducted his own experiments to test their claims. He recounted his original experiments on five cholera cases, performed in Paris in November 1884, at the *Quinze-vingt Ophthalmogiqué Hôpital*. In accord with the methods of Lister and of Koch, in the dejecta, blood, and intestinal tissue of the three cholera decedents, Cheyne consistently found Koch's cholera-bacilli. Using glass-plate, nutrient jelly cultivations, sterilized pipettes, and other instruments, he meticulously avoided contamination. With pure cultures, he could count each colony and from this figure estimate the quantities and types of bacteria present. He began this process by introducing specimens into sterile flasks of distilled water, placing a sample of the dejecta into each flask with a sterilized platinum wire, stoppering the flasks, shaking the containers to diffuse the specimens in the fluid, and applying drops into tubes of liquefied jelly. Once they had been shaken, specimens were then deposited onto glass plates where they would solidify at a temperature of 20 degrees Centigrade (68 degrees Fahrenheit).

Cheyne estimated the number and identified the kinds of bacteria present in the choleraic sample through cultivations of human specimens (823). Case #2 exemplified how cultivating in this way could reveal much more than ocular examination alone. This specimen was presumed to have been a pure cultivation of cholera-bacilli, but the microscopic examination could not confirm that, since less than half of the micro-organisms were immediately recognizable as comma-bacilli (823). Visual evidence, especially with respect to these organisms, was notoriously untrustworthy; in case #3, for example, less than 20 percent looked choleraic, while in case #5, 50 percent fit the visual description. To enhance visual acuity, Cheyne used Weigert's and Koch's staining innovations. Yet, when he examined sections of the intestinal walls stained with aniline dyes, he saw very few comma-shaped bacilli (823). This anomaly, Emanuel Edward Klein would use to refute Koch's causality hypothesis. Cheyne had to figure out why the course of the disease did not appear to correspond directly to the life-cycle of the colony.

In the second issue of the "Report," published on 2 May 1885, Cheyne definitively established that the comma-bacilli had a growth pattern. As the bacilli progressed through phases, the colony changed shape. The immature, rod-shaped forms developed curvatures. As the colony increased in size, it disaggregated, and gelatin in the immediate vicinity liquefied. Cheyne noted that the average area of a colony, conforming to Koch's estimation, had been about one millimeter (879). Microscopic examination at 120x revealed "an irregular shaped mass of highly refracting granules," centered in an area of fluid jelly; and similar masses floated in the fluid. These aggregations or zoöglœal masses contained an abundance of S-shaped and spiral organisms moving in a "corkscrew manner" (878). As the organisms migrated towards the arc of the zoöglœa, presumably to attach themselves to it, they vibrated and floated backwards and forwards, a characteristic that Cheyne attributed to their "flagella"— cellular appendages of locomotion ("zoöglœa," *OED*. **II**: 3781; "flagella," Campbell, Reece, et al.: G-15). Cheyne logically inferred that the rapidity of colonial growth varied directly to the amount of peptone (the food source) in the culture medium. For him, the most significant aspect of the experiment was that, formally and behaviorally, the colonies of cholera-bacilli appeared to be quite unique; and, in their mature phase, they were unmistakable (879).

Test-tube cultivations performed alongside the glass-plate experiments also revealed comma-bacilli, even though optical resolution was lower with the latter than with the former method ("Report" [2 May 1885]: 879). In tubes, the inoculated specimen underwent a typical cycle: the colony consolidated, an air-bubble domed the surface; and the jelly liquefied as the colony expanded, became funnel-shaped, reddened, and sank to the bottom of the tube, the entire colony dying in six or seven weeks. The comma-shape was present in all mature bacilli. Cheyne confirmed Koch's observation that dehydrated bacilli died in three hours (879).

Recalling Koch's claim that gastric acid inhibited growth, Cheyne retested its effects on ingested cholera bacilli ("Report" [9 August 1885]: 931). The results were positive: ten minutes of exposure to gastric acid killed the bacilli (931). Repeated experiments of this kind demonstrated that "at certain stages of their existence these bacilli are more easily killed than at others, as probably due to difference in vitality in different bacilli" (931). Even when the solution was greatly diluted—1–1,850 parts hydrochloric acid to water—bacilli died in 25 minutes (931).

Cheyne's experiments with Koch's cholera bacilli showed (1) that large numbers of bacilli were present in all cholera specimens; (2) that no other organism was *consistently* present in these secretions; (3) that blood and tissue samples of internal organs, in one case, revealed no bacteria of any kind; (4) that Koch's comma-bacillus could be differentiated from other bacilli, even from similar forms; and (5) that the form, mode of growth, and general characteristics of the cholera bacillus corroborated Koch's 1884 findings ("Report" [9 August 1885]: 932). Cultivation techniques made it possible for Koch, Cheyne, and others to distinguish between different species of vibratory bacilli (932).

By 1885, Vibrio classification had come a long way from the observations of O. F. Müller and others. Cheyne's analysis indicated that cholera bacilli were comma-shaped, except for immature rods. An observer could distinguish similarly-curved Vibrios from helical forms (the cholera bacillus' mature stage) on the bases of size and number of rungs per helix; thus, the cholera helix had from eight to 30 turns, whereas that of the Finkler's bacillus was larger, not markedly curved, and had from three to six turns. Flügge's comma-bacillus, on the other hand, was smaller than Finkler's, and it had more rungs than did the helical stage of *Bacillus cholerae*.

Physiologically, the bacilli species were more easily differentiated from one another, whether isolated or en masse. In nutrient jelly/glass-plate cultivations, cholera bacilli, for example, gathered in small irregular colonies rather than in round granular ones; they slowly liquefied gelatin; and their colonies widened to an area of 1.0-square millimeter. On the other hand, Finkler's colonies, which were flat and contoured, liquefied gelatin, and in 30 hours occupied a 1.0-square-centimeter area; and lastly, Flügge's bacilli, which differed significantly from both Koch's and Finkler's, produced very dark, curved, and granular surfaces; the Flügge colony liquefied the medium more rapidly than had Koch's, but not as rapidly as had Finkler's; and Flügge's earliest growth-stage corresponded to Koch's latest. Thus, on the bases of form and of function, the three species of comma-bacilli could be differentiated from each other more accurately ("Report" [9 May 1885]: 931–934).

With pure cultivations in hand, Cheyne moved on to animal experimentation in search of the elusive toxin, a critical phase of the investigation. New questions arose at every major turn. One was that the bacillus did not affect all species. Another, as the result of extensive guinea-pig experiments, was that it neither damaged the intestines nor appeared to have entered the bloodstream; however, the toxin (not the germ) likely had entered the circulation (933).

Murine experiments with pure cholera cultivations, recounted in the 16 May 1885 installment of the "Report," had surprising and sometimes inconsistent results. Of four mice inoculated, two that had received a nine-day-old cultivation died in six hours; a third died after receiving a three-week-old culture specimen; but a fourth was unaffected. Dissections of the dead mice provided an interesting lead: murine and human pathology in certain respects were comparable. Fecal and blood cultivations were then made to see if the bacillus was present in each species: the results were again positive, as "large numbers of cholera-bacilli developed" ("Report" [16 May 1885]: 975). On the basis of these data, Cheyne became more convinced that a "chemical poison" was responsible for the disease in the first two mice; but, contrary to expectations, the third died from the growth of the organism in the blood (975), findings inconsistent with the guinea-pig studies of 16 May.

Inconsistencies notwithstanding, Cheyne connected the majority of these results to those of S. V. Richards who, in 1884, had blamed chemical poisoning for the rapid deaths of pigs receiving large doses of comma bacilli (cited in Koch, "An Address" [6 September 1884]: 455–456). However, Richards' five tests, because they lacked controls, were inconclusive: three experiments were negative (two of three employed human cholera secretions), and the two remaining animals died. Since porcine lethality in the former's experiments could not be definitively blamed on a toxin, Cheyne attempted to reprise Richards' investigation. Aware that stomach acid could kill the bacilli, he therefore fed cholera cultivations to a pig on two occasions but with no adverse effect on the animal. The gastric-acid line of inquiry was discontinued when it was learned that the experiments of Nicati and Rietsch inducing mortality in an undisclosed number of animals had raised the possibility that the age of the culture affected its level of toxicity ("Report" [16 May 1885]: 975; *Comptes Rendus,* xcix: 928–929, respectively). Comparing the Nicati and Rietsch experiment to his own on mice and pigs, Cheyne conjectured that both the age of the cultivation (old cultures becoming attenuated or devitalized) and the species of the test animals might have been determining factors: older colonies, for example, had no effect on the pigs, whereas a fresh, eight-day-old culture killed mice (Nicati and Reitsch, "Expériences"; "Recherches").

Cheyne summarized the results of his choleraic experimentation to date. Two guinea-pigs succumbed to the cholera bacilli. Though glass-plate cultures grown from their secretions teemed with the bacilli, their intestinal tissues and not their blood (an important distinction) were populated. As for the guinea-pigs that had survived the challenge, Cheyne again detected no micro-organisms in the blood. He attributed the survival of the pigs that ingested culture fluid to the neutralizing effects of gastric acid; he conjectured that the dead mice that had received subcutaneous doses of culture fluid succumbed to choleraic toxin; a third mouse had presumably died of cholera bacilli in its blood and intestines ("Report" [16 May 1885]: 975).

In order to derive diagnostic value from these experiments, Cheyne needed to induce cholera consistently in laboratory animals, to extract the probable cause from the sickened animals and, re-growing it in pure cultivation, to induce the disease in healthy animals. Despite the variability of the results, he knew he was on the right physiological track: "microscopic appearance is ... of no consequence; the essential fact being that, on cultivation ... almost all the colonies which developed on the plate were colonies of cholera-bacilli" ("Report" [16 May 1885]: 975). Without the use of pure cultures throughout, the investigator had little hope of demonstrating, unequivocally, that an organism caused a disease. On the basis of accumulated evidence, Cheyne inferred that the complete life-cycle of the comma-bacillus progressed through three stages: from the linear stage of the small bacillus; to the mature, curved stage; to the helical stage (976). He had learned about other features of the organism, as well: though virulent, it had no spore-forming capability, was fragile, and was sensitive to the external environment. It died rapidly when dried, tended not to occur in droughts, was susceptible to gastric acid, and decomposed when nutrients were either low or absent (977).

In the 23 May 1885 Appendix to the "Report" series, Cheyne summarized in seven parts the physiology of the comma-bacillus. The cholera-bacillus was distinguishable from other bacteria, as Koch had said in the 6 September 1884 "Address." Agreeing with Koch, Cheyne restates that "the microscope alone will not suffice for their detection in most cases, but that the culture-test must be employed as well." Concurring with Koch, he asserted that the bacillus was specific to choleraic disease. In acute and uncomplicated cholera cases, it was found in great numbers and was present in diseased tissue; however, it could neither be detected in healthy persons or animals, nor was it extraneous to the body during an outbreak (with the exception of moist, contaminated surfaces and fabric, as Koch had proven in September 1884). Although direct evidence from animal experiments fell short of confirming the causal linkage, and although human trials were not possible, Cheyne recorded important observations on the role of persons, food, and water in choleraic infection. Moreover, his work with pure cultures and animal models was "almost as good as experiments on man" ("Report" [23 May 1885]: 1027).

As the cholera bacillus had been "well characterized" and had become easier to identify, Koch began to explore in greater depth the etiology of cholera: its origin, cause, and the connection between the putative microorganism and the disease ("An Address" [6 September 1884]: 453). From *in vivo* experiments on lab animals, he turned to human post-mortem specimens. A central problem to be solved, in 1884 (as we have seen), was whether the presence of the comma-bacilli always, or only occasionally, coincided with the disease. To find the answer, tissue samples from ten post-mortem Egyptian decedents were examined. Koch's Commission in India had conducted a total of 42 autopsies and extracted specimens that were examined and grown in food-gelatin. In each specimen, comma-bacilli were found (453). Similar findings were confirmed through laboratory analyses of choleraic autopsies performed in Toulon, France. In this sampling, the bacilli had been identified in nearly 100 cholera patients. Since the concentration of the micro-organisms always stood "in exact proportion to the cholera-process itself," and since healthy persons showed no trace of the germ, Koch asserted that the bacilli were specific to cholera (453). Both negative and positive evidence buttressed the causality hypothesis: dejecta from dysentery patients, for example, showed no sign of comma-bacillus (455). On the basis of the accumulating evidence, and although the etiological cycle had not yet been duplicated in an animal model, Koch felt confident, in 1884, that comma-bacilli caused the disease (454).

In place of human experiments using live organisms, Koch developed an alternative explanation, one that Cheyne would later endorse, to account for the transmissibility of comma-bacillus from person to person and via inanimate objects. The disease, for example, could have been transmitted from soiled linen to a laundress' hands and, then, from her hands to her mouth or eyes. Koch remarked that the only instance in which he had ever found the bacilli outside of the human body was on moist linen. Transmission of the infectious agent through physical contact was therefore a viable theory, as the comma-bacilli could persist for some time on fabric, in drinking water, or on moist food ("Address" [6 September 1884]: 456). The linen-transference discovery was comparable to an experiment "in which a human being is fed with a small quantity of a pure cultivation of comma bacilli." Thus, "a human being unconsciously performs [an experiment] on himself, and the same demonstrative power lies in it as if it had been intentionally made" (455). What occurs in cultivation is therefore imagined as happening inside the human body; hence, the comma-bacilli's proliferation and displacement of competitive bacteria was analogous to what transpires in the intestines. The colony multiplies in the intestinal canal, irritating mucous membranes and causing diarrhea, loss of fluids, and toxic shock. At the maximum level of growth, the infection produced "the peculiar complex of symptoms that constitute the real attack of cholera" (455). Koch believed that learning about the real cause of the disease and about the pathogen would allow one to construct "the etiology of cholera ... on definite and fixed lines"; contradictions and anomalies could thereby be explained (459).

Cheyne agreed with Koch that causality could be based on circumstantial evidence or on the linen analogy, for a preponderance of indirect evidence and similarities of various kinds routinely led to experimental proofs. This was a reasonable expectation wherever a definite organism was discovered in large numbers in the diseased tissue of an animal, when the organism was known to be associated with no other diseases, and when it was not present in that animal normally. By the spring of 1885, Cheyne was reasonably certain that Koch was correct, although he remained circumspect. The possibility existed that, under certain circumstances, a disorder similar to cholera could be induced in animals ("Report" [25 April 1885]: 821). Conversely, the failure to reproduce the disease in animals did not automatically invalidate the causality hypothesis. Koch had also encountered anomalies. Cheyne referred to the latter's negative results with animals when the bacillus was directly injected into intestines and veins or when fed to them in contaminated food. The Nicati and Rietsch experiments on dogs and guinea-pigs, alluded to above in the discussion on physiology, indicated that the excretion cultures, despite irregularities noted earlier, could possibly communicate cholera; and Koch had had some positive results as well with guinea-pigs (822–823).

Upon his return to Paris, Cheyne contacted Koch to inquire about the operational details involved in guinea-pig experiments. Promptly forwarding directions, Koch stressed that strict antiseptic precautions had to be observed and that the cholera strain was to be injected into the duodenum, between the pyloric end of the stomach and the entrance of the bile-duct ("Report" [9 May 1885]: 933). Following these directions carefully, Cheyne replicated Koch's experiment (just as Lister had routinely replicated Pasteur's). The project involved 17 guinea-pigs, three of which had become symptomatic and died of cholera in 2–4 days. He was not afraid to admit to the inconclusiveness of the experiment and that his inoculations were not performed to his complete satisfaction; however, he continued to replicate the work of Koch and of Nicati and Rietsch (933–934).

Based on two observations, Cheyne implicated the bacillus in choleraic disease. Accumulating, negative evidence strengthened his argument for causation. Undetectable in similar intestinal disorders, such as typhoid, the comma-bacillus had been sufficiently proven to be exclusive to Asiatic cholera ("Report" [16 May 1885]: 976). But coextension was only one aspect of the investigation, as Koch had realized. Other possibilities had to be systematically tested. In this regard, Cheyne deferred to Koch's etiological narrative: in the earliest stage of the disease, the comma-bacilli proliferated, as one-day-old human dejecta showed, and numerically diminished over a short period of time; animal experiments that Cheyne had contributed verified that the bacilli grew in the intestine and set up morbidity in lab animals, consistent with cholera pathology in humans ("Report" [16 May 1885]: 977).

1884–1892: Epidemiology

Koch's Calcutta inspections, documented in the 6 September 1884 "Address," revealed the presence of comma-bacilli in communal water tanks, situated near habitations where residents had died of cholera (453–459). It was later determined that the linen of a fatal cholera victim had been washed in the tank. This was an enlightening discovery since it was the first time Koch had found the comma-bacilli outside of the human body; moreover, this insight was as significant as John Snow's discovery that the Broad Street water pump, in London, was the focus of the 1854 cholera outbreak.

The topography of the Calcutta village where the latest outbreak had occurred bore directly on Koch's investigation. On the bank where the receptacle had been located was a village of 200 to 300 people, occupying some 30 to 40 huts; thus, the place was densely populated, and the tank was central to everyone's needs ("Address" [6 September 1884]: 453–459). Between 6 percent and 8 percent of the population had died of cholera; morbidity figures, likely higher than those for mortality, were unavailable. The tank where water collected also received the household refuse. The people bathed and washed utensils in the tank daily. Human waste, routinely deposited on the dry banks, leached into the water-table; and the contents of hut-cesspools, a direct source of contamination, also drained into the tank. When the epidemic was at its worst, Koch and his assistants detected high concentrations of comma-bacilli at different points on the bank; conversely, as the epidemic abated, lower numbers of microbes were found.

A familiar anomaly arose during a secondary investigation. Since cholera morbidity was high, it was expected that the microbes would always be at a correspondingly high level in the affected tissues. The opposite appeared to be true: the bacilli, as far as Koch could see, were fewer in number. This set of observations confounded him. On the one hand, the life-cycle of the bacilli was brief. On the other hand, the subsequent damage to the community continued, even as these fragile organisms died and deteriorated. From a modern perspective, this could make a case for the existence of cholera endotoxin (poisons resulting from cell-membrane deterioration), as a concomitant to the exotoxins modern scientists would find. Speculating, in the 6 September 1884 "Address," about what had caused morbidity, Koch supposed that comma-bacilli produced "a special poison" (i.e., exotoxins) that, once absorbed into the vertebrate system, brought prostration and death (456).

Undoubtedly, cholera was spread by human beings, and the bacilli could grow outside the human body in well- or river-water. Though a tank or reservoir lacked sufficient nutrients to support luxuriant colonies, stagnant areas, such as gutters and cesspool outlets where animal refuse and vegetation collected, were environments conducive to choleraic development (von Pettenkofer was right about this). Wherever a water course was not stagnant, however, "the formation of a local concentration of nutritive substances" and of luxuriant microbial growth was less likely ("Address" [6 September 1884]: 456).

According to the historical record, no cholera epidemics had ever broken out spontaneously outside of India. The presumption, according to a broad consensus, was that a specific organism caused cholera, the origin and habitat of which was the Delta of the Ganges River (Koch, "Address" [6 September 1884]: 457). In fact, the entire region, from the banks of the Ganges as far up as Benares, was a broad choleraic focus. Koch had surveyed the geography. Where the Rivers Ganges and Brahmaputra ramified into streams, seawater mixed with river water, ebbed and flowed, and at high tide flooded the uninhabited southern district of the Sundarbans region. This alluvial region supported lush vegetation, wildlife, and microbes inimical to man. As refuse from the densely-populated upper Ganges floated downstream, it mingled with the brackish Sundarbans' water, already saturated with putrefying organic matter. Koch postulated that cholera originated in this frontier territory, for the epidemics had erupted south of Bengal.

To Jessore and Calcutta, both of which border on the Sundarbans, cholera was endemic ("Address" [6 September 1884]: 457). Cheyne described the topography at length. Adjoining districts were also ecologically and topographically conducive to cholera outbreaks and transmission. Because tropical rains inundated the lower Bengal's flat country, huts had to be built on raised ground. All the Delta villages were built this way, even in Calcutta. Excavated earth raised the houses above the waterline, but the resulting gullies and troughs inevitably filled with water. Earthen water tanks which were everywhere served the needs of small communities. Calcutta had 800 such tanks. The geography and living conditions of the indigenous people, therefore, combined to make cholera an endemic danger ("Address" [6 September 1884]: 457). The disease gradually waned after the construction of aqueducts that provided clean drinking water. However, when the Calcutta population reverted to the impure water of the Hooghly River and from the local tanks, cholera returned (457).

Koch knew that human migrations were an epidemiological factor during the pandemic. Pilgrimages to Hurdwan and Puri contributed directly to the spread of cholera. Hundreds of thousands of people crowded together in these areas and, under miserable conditions, were "penned together in a confined space." Thousands used available tanks in unsanitary conditions, and the consequences were predictable: "the disease, when it does break out amongst the flocks of pilgrims, is speedily scattered over the whole of India, and reaches every place" ("Address" [6 September 1884]: 458). Cheyne rightly suspected that overseas travel routes widely-disseminated the disease, intensifying the health crisis. Travelling from India to Europe via the Red Sea, especially from the main harbor of Bombay where cholera was endemic, could bring an infected person to Egypt in 11 days, to Italy in 16, and in 18 to 20 days to France (458); thus, the danger of cholera spreading from India to Europe increased significantly (458). The danger of cholera outbreaks on ships was variable. Merchant ships with small crews had had cholera cases on board, but epidemics tended not to occur. The direct linkage between cholera and dense populations held true for large vessels. If a transport-ship were crowded with pilgrims, emigrants, or migrant workers, and if cholera were to break out on board, it could spread quickly (458). Cheyne pointed out that Koch was interested in another factor integral to cholera epidemiology: that of adaptive immunity. Having one bout with the disease, thought Koch, conferred short-term immunity. The allusion was to anecdotal information about someone who, on two occasions, had experienced the disease. Contracting the disease in an epidemic conferred a degree of immunity to secondary exposure ("Address" [6 September 1884]: 458).

The value of Koch's work on cholera in British India was put to the test in 1892. The City of Hamburg initially resisted his etiological and epidemiological advice but realized its value the hard way. In August, according to one source, nearly 9,000 of the 18,000 people stricken with cholera died of the disease (Snodgrass: 232). Koch had been appointed Health Inspector in 1890 and, as author of the book *German Bacteriology*, was unquestionably the resident expert. The city had a modern water and sewer system, but it lacked a sand-filtration process that would have prevented infectious sewage from entering the water supply (J. N. Hays: 150–151). Koch's twofold remedy was to kill the organisms in the water and, through isolation and quarantine, to break the infectious cycle. Emigrants from Russia to North America were identified as carriers. Instead of initiating Koch's measures immediately, to its detriment the Hamburg government delayed. When Koch arrived in Hamburg, the strategy was finally activated: public health brochures were distributed in impoverished sections, water-boiling stations were created, and pure water was imported (Snodgrass: 232). Wherever strict quarantine and public health measures were enacted, outbreaks were contained.

1884–1885: The Koch-Klein Debate

This section describes the Koch-Klein debate over the origin, characteristics, and disease-causing properties of the cholera germ. Cheyne played a significant role in the controversy, experimentally verifying Koch's claims, and in published documents soundly refuting Klein's, which were aligned with von Pettenkofer. In July 1884, the British government had commissioned Klein and the pathologist Heneage Gibbes to research the latest cholera epidemic in India. Their findings were published, in 1885, under the title, *An Inquiry into the Etiology of Asiatic Cholera*. Upon his return to London, Klein delivered the Report to the Royal Society. In it he questioned the validity of Koch's views on cholera and, in the process, revealed his agreement with von Pettenkofer (Worboys, *Spreading Germs*: 250).

Klein who published 264 scientific papers was a distinguished and prolific contributor to British bacteriologist (Atalić et al., "Emanuel Edward Klein": 114). Experimentally, he abided by the Henle-Koch method on the necessity of microscopic identification, of cultural isolation, and of animal inoculations, the process linking germ to disease (114). His partiality towards the soil-borne theory, however, resulted in eclectic incongruities, reflecting a limited understanding of cholera etiology and pathogenesis.

Klein's critique, in a "Report" delivered to The Royal Society, 28 January 1885, highlights five investigative differences between Koch and himself. The first was legitimate: choosing vibratility over shape as a defining feature, Klein argued that the physiological reference to Vibrio improved upon Koch's empirical phrase, "*comma* bacilli" (156). A second area of contention was pathogenesis. Post-mortems of acute cases had led Koch to believe that the ileum, the lowest part of the small intestine extending from the duodenum, contained the bacilli in a nearly pure state (Klein: 156; Henry Gray, *Anatomy*: 117). Klein rejected this claim, along with Koch's assertion that infected mucous membranes were the source of toxins. A third point of disagreement was taxonomic. Von Pettenkofer's soil-based theory guided Klein's analysis of growth patterns in cholera cells, as he misidentified the microbe as a soil-bound saprophyte (it would prove to be an obligate, water-borne bacillus) (156). Although concurring with Koch that cholera infection depended on the quantity of ingested micro-organisms and on individual susceptibility, Klein nevertheless attempted to subvert the German physician's basic causality premise and, as an anti–Contagionist, even recklessly ingested the bacillus to prove his point (Worboys, *Spreading Germs*: 250).

In regard to the microscopic study of tissues, Klein was opposed to Koch. According to the former's post-mortems of acute cholera cases, small, immotile, straight bacilli were only minimally present, and, upon further study, microscopically undetectable in the blood, in mucous membranes of the intestine, and in other tissues ("Report to The Royal Society": 157). Recounting experiments performed from October 1886 to May 1887, Klein realized that he had to clarify what had become his self-contradictory opinion. On the one hand, he fully agreed with Koch, "as to the constant presence of the comma-bacilli in the cholera intestine and cholera discharges *during the early stages*" (italics added). On the other hand, in 1887, Klein contended that, since the putative bacillus was absent in the late stage of the disease and at post-mortem, he could not accept the comma-bacilli of Koch "as the proved cause of cholera" (*The Bacteria of Asiatic Cholera*: viii; *Micro-Organisms*: 416). Klein's experiments on small animals using cholera cultures, for example, had not found the microbes at dissection ("Report to the Royal Society": 157; "The Etiology of Tuberculosis" [Koch's Postulates]: 116). In the 24 January 1885 issue of the *British Medical Journal*, Klein restated the third point that Koch's comma-bacilli were not specific to choleraic disease and that they were related to saprophytic germs ("The Organisms of Cholera": 200). He would be proven wrong on both counts. In 1896, Klein reversed himself: in almost all cases of cholera and at every stage of the disease, Koch's comma-bacilli *were* present. Furthermore, cholera excreta contained Koch's Vibrio and **S**-shaped Spirilla, of varying lengths and numbers of curvatures, indicating that there were different strains and subtypes of the species (*Micro-Organisms*: 416). Irrefutable evidence eventually forced him to reconsider the possibility that the comma-bacilli caused the disease (*The Bacteria of Asiatic Cholera*: viii; *Micro-Organisms*: 416).

Pathology was another area of lively discussion. During the 24 March 1885 meeting of the Royal Medical and Chirurgical Society (R.M.C.S.), Klein insisted that clinical and post-mortem investigations did not support Koch's conclusion that low levels of bacilli in the ileum could be consistent with severe pathology. Cheyne's six-fold explanation, recorded in the "Report," attempted to account for the reputed contradictions ("Report": 655–656). First, if microscopic examination of the contents and tissues of the ileum were to reveal a low count of bacilli, such a finding did not contradict what was known about *Vibrio cholerae*'s physiology since the bacilli grew rapidly and died quickly, and since tissue damage was the aftermath of the infection. Second, even if one did not accept the aforementioned premise, one could not accurately quantify how many bacilli were present in the acute phase of the disease from the diminished numbers found at its aftermath. Third, Koch had discovered, in the early stage of growth in acute cases, that immature comma-bacilli appeared to be "straight" rather than curved; thus, the number of comma-bacilli could have been underestimated because undeveloped cells had not been counted. Fourth, microorganisms might have eluded detection because, having penetrated and grown *within* the intestinal wall, their colonies could not be fully observed ("Report": 655). Fifth, Cheyne answered a crucial question: were the putative bacilli detectable in conditions *other* than in Asiatic cholera? He cited evidence from Koch's samples that, in every case of cholera seen in India, Egypt, France, Italy, and Spain, the comma-bacilli had been present in some form and at some stage of the disease. When the epidemic had abated, in the summer of 1884, a thorough microscopic and physiological investigation of saliva, of feces, and of decomposing matter turned up no evidence of the germ. This information countered Klein's assertion that it was a saprophyte or decomposer of organic material. And sixth, Koch also knew of other comma-form bacilli, distinguished by size, shape, and vibratory capacity, that were non-choleraic. These specimens had been presented to the R.M.C.S., and Koch and Cheyne could distinguish one from the other ("Report": 655).

On 4 April 1885, Klein reacted to Cheyne's cogent defense of Koch ("Some Remarks" [4 April 1885]: 693). Instead of refuting the latter's data and Cheyne's six-part defense enumerated above, with animus Klein questioned his German opponent's laboratory proficiency. While in Egypt examining cholera dejecta microscopically, Koch had admitted being unfamiliar with cholera cultivations grown on various media. Seizing the opportunity, Klein suggested that, if this was true, Koch might have been relying on dubious microscopic evidence to support his claim that comma-bacilli were consistently present in Egyptian cases (Klein, "Some Reasons": 693). As a bacteriologist, Klein could not accept the premise that a low level of bacilli could be consistent with extensive disease of the ileum, even though Cheyne had gone to great lengths to explain the extenuating circumstances. Klein, it is important to recognize, was following his own variation of the Postulates, the second criterion of which stipulated that causation depended on the presence of these microbes in the tissues of the small intestine during the acute phase of the illness; furthermore, an enormous number of microbes were presumably required to produce a high concentration of toxin (Atalić, "1885 Cholera Controversy"). In addition, Klein rejected Cheyne's counterclaims that in acute cases immature comma-bacilli were linear, not curved, and therefore difficult to see and quantify; that they invaded the intestinal epithelia; and, though they were ephemeral, exuded powerful toxins into the body. Klein insisted that, if the comma-bacilli caused the disease, then they would have to be amply present in devitalized tissues, in the intestinal fluid and mucous, and at high levels in the lower part of the ileum.

In the 1884 "Lecture at the First Conference for Discussion of the Cholera Question," Koch had reported on the findings of choleraic post-mortems. He had discovered that cholera had destroyed extensive areas of tissue, leaving a nidus for comma bacilli and for opportunistic micro-organisms. Initially, he saw that the bacilli had indeed penetrated the tubular glands and the interstices between the epithelium and basal membrane. Following the choleraic invasion, other bacterial flora had penetrated the intestinal wall. The presence of multiple species and of widespread disease were unmistakable: "there are necrotic diphtheria-like formations in the mucous membrane of the intestine and typhoid-like abscesses where the tissues, which are killed by pathogenic bacteria, are also penetrated by other non-pathogenic bacteria" ("Lecture at the First Conference," *Essays of Robert Koch*: 152).

Cheyne corroborated Koch's observations. In an 11 April 1885 Correspondence to *The British Medical Journal*, he countercharged, on the basis of persuasive histological evidence, that Klein had misinterpreted Koch's bacteriological analysis of cholera-bacilli. Furthermore, he swept aside Klein's unfounded assertion that Koch's causation-argument was not based on morphological *and* on physiological evidence. To the contrary, Cheyne vouched for the thoroughness of Koch's method: gelatin cultures were seeded after the trip to Calcutta; and the Indian examinations of 42 fatal cases and of the dejecta of 32 others, performed both microscopically and physiologically, proved that the organisms were invariably present ("Correspondence": 756). Even before Koch learned that several forms of comma-bacilli existed, he had known that progress towards more accurate identification of the pathogen depended on its physiological characterization (757).

With respect to sanitation, Koch and Cheyne were closer to von Pettenkofer's thinking. On the incidence, distribution, and control of cholera, Cheyne pointed out, in the 31 July 1886 "Report on a Study of Certain of the Conditions of Infection," that particular conditions had to be in place if choleraic disease were to gain a regional foothold, and a substantial quantity of bacilli had to be active (197–207). A quantity of comma-bacillus deposited on the soil and in the water of a locale would flourish or diminish in accordance with local conditions; hence, in a sanitary region, environmentally inhospitable to choleraic growth, colonies would "either die out or grow slowly and with difficulty" ("Report" [31 July 1886]: 207). Where the environment supported minimal growth, small quantities could still make their way into the water and food supply; but, because of diffusion and low numbers, the bacilli would be unable to cause virulent disease, would either have no effect or only a mild one (207). Under sanitary conditions, endemic cholera was, therefore, a minor concern. Under the best circumstances, the infectious organism resulted in a few mild attacks and, rarely, in severe ones.

Immunology was also discussed ("Report" [31 July 1886]: 206). Cheyne considered the interesting possibility that recurrent, mild episodes of cholera could stimulate immunity, attenuating the effects of a subsequent infection. Here, he was indebted to Koch's 6 September 1884 "Address." In the interests of public health, Cheyne correctly believed that early recognition of, and swift reaction to, an outbreak was absolutely necessary if an epidemic were to be prevented. According to epidemiological data, before a full-scale epidemic occurred, a gradual increase in morbidity and mortality was an obvious warning sign. Worsening conditions meant that the bacilli had penetrated the human food chain and water supply and could overcome acquired immunity in some and devastate the nonimmune population. Although large numbers of bacilli had to be consumed for the disease to develop, each victim was host, reservoir, vector, and a danger to many.

For Cheyne, the control of cholera in British India required a multidisciplinary approach. The bacillus proliferated in soil and in bodies of water, entered the human body through the alimentary route, propagated through infectious secretions, and was reintroduced into the environment through human evacuations. This pattern, in Cheyne's view, confirmed "the causal connection between Koch's cholera bacillus and Asiatic cholera" ("Report" [31 July 1886]: 207).

9

1870–1902
Listerism on the Battlefield

Cheyne believed that the systematic use of antisepsis during the conflicts of this period could have considerably improved care at the point of injury, during transportation from battlefield to base hospital, and during recovery. Although the published "field tests" on Listerism were inconclusive, knowledge accrued from them would influence the development of military medicine in the twentieth century (Fisher: 153). As historiographer and battlefield surgeon, Cheyne contributed to antiseptic medicine in the tradition of Lister.

1870–1902: The Military Use of Listerism

In 1870, an official request prompted Lister to write "A Method of Antiseptic Treatment Applicable to the Present War" (1870) (*CP*. **II**: 161–164).[1] This groundbreaking essay set forth a "plan" for maintaining post-operative asepsis. It was intended, in Lister's words, to combine "efficiency with the simplicity and facility of execution essential under such circumstances" (*CP*. **II**: 161). A number of factors had to be considered in order to devise a treatment protocol for the battlefield. Among the factors one had to consider was the variety of wounds and pathogens, the possibility of mass casualties, and the logistical uncertainties of rescue and transport under fire and over bad terrain. Although, in this text, Lister makes passing references to the battlefield, in the opinion of his contemporaries, his "plan" was best suited to the care of one or more uncomplicated gunshot cases rather than to mass casualties with serious burns, multiple bullet wounds, and shrapnel. I would like to outline Lister's system and review how he envisioned its use in armed conflicts.

Lister's 1870 plan, which had two phases, was best designed for the field hospital. His overriding concern was with the management of wounds and control of infection. The first phase comprised three steps: (1) the wound was to be thoroughly and repeatedly saturated with 1–20 maximum-strength solution of carbolic (acid-to-water), applied with a syringe and manipulated into the recesses of the wound; (2) bleeding blood vessels were to be sutured with aseptic catgut and arteries secured by torsion; silk or linen thread were to be steeped in an oily solution of carbolic acid; ligatured ends to be left projecting from the knot; and (3), with the acid in the wound, visible foreign material (bullets, shrapnel, dirt, and clothing) and bone fragments were to be extracted.

The second phase was concerned with protecting the wound from external micro-organisms and with promoting healing. It had five consecutive steps, all involving the dressing of the wound: (1) it was to be covered with two or three layers of oiled silk, both sides coated with carbolic acid solution and fixed oil (olive, almond, linseed, or equivalent, in 1–5 aqueous solution); the raw surface was to be completely covered, with the bandage overlapping the skin; (2) cotton or lint cloth, steeping in the oily acid, was then to be applied; the cloth, folded to ¼ inch thickness, had to have a three-inch-square border extending over the primary silk bandage; the margin of the cloth was to be thinner than that which had been superimposed on the silk; (3) this second layer was then covered with thin gutta-percha tissue (Malaysian tree latex), overlapping on all sides by one inch and retained in place by an acid-soaked roller towel; (4) a larger folded towel steeping in oily acid was then wrapped around the layered bandages; and (5) this multiple-bandaging system was completed with the superimposition of oiled silk or of gutta-percha (*CP*. **II**: 161–162).

Lister accounted for each aspect of the treatment—every solution, every layer having a systematic function. The watery solution of carbolic acid destroyed septic organisms in the wound while barring epidermal bacteria. The permeable, oiled silk protecting the skin from acid irritation allowed the wound to scab over, and the carbolated oil, with which the silk had been imbued, diffused into the wound. The gutta-percha bandage also had a multiple purpose: it prevented blood and serum discharges from soaking through and diluting the acid in the fabric, it stopped bacteria, but it permitted the antiseptic to soak through from the outer cloth. While the outermost cloth was regularly changed and treated with fresh acid, the gutta-percha and all lower layers of fabric were designed to guard against external contaminants from ambulance bedding and other sources; the circumferential cloth resting on the raw wound would remain completely aseptic, the effect of carbolic acid having permeated the gutta-percha (*CP*. **II**: 162–163).

Since the outermost cloth had to be changed regularly or whenever the discharge had saturated it, care had to be taken not to displace the surface dressing. To perform this delicate procedure, the edge of the gutta-percha had to be gently raised as the dressing was changed. Wary of the intrusion of air-borne bacteria during this procedure (later deemed an exaggerated concern), Lister promoted the use of bilateral cloths, each overlaying one-half of the gutta-percha. As one half was raised during the redressing, a freshly-oiled cloth was immediately applied just as the portion of old cloth was being lifted off.

Lister had not ignored the prospect of serious casualties arriving at an understaffed field hospital. He therefore stressed the need for cleanliness in an overcrowded ward to avoid dreaded hospital diseases. All wounds had to be kept aseptic with the regular use of dressings and carbolic acid solution, which promoted healing (*CP.* **II**: 164). He, no doubt, understood that the conditions of civilian and military medicine differed greatly from each other, but his method and materials "were likely accessible to the surgeons of both armies," and with reasonable adjustments could be profitably used under the worst conditions. As he would soon learn from Dr. William MacCormac (1836–1901) and others, having the material and understanding the intricacies of the method did not guarantee success (*CP.* **II**: 164).

During the Franco-Prussian War (19 July 1870–10 May 1871), Lister's former student and colleague, William MacCormac, served as Surgeon-in-Chief to the Anglo-American Ambulance Corps; his staff included the American surgeon, J. Marion Sims (1813–1883), and 14 medical personnel. Assigned to the French army, they attempted wherever possible to apply Lister's antiseptic system to battlefield wounds. As a humanitarian gesture, they presented the system to both armies. The Germans, though interested, had not yet acquired practical use of Lister's techniques, while the French rebuffed the offer (Fisher: 182). Thus, the opportunity to test the medical use of carbolic acid systematically and under extreme conditions was lost. In letters of 1 September and of 21 November 1870, Lister was confident that, if applied to battle injuries, the acid would control infection. The official French decision not to cooperate perplexed him (Godlee: 354–355).

When MacCormac arrived in Metz (northeastern France), prior to the German siege of the town (3 September-23 October 1870), he was assigned to a main military hospital. Of the large number of wounded, he was surprised to find that few serious cases were among them. The implicit reason was that, despite army policy, carbolic acid and other agents were unofficially being employed. This is evidenced passingly in his reminiscences. One entry reads that the surgical treatment had been greatly simplified, wounds were bandaged, and a surgeon made considerable use of "*l'acide phénique*" or carbolic acid. Friedlieb Ferdinand Runge (1795–1867) had discovered this chemical in 1834, and, in 1841, Auguste Laurent (1808–1853), associating it with alcohol, named it, *phénique* (phenol) (E. V. Howell: 26; M. Blondel-Megrelis: 303–314; B. Anft: 566–574). At Metz, sores were bathed in phenol. Lister's ideas appear to have been in practice, but it is more likely that the French medical corps had been influenced by the earlier pharmacological work of Jules Lemaire (1814–1886) and of Gilbert Déclat (1827–1896), who had tested the disinfectant property of the acid, in the 1860s, on gangrene and on other infectious disorders (Lemaire, *De L'Acide Phénique* [1863]; Déclat, *Nouvelle Applications* [1865]).

Overall, the patients whom MacCormac met looked well: "There was hardly any inflammation or general fever to be seen in any of [them]" (*Notes*: 9, 11). At Caserne d'Asfeld, a barrack was transformed into a hospital (*Notes*: 27). Even though MacCormac was disheartened by supply shortages, notably of chloroform and of carbolic acid, at the Battle of Sedan where 12,500 French soldiers were wounded overcrowded hospitals were kept as sanitary as possible, "asphalt floors [being] washed twice daily with a solution of carbolic acid" (*Notes*: 27, 32). On an ambulance-run to beleaguered patients in the village of Balan, Sims was reported to have loaded up "surgical instruments, dressings of all kinds, chloroform, carbolic acid, and some provisions" (*Notes*: 34). Listerism, under MacCormac's direction, was applied to the most horrendous injuries. Even in the most difficult cases, the antiseptic system supported surgical care and natural healing. To a soldier brought in with a severe facial wound, soft tissue lacerated, bone comminuted, and masseter muscle torn off, MacCormac applied carbolic dressings (*Notes*: 49–50).

MacCormac's efficient work impressed observers, one of whom was a British military surgeon who had inspected the hospitals near Sedan ("The Wounded": 511). Another was a correspondent for *The Lancet* who reported that surgeries in these hospitals generally had positive outcomes. This was true for amputations, although the results of resections were not as favorable. Postoperative infections, particularly gangrene and erysipelas, were rare; however, pyemia which required close attention often occurred despite scrupulous measures and efforts to limit overcrowding. Powerful weapons, however, made wound care very difficult. Both the French and the Germans employed effective small arms resulting in horrific internal injuries. The Prussian Dreyse needle-gun, a breechloading weapon, the pointed striker of which ignited the bullet's powder, had a high rate of fire and high muzzle-velocity. The French had two systems: the chassepôt breechloader, having twice the range of the Dreyse, and a 25-barrel machine gun (McCallum: 125). The chassepôt bullet comminuted bone, but the orthopedic injuries produced by the bone-splintering needle-gun were far worse (512). Greater fire-power meant more casualties and more complicated wounds.

Antisepsis was not a panacea for joint and bone injuries, even though the

1870 paper included directions on dressings and splints for compound fractures ("A Method of Antiseptic Treatment," *CP*. **II**: 164). At Asfeld, gunshot wounds involving joints and limb-amputations, unlike wounds of the face and neck, were disastrous because secondary infections were difficult to manage under the circumstances. It was not until the War's end, in 1871, that Lister would again discuss the antiseptic treatment of joints, using carbolic acid and oil directly on the wound (*CP*. **II**: 194–195). It is not clear from MacCormac's early recollections whether the antiseptic had been applied to gunshot wounds of the extremities. The inference is that Lister's antiseptic system, as discussed in the papers of 1867–1868, could not be employed in those cases because injuries were complicated, often very severe, and the casualty count too high. MacCormac's mortality statistics included orthopedic injuries, especially amputations at the thigh (85%) and of the leg (40%), and gun-shot fractures of the femur (>70%) (*Notes*: 118–121). He ascribed the bad results to privation, to the exhausting influences of the wounds, and to the unhealthy conditions of the hospital at the time operations were performed (*Notes*: 123). Secondary hospital infections in an overcrowded facility, along with the dearth of anti-septics, were the likely causes of high mortality.

As a medical historian, summarizing the content of Lister's 1870 paper in the 1885 *Manual of the Antiseptic Treatment of Wounds*, Cheyne compared Listerism to another plan that had shown promise in 1870–1871 (106–111). Friedrich Johann von Esmarch's (1823–1908) treatment plan was largely based on his considerable experience as a combat surgeon during the first (1848) and the second (1864) Schleswig-Holstein War. While stationed in the field hospital at Flensberg, he had worked alongside renowned surgeons, Bernard von Langenbeck, of Berlin (1810–1887), and Georg Friedrich Louis Stromeyer, of Hanover (1804–1876). Von Esmarch gained further experience, in 1870, during the Franco-Prussian War, where he served as Surgeon-General to the army and then as consulting surgeon at the military hospital in Berlin (J. E. Herzenberg; Manring, et al.: 2170).

Von Esmarch learned about Lister's system when visiting England and Scotland. Later, at the 19 April 1876 Fifth Congress of the Society of German Surgeons, von Esmarch was reported to have extolled the value of Lister's war-wound system. Testing it under peaceful conditions could not simulate war conditions, but he extrapolated, nonetheless, based on considerable surgical experience ("Fifth Congress": 273). Accordingly, he stressed that military hos-pitals should emulate the system's implementation in civilian practice (273). Although von Esmarch was skeptical about the success of Lister's 1870 system under battlefield conditions, he did not decline the use of carbolic acid. In light of its documented successes, during the Revolution of 1848 in the German

States, he treated perforating gunshot wounds and infection with the acid. Von Esmarch raised the possibility of modifying Lister's system for use in war where debris, microbial-rich soil, and the manual exploration of wounds were unavoidable sources of infection (273). In the 1878 edition of the *Surgeon's Handbook on the Treatment of Wounded in War,* he described amputations that he had performed under carbolic acid spray during the Franco-Prussian War. The sprayers that Lister used had developed from Benjamin Ward Richardson's small, hand-held aerosolizer, to a clumsy tripodal mechanism, and, finally, to steam-powered devices, the noxious chemical fumes of which sickened operators and patients alike, leading to the disuse of this modality (Lister, "On a Case Illustrating the Present Aspect of the Antiseptic Treatment in Surgery," *CL.* II: 166; Godlee: 282–284).

Von Esmarch's extensive experience in war had led him to question, not the validity Lister's antiseptic principle, but rather its practical application on the contemporary battlefield. In his view, the multi-layered procedure outlined earlier in this chapter needed modifications. Von Esmarch proposed that disinfection had to begin at the point of injury. As Surgeon-General, he therefore required every German soldier to carry an antiseptic dressing package, containing a 40 cm. × 20 cm. antiseptic muslin compressor, a 300 cm. × 5 cm. cambric bandage, one safety pin, and as 28 cm. × 18 cm. waterproof material for covering (J. E. Herzenberg).

According to Cheyne's 1885 survey, von Esmarch also commented on the medical implications of a new high-velocity bullet. Issued in 1870, this bullet passed quickly through clothes and tended not to carry dirty fabric and organic material deep into the wound. Consequently, if the wound were not handled in the field with dirty fingers and contaminated instruments, and if it were treated as quickly as possible, "it may in most cases be regarded as aseptic" ("Fifth Congress": 274; Cheyne, *Manual:* 125). With this idea in mind, he called for each soldier to be provided with the bandaging kit described in Figure 3, along with tampons of salicylic cotton (a mild antiseptic) wrapped in salicylic-treated gauze. During action, if there was a reasonable chance to save an injured limb, these tampons could be introduced into the wound and bandaged on, "without any preliminary probing or examination" (125). Only necessary apparatuses should be applied as the soldier was transported to the rear. Once away from the Front, doctors and attendants would immediately cleanse the wound with antiseptic lotion. If the surgeon saw that the wound had to be explored for bullets, splinters, or debris, the tampon was removed under an antiseptic spray, the wound cleansed, and an antiseptic dressing applied. If the injury did not need exploration, the surgeon cleaned the skin and then added salicylic jute or other antiseptic material without disturbing the tampon. And,

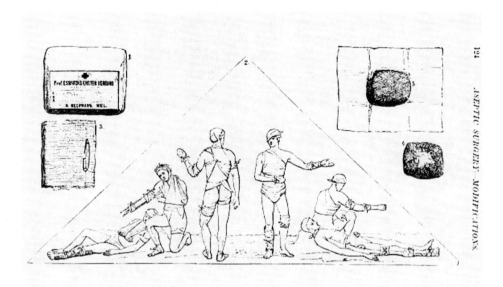

Figure 3: von Esmarch' first dressing for the wounded in battle: (1) Packet folded up. (2) Triangular bandage. (3) Gauze bandage. (4) Antiseptic tampon. (5) Tampon and square of oiled paper. From William Watson Cheyne (*Antiseptic Surgery, Its Principles, History and Results*, London: Smith, Elder & Company, 1882, p. 124, http://www.archive.org).

if infection ensued, the surgeon had then to enlarge the opening to disinfect it (125). If aseptic conditions were maintained, the chance of amputation was lowered; however, when ball-shaped projectiles shattered bone, amputation could be the only option (125). Von Esmarch understandably called for a supply of antiseptic in every ambulance (125).

A third physician renowned in military medicine up to 1882 was the Russian surgeon, Carl von Rehyer (1846–1890). In 1874, he had spent one month in Edinburgh and, at the German Surgical Association, read one of Lister's papers on wound treatment. After visiting Lister's clinic and becoming familiar with the antiseptic system, he presented a controlled study of infected gunshot wounds combining debridement with antisepsis. The results were statistically significant: the mortality rate had been lowered by 43 percent (Murray, et al.: S223; Lehnhardt, et al.). In 1874, von Reyher who had been appointed a consulting surgeon to the Russian army of the Caucasus during the Russo-Turkish War (1877–1878) treated many cases successfully (Godlee: 350). Agreeing in part with von Esmarch, von Reyher contended that the probability of wound infection was at its highest if medical treatment were begun in the midst of the battle. Accordingly, he categorized wounds either as those that had been handled and therefore contaminated or as those that had never been touched. The influence of Lister in this regard is obvious, as von Reyher brought a mechanism along with him to make carbolic-acid gauze at the field hospital (Godlee: 351).

Von Reyher's aseptic system in the field hospital offered two surgical alternatives. The surgeon could either close the wound and disinfect the skin or clean out and purify the track of the bullet, and then set up aseptic drainage. In the first instance, healing was expected to occur without treatment under a crust; in the second, under an antiseptic dressing (Cheyne, *Manual*: 109). The first method, "treatment by crust," was reserved for uncontaminated small wounds. Von Reyher's approach to war-wound management, as Cheyne's remarks, depended on shutting out the outer world as soon as possible (109).

For wounds that did not heal naturally, von Reyher had a contingency plan. In the more complicated cases where crusting might not occur, the bullet track was to be irrigated, and, with unimpeded drainage ensured, the wound was then treated as one would a compound fracture. Since low-velocity bullets tended to make larger wounds, and since the chance for debris entrance was greater in this case, the second method was indicated. If a wound had been manipulated before arrival in the field hospital, the second method was to be used (Cheyne, *Manual*: 109).

To some extent, the British military, unlike the French and, to a lesser extent, the German, did not waste the opportunity of testing and evaluating Lister's system under fire. Four surgeons—Drs. William MacCormac, T. B. Moriarity, Edgar M. Crookshank, and Alexander Ogston—favored its use, reported on their experiences, and provided assessments. Cheyne contributed his opinion in 1900, and again in 1902, and we shall look at his firsthand experiences as well.

In his 1880 monograph, *Antiseptic Surgery*, MacCormac dispelled any doubt that he favored the use of Lister's system in war. His experiences as a combat surgeon in the Franco-Prussian (1870–1871) and Serbian-Ottoman (1876–1878) Wars had evidenced the value of carbolic acid. Yet MacCormac, like von Esmarch, needed to evaluate why the system, as presented in 1870, appeared to be impractical on or near the battlefield (267). The problems besetting antisepsis in war, he realized, were complicated and manifold: numerous wounded, long evacuation periods, poor transportation, widespread confusion, and the "insufficiency of aid of all kinds" (*Antiseptic Surgery*: 267). Immediate care was needed at the first dressing lines. Medical professionals were advised only to deliver patients after applying antiseptic occlusion; permanent dressings would be applied later (267–268). MacCormac's reasoning was that "the primary antiseptic occlusion should be regarded as an expedient to gain time, as being provisional to a more complete and perfect dressing which can be made later when ... it appears desirable" (270). In effect, he endorsed von Esmarch's procedures for the field hospital, to which carbolic acid was essential (271).

Brigade Surgeon T. B. Moriarity, the fourth physician surveyed here, served during the Second Afghan War (1878–1880). In "Listerism—Simple Application of, to Recent Injuries" (1881), he recorded cases encountered while in charge of wounded at the Peshawar base hospital. For the wounded coming under his care, Moriairty used Lister's system. Two case studies showed promising results, primarily in the management of infection, and he acquired valuable experience in the treatment of complicated surgeries and in the management of long-term rehabilitations. Of the two reported successes, the first was of a soldier whose arm had been shattered by rifle-fire. Moriarity had filled the bullet hole with antiseptic solution (1–20 acid to water), splinted the arm, and maintained asepsis with syringed carbolic solution. The soldier was well on the road to recovery over a period of five months, largely due to Listerism: re-dressing, antisepsis, and splints ensured bone healing. The fracture united in August, although the soldier had chronic discharges and a difficult readjustment to civilian life. He definitely owed his survival to carbolic acid, the use of which had allowed Moriarity to treat the case conservatively and in accordance with Lister's system. Most important was the fact that having the means and opportunity to control infection obviated amputation (411–412). Moriarity's experience, though anecdotal, illustrated the value of Listerism as it was applied to a single individual in a controlled and well-supplied setting.

Edgar March Crookshank, physician and bacteriologist, obtained his medical degree in 1881, at King's College, London, and he served as assistant to Lister who had arranged for Crookshank to go to Egypt in order to test the system in combat (Mortimer: 581). Accordingly, the army to which Crookshank had been assigned had been stocked with antiseptic supplies: carbolic acid, iodoform powder, lint, bandages, and chloroform. What appeared as an opportunity to test the value of Lister's system under combat conditions was lost through a fortuitous irony when a British naval bombardment achieved a swift victory, destroying Egyptian rebels on 13 September 1881 and thereby limiting the number of British casualties (581). As Civil Surgeon during the 1882 Egyptian campaign, Crookshank was decorated for his service at the Battle of Tel-el-Kebir (O'Connor: 192).

Crookshank's "Remarks on the Antiseptic Treatment of the Wounded on the Battle-Field" (8 March 1884) advocated the use of iodoform (a volatile crystalline compound) and boracic dressings for patients transported from the battlefield to the field hospital; but he found the use of sprays impractical. A better alternative, congruent with von Esmarch's thinking, was for every soldier to carry materials for wound dressing in the field. Crookshank, therefore, recommended that every service kit be equipped with a waterproof inside-breast pocket, one large enough to hold "a fold of antiseptic lint or a pad of compressed antiseptic wool, with a few yards of gauze bandage, and a miniature canister containing iodoform" (422). A carbolized gauze bandage was his choice since it could be folded up tightly, and since it contained an effective antiseptic; an infusion of carbolic acid made the bandage moldable to the limb, securely fixing the dressing. In Egypt, Crookshank improvised iodoform dusters out of perforated tin cans, an idea that stayed with Cheyne in 1900–1902 during the Second Anglo-Boer War. The duster had to be moisture-resistant if the powder was to suppress bacterial growth. The re-designed field jacket would also include a small canister or paste-tube of iodoform and eucalyptus oil, along with a gauze bandage and "a packet of compressed carbolized rope and a strip of protective, in lieu of iodoform, or some more expensive antiseptic" (422). Obviously an elaboration of von Esmarch's scheme, Crookshank's kit would be made available to all frontline personnel (422).

Crookshank, like von Reyher, thought "Meddlesome surgery" compromised the patient, although it is not clear if he was proscribing wound handling and probing on the battlefield, in the field hospital only, or in both locations. Like von Esmarch, Crookshank embraced the concept of self-dressing simple wounds. This practice would aid an ambulance system, critically overloaded with mixed injuries and non-triaged patients. Soldiers who were not incapacitated could help themselves and their comrades if wounds were minor, just as in civilian practice where out-patients are prescribed small supplies of iodoform and lint to dress their ulcers and soft chancres (422).

As far as pharmacology was concerned, Crookshank thought antiseptic lotions unstable (evaporation, spillage, lack of adhesion). For this reason, he favored dry, adherent dressings, relegating lotions to field-hospital use where Lister's corrosive sublimate was recommended. Chloride of zinc, Crookshank found to be most effective. Its value had been established at King's College Hospital as a hemostat and antiseptic, administered in 40-grain solution to swab amputation flaps and suppurating bullet channels (422).

The Lister-Esmarch system, a variant of which Crookshank developed, was practical and potentially an asset to later conflicts. Portable dressings would be important in wars where casualties were likely to be in the thousands; and where wounds, on average, more severe. The plan was that minor wounds could be attended to on site. Furthermore, if surgeons should be killed or wounded along with combatants, if units were isolated from field hospitals, or if the wounded could not be quickly retrieved in difficult terrain, then at least the injured man, assisted by his fellow soldiers, could manage injuries temporarily. Crookshank imagined that, with the development of powerful weapons and with the possibility of large-scale regional warfare, Lister's complex system, in principle, had value; but it would have to be adapted to combat conditions.

Another prominent Listerian, Alexander Ogston, whose work on *Staphylococcus* is discussed in chapter 4, was a decorated military surgeon who saw action in Egypt in 1885, served on the hospital ship S.S. *Ganges,* and treated the wounded at Hasheen and Suakin. Upon Ogston's return from his Egyptian tour, he became Vice-President of the surgical section of the British Medical Association. In that post, he criticized the poor medical care British forces received in Egypt and, in so doing, offended the medical establishment. As a consequence, he was banned from serving in the Second Anglo-Boer War; but Queen Victoria's intercession on his behalf reversed the decision, and he was deployed to South Africa (A. G. Ogston).

Ogston assessed Lister's 1870 plan in a 1902 paper against the background of the Second Anglo-Boer War. In this article, Ogston misleadingly claimed that surgeons of the Franco-Prussian War were "stimulated ... to secure the benefits accruing from the system to the wounded on the field, among whom the mortality had under older methods been very heavy" (1837). Most surgeons realized that the mortality rate due to infection had to be reduced, but most were not as enthusiastic about carbolic regimens and about Lister's plan as Ogston would have it. The Germans became more inclined to use Lister's system but only after the War of 1870–1871, when it became clear that bacterial complications plagued convalescents. Few combatants had received the benefits of antisepsis during the War, perhaps with the French exceptions that Mac-Cormac had cited in 1870.

Ogston alluded briefly to Richard von Volkmann's (1830–1889) experiences in the Austro-Prussian War of 1866. Von Volkmann had innovated surgically to reduce mortality through an open treatment relying on scabbing rather than on surgical closure of wounds; and he would benefit greatly from Lister's antiseptic system (Godlee: 336). After spending eight months at Sedan and Paris during the Franco-Prussian War, von Volkmann returned to Germany only to find that serious combat injuries, coupled with overcrowded and unsanitary conditions in hospitals, had led to widespread bacterial infections, and post-operative mortality figures had increased (336–337). In the winter of 1871–1872, despite efforts to improve the situation, he desperately considered closing the hospital (337). In November 1872, he studied and began to practice Lister's system, without having visited Edinburgh for instruction in its use. The results were astounding: post-operative infections were virtually eliminated (337).

Ogston surveyed the influence of antisepsis on military surgery from 1870 to 1902, against the background of Lister's treatment of traumatic injuries. The use of carbolic acid in varying strengths, Lister wrote in "On the Principles of Antiseptic Surgery" (1891), had initiated "a complete revolution in the practice of surgery." Antisepsis had dramatically reduced post-operative infections and had permitted surgeons to undertake longer and more complicated procedures without having to worry as much about the danger of *hospitalism* (*CP*. **II**: 341). Ogston then referred to the progress made by the veteran surgeon, Bernhard von Langenbeck (1810–1887), who had become an advocate of conservative surgery in warfare, presumably during the Franco-Prussian War. According to Ogston, antiseptic re-sectioning and other conservative procedures, despite opposition, had become applicable to gunfire-damaged joints ("Lister and Military Surgery": 1837). If Ogston was correct about von Langenbeck having benefited from Lister's system in 1870–1871, then it is surprising to learn that, in 1875, the former was somewhat hesitant about using the method in civilian practice; however, when Lister visited him in Berlin, von Langenbeck performed his first operation with a carbolic sprayer and invited Lister to dress the wound ("An Address on the Effect of the Antiseptic Treatment," *CP*. **II**: 252; Godlee: 342, 370).

Ogston also traced the genesis of von Esmarch's kit back to Lister. In place of carbolic acid dressings, however, von Esmarch's understocked kit included two small balls of salicylic wool and gauze to plug entrance and exit wounds, materials intended to stabilize small-caliber gunshot injuries. In an August 1881 Address delivered before the International Medical Congress, London, Lister congratulated von Esmarch for having attained surgical asepsis with the limited use of antiseptic chemicals. In certain cases, asepsis had been achieved with a *Dauer-Verband* or permanent dressing. Lister observes, however, that blood-encrusted wounds covered with old dressings did not necessarily breed harmful micro-organisms. He attributed the surprising outcome to von Esmarch's practice of keeping the dressing dry and of inhibiting bacterial growth in this way ("An Address on the Treatment of Wounds": *CP*. **II**: 291).

Irrigation was called for in larger wounds. To supplement the principle of the freely-draining wound, as Cheyne explained in 1882, von Esmarch designed an irrigator, consisting of a cylindrical leaden or zinc vessel fixed to the wall. At the bottom of the vessel, an India-rubber tube with a nozzle was attached. Its flexibility permitted the physician to work it into the deeper parts of the wounds to flush out organic debris with diluted antiseptic fluid (Cheyne, *Manual of the Antiseptic Treatment of Wounds*: 131; Ogston, "The Influence of Lister": 1837). The management of traumatic wounds in battle, as physicians quickly learned, demanded flexibility and reasonable improvisations.

Ogston credited continental military medicine for advancing "along Listerian lines" ("The Influence of Lister": 1837). This advancement occurred despite an overabundance of caution—of doing too much or the wrong thing, believing it would save or prolong life. "The principle of non-intervention" or "the Abolition of Probing" meant that, beyond basic dressing techniques, war wounds should not be tampered with on the field. This concept was related to the theory that the high-velocity bullet was aseptic. Small-bore bullets, such as those used with the Lee-Metford or Mauser rifles, tended to produce entrance wounds that healed naturally without infection. Ogston speculated that the sterility of a projectile was caused by its rotation: 2500 revolutions per second in the Lee-Metford; and 3600, in the 5-millimeter round. In flight, centrifugal force supposedly burned off adherent impurities (1837).

1899–1902: The Second Anglo-Boer War

During the Second Anglo-Boer War, which had broken out on 11 October 1899, Cheyne served as a Consulting Surgeon. Five hundred civilian surgeons were appointed to the army, among whom were Anthony Bowlby, George Makins, Cuthbert Wallace, and Frederick Treves (de Villiers, "The Medical Aspect": Part II). His experience there, recorded in three articles, was intense and far removed from civilian practice.

One of Cheyne's earliest battlefield experiences occurred on 27 February 1900 when over 4,000 embattled Boers and 50 families, under the command of General Pieter Arnoldus Cronjé (1836–1911), had camped along the banks of the Modder ("Mud") River. A British artillery and mortar bombardment of the position over a ten-day period led to a health crisis in the Boer laager.

Cheyne, along with other members of the R.A.M.C., was dispatched to the scene after hostilities had ceased. Dr. Arthur Conan Doyle, who also was on duty at Paardeberg, recalled how British troops had reacted after visiting the laager:

A visit to the laager showed that the horrible smells which had been carried across to the British lines, and the swollen carcasses which had swirled down the muddy river were the portents of its condition. Strong-nerved men came back white and sick from a contemplation of the place in which women and children had for ten days been living. From end to end it was a festering mass of corruption, overshadowed by incredible swarms of flies [The Great Boer War: 340].

Cheyne's description of conditions at the laager were published in The British Medical Journal, of 5 May 1900.[2] The river had grown thick with mud and rotting horseflesh, but even that had not dissuaded British soldiers from filling their canteens and risking enteric fever (1094). The overcrowded and static camp at the Modder River and at Paardeberg had created ideal conditions for the outbreak of typhoid (J. C. de Villiers, "The Medical Aspect": Part II). Although the British were able to acquire fresh water from a farm some five miles away, there was not enough for surgical needs. With no other choice, Cheyne and the medical staff had to use the polluted water for that purpose. To make conditions worse on the British and Canadian side, a shortage of kettles meant that the polluted water could not be boiled, implying that, contrary to the principles of asepsis, pathogens were routinely being introduced into wounds during treatment (1094). The environment was harsh and given to extremes. Along with the oppressive heat, dust storms forced surgeons to improvise when the numbers of wounded increased; thus, for lack of shelter, "operations had to be undertaken in the open" (1094). Heavy rain was not refreshing. The torrential downpour on 24 February, for example, caused the river to rise, increasing the downstream flow of dead horses from the Boer camp. Hundreds of carcasses washed onto the banks and clogged the river when snared in tree branches. As many as 15 to 20 at a time had to be disentangled and removed by sanitation units at great risk of contracting diseases. After the surrender, the camp had to be leveled because it was a focus of environmental pollution (1094).

On the morning of 27 February, after the official surrender, Cheyne and officers of the Medical Corps rode to the laager, "to see what assistance was necessary for the Boer wounded" (1094). The utter devastation and the odor emanating from the mangled horses was unendurable. The smashed and burned wagons reflected the intensity of the barrage, as did the cratered and debris-strewn ground (1094–1095). The British medical team had to find wounded Boers in the wreckage and uneven landscape who were in need of urgent care. Wounded combatants who had been stationed in deeply-burrowed firing positions along the river bank had been left to their own devices, Boer doctors and ambulances having left the laager for Jacobsdal during the retreat from Magersfontein. According to Cheyne,

> The wounded had been lying in the foul laager without medical attendance or dressings of any kind, and as a consequence the wounds were very foul, and many of the patients were in a deplorable condition from sepsis. The septic state of the wounds is very well illustrated by the fact that next day, while we were again out at the place where the Boer wounded had been collected, secondary hemorrhage occurred in two cases, and would have proved fatal had we not happened to be on the spot [1095].

Overall, the rescue and care of 159 Boers found at the laager presented an opportunity for Cheyne to reaffirm the value of antiseptic surgery (1095). Lister had earlier provided Cheyne with "double cyanide powder," with which to impregnate bandages. In the 1907 "Note on the Double Cyanide of Mercury and Zinc as an Antiseptic Dressing," Lister would reflect on the usefulness of this antiseptic compound to "military practice as a first dressing, by dusting it over the wound with a pepper box, and covering with any absorbent material that might be at hand" (*CP.* **II**: 330). He recommended that the salt could be used, "with the utmost freedom," alleging that it had no toxic effects. Surgeons who tried to use cyanide as a powder were not always able to do so at the Front (*CP.* **II**: 330–331). Lister cited the experiences of his former student and colleague, the surgeon Lenthal G. Cheatle (1865–1951), who had found the cyanide more effective and less odorous than iodoform when treating granulating wounds (*CP.* **II**: 331). Cheatle, in an 8 September 1900 paper, "A First Field Dressing," described the kit's contents and deficiencies: stitched into the jackets of soldiers, it consisted of antiseptic gauze and wool with a strip of bandage, in a parchment or macintosh bag. It had been designed for emergency use; thus, the patient or anyone else in the vicinity, with or without medical training, could apply it to a wound. The main problem, in Cheatle's view, was that it was almost always put on without "purifying lotion." As a result, the patient's skin, fingers, and clothes, or the fingers of the dresser on the scene, were sources of infection. To solve the problem, Cheatle called for an additional antimicrobial chemical to be part of the first-aid gear. Surprisingly, he found 1–20 water solution of carbolic acid, used alone, to be unsatisfactory. He opted instead for double cyanide of zinc and mercury salt in powder form, an ample supply of which Lister had given Cheatle before the latter had embarked for South Africa.

The powdered antiseptic had practical disadvantages. In the field, Cheatle was frustrated when the powder blew away in the wind or proved difficult to apply to posterior aspects of the trunk and limbs, as an injured patient had to be rolled over on his stomach ("A First Field Dressing": 668). The obvious solutions were, first, to create a stronger agent and, second, a more practical delivery system; therefore, they turned to a paste, combining cyanide of zinc and mercury powder with a watery solution of carbolic acid. Not only did this formula adhere to wound surfaces, but it was germicidal on contact. Cheatle, too, thought the cyanide neither toxic nor irritative. The paste was easy to dispense, even under the worst conditions: it could be manually rubbed on healthy skin and on wounds, disinfecting the dresser's fingers in the process. He calculated the amount necessary for each field dressing and delineated a method for administering the compound. As Crookshank had found in 1884, collapsible tin tubes proved the best containers: the chemicals did not react with the tin, and the paste could be squeezed out like paint; additionally, a coat of petroleum jelly on the tube-outlets prevented evaporation. The Cheatle compound was composed of double cyanide of zinc and mercury powder; tragacanth (a plant gum to keep the powder in suspension); carbolic acid; and distilled, sterilized water. Cheatle attested to its special value in South African action: the paste was ready to use because it did not require water. The dresser could apply it when needed, then place the outer dressing and bandage on.

The Cheatle kit, carried in a macintosh pouch, contained 12 collapsible tubes, cyanide wool sufficient for two wounds, and a bandage with four safety pins. It was a great advance over the specifications of the 1891 First Field Dressing. William Flack Stevenson (1844–1922) itemized its contents: a pad of gauze; a pad of compressed flax charpie between layers of gauze; a loose-wove bandage, 4½ yards long; a piece of water-proof jaconet (light-weight cotton cloth), to be applied outside the dressings; and two safety pins. All of these dressings were imbued with a solution of 1–1,000 corrosive sublimate. Towards the end of the Second Anglo-Boer War, however, it was determined that the 1891 specifications allowed for neither sufficient drainage nor for drying; hence, scabbing could not occur, and the dressings became a nidus for bacterial regrowth. To solve this problem, the jaconet was discontinued, and infection was checked (270–271)

Equipped with double-cyanide dressings and other preparations, Cheyne who had helped to rescue the infirm Boers at the laager could have reasserted the effectiveness of Listerism. That he and his R.A.M.C. colleagues had reversed late-stage sepsis in critically-ill patients had astonishing implications: two possibly exsanguinating patients, in the throes of hypovolemic shock, and with systemic infections, had been saved by rapid intervention. Cheyne, Major W. F. Stevenson, and Colonel Sylvester had, doubtless, accomplished the unexpected. It is very likely that, in the tradition of Listerism, they had employed the Cheatle formula.

Bloemfontein, South Africa, where Cheyne was stationed in the spring of 1900, had become "the great centre to which the sick and wounded were conveyed during the operations that followed to the north and east" (*The Times History*, **VI**: 523). At this point in the War, facing superior British armaments, the Boers had shifted from full-scale, frontal engagements to guerrilla tactics (Pretorius: 111). The Republican forces were armed primarily with small arms, notably the 7 mm 96 Mauser (Benton: 278). Ensuing skirmishes and ambushes took a toll. Cheyne was on call when, on 29 March, an engagement took place at Karee Siding, leaving 149 British wounded. On 31 March 1900, at the Battle of Sannah's Post, Bloemfontein, 129 more were wounded. In total, between Driefontein and the Modder River, the British suffered 184 killed in action and 723 wounded (*The Times History*, Appendix III; **VI**: 24). To make matters worse, troops occupying Bloemfontein, on 13 March 1900, were already worn out, short on rations, and stricken with enteric fever. The situation worsened, on 13 March, when Frederick Sleigh Lord Roberts (1832–1914) arrived with an additional 200 sick and wounded. By 16 March, 327 were under treatment and, by the end of May, the epidemic and the war had hospitalized 4,000 men. All available buildings had to be converted to hospitals in order to supplement the overcrowded civil institutions. Stationary hospitals were located along arterial railroad lines that traversed South Africa, connecting field and base hospitals (Benton: 277). When a new stationary hospital was opened on 29 March, personnel had to be augmented quickly. One hundred and twenty-two nurses arrived between March and April (*The Times History;* **VI**: 522–523). Stretcher-bearer companies staffed medical stations: two bearers per company, responsible for taking the wounded to the battalion aid post, were in the action; from there, stretcher-bearer companies, under the care of a Medical Officer and two N.C.O.s, transported the injured to collecting stations; and then, by ambulance, patients were sent to field hospitals which were prepared to accept the overflow of patients (J. C. de Villiers, "The Medical Aspects": Part II). Overall, Medical conditions at Bloemfontein were considered to be unsatisfactory (**VI**: 523).

Cheyne improvised his own antiseptic delivery system: a handy dusting canister made out of perforated pepper-boxes, probably Lister's idea, the dispenser resembling Crookshank's apparatus for the use of stretcher-bearers on the front lines. Cheyne's cans also contained the Cheatle antiseptic compound: cyanide of zinc, mercury, and, in various proportions, drying additives such as chinosol and kaolin (Eisler: 410; "Chinosol"). He recommended that scabbing could be promoted if wounds were dusted with the powdery compound before the first field-dressings were applied. For large skin wounds, he instructed that the dressing be removed, the injury thoroughly cleaned, and a fresh mercuric cyanide dressing applied (1193).

Cheyne's report, "The War in South Africa: The Wounded from the Actions between Modder and Driefontein," was written in April 1900, while he was stationed at Bloemfontein. It contains perceptive observations on ballistics and on how battlefield conditions affected wounds. Although asepsis was ordinarily difficult to maintain on the field or in war-zone hospitals, he was surprised that most of the bullet wounds he had treated healed "without any trouble." Cheyne attributed this unexpected outcome, early in the conflict, to the small entrance wound of the Boers' Mauser bullets. This small, ogival bullet tended to split the khaki uniform cleanly, rarely driving contaminated fabric into the wound channel. Cheyne envisioned the standard Mauser's clean entrance wounds as if the projectile were being viewed transversely and in slow motion: "I can only explain them by supposing that a wave precedes the point of the bullet in the soft fat, which drives the harder structures to each side, and thus permits the passage of the small bullet" (1193). Such wounds barely oozed, and the dryness of the South African climate often allowed for rapid evaporation of fluids on the wound surface, promoting the formation of scabs and rapid healing beneath. Patients who had been wounded early in the daylight, and who had lain for hours in the sun actually had the best results. All of these factors contributed to asepsis at the surface and interior of the injury, as well as underneath the crust. Reputedly, if the uncomplicated Mauser wound were left untouched, as long as it was not close range, the chance of healing without medical intervention was possible (1193).

Many factors complicated prognoses. Even though fighting was relatively light on the Modder River-to-Driefontein march, sepsis occurred. This was attributed to prolonged mobility in ox-carts which inhibited natural healing. Because it took a march of two-to-three days to get to the field hospital, casualties who were jostled in transit had limited exposure to the benefits of dry climate. Large bullet wounds and compound fractures posed the most difficult challenge. Amputation had to be performed in some cases, to stop spreading gangrene. A superficial crust which signaled natural healing could only be expected under dry conditions and if the patient were immobile.

The survival rate for the infantry amputees was poor. Cheyne recorded the loss of five of ten Karee Siding patients under his care, and the convalescents were transferred to the field hospitals, which indicated that Cheyne was a Medical Officer at the collecting station. Many of these fatalities were the result, not of sepsis but of traumatic internal injuries. Of the Paardeberg and Driefontein cases he records, 11 were untreatable. Of the 11, seven died, and four recovered enough to be sent to the base hospital. Of 25 inoperable cases, 16 died, and nine seemed to be recovering, but their outcomes at the base were unknown (1194). These dismal statistics had to have dismayed Cheyne who could do little more than to stabilize patients and to watch some of them bleed to death internally (1194). To provide critical care rapidly, Cheyne acquired a cart that allowed him to travel about with a complete surgical kit, including an acetylene tent lamp, and plenty of supplies (Benton: 176–177).

Cheyne's Case reports #84 and #87, recorded in June 1900 below, involved cleanly traversing bullets: one struck a vital organ; and the other just missed the spine, respectively. Case #84 is a grim clinical description of the damage a single, small-caliber bullet could cause if it hit a vital area, and if evacuation from the battlefield was delayed:

> Case #84-projectile entered in middle of right buttock, a little above the trochanter [flat process on femur]-exited through abdominal wall in right semi-lunar line at level of umbilicus. Abdomen distended-gas escaping from exit wound on pressure. Laparotomy; perforation in anterior wall of ascending colon sutured; posterior opening 2 inches lower,

firmly closed by adherent omentum; abdominal cavity cleaned and drained; died after 24 hours [Cheyne, Case #84, in Stevenson, *Report*: 92].

According to Cheyne's description, this projectile had entered the right buttock, moving transversely across the lower torso and through the ascending colon, exiting at the level of the umbilicus. Cheyne's sectioning of the abdominal wall revealed that the bullet had gone through the colon, leaking contents for some undisclosed time into the abdomen. By the time the soldier had reached the operating table, peritonitis had developed, and it was too late. It is difficult to estimate how long it took to get the wounded patient to treatment. Although, as Benton observes, small-caliber bullet wounds left the possibility of recovery "through conservative wound management." Surgery of this kind, because it was high risk, was not advisable. But Cheyne knew that a laparotomy was the only chance the patient had (283).

Case #87, although it sounds like a more serious injury, had a much different outcome. According to Cheyne's record:

> Major H., wounded on 14th, June 1900, carried to hospital three miles distant. Entrance in mid lumbar region, slightly to the right of spine; exits same side, midway between ribs and anterior [lower back: proximity to 5 lumbar vertebrae] superior iliac spine [bony projection of the pelvis].
>
> Suffered much pain, but no collapse; general condition remained good—*no evidence of peritonitis.* Starvation treatment, with absolute rest; recovered [italics added; Cheyne, Case #87, in Stevenson, *Report*: 92].

To this soldier's advantage was that the clean wound had neither struck the spine, nor a vital organ, nor caused peritonitis. His abdomen had been injured—hence, the starvation treatment. Although radiographic imagery had been available at Deelfontein and later in Pretoria, and had proved effective in finding bullet fragments, Cheyne was unlikely to have had this capability (de Villiers, "The Medical Aspect": Part II).

Unlike in the two cases noted above, the expanding bullet was responsible for large projectile wounds. The Boers began to rely on this kind of projectile once the fire-power advantage had shifted to the British. These bullets were being found in the wounded and dead, as well as in Boer bandoliers and wallets retrieved from captured camps (Cheyne, "The War in South Africa": 1193). Cheyne extracted firsthand evidence during surgery: "one was soft-nosed, one a Jeffery's sporting bullet and one an ordinary Mauser with the end filed off" (1193). A brief historical digression will help to put this important turn of events into perspective.

At The Hague Conference, legislation was passed on 29 July 1899, outlawing the so-called Dum-Dum bullet. Co-signatories to Declaration IV.3 agreed "to abstain from the use of bullets which expand or flatten easily in the human body, such as bullets with a hard envelope which does not entirely cover the core or is pierced with incisions" ("Declaration on the Use of Bullets"; Waldren: 17–18). The British had developed a soft-point projectile by removing one millimeter of the copper-alloy jacket from the nose of the standard Mark II bullet, exposing the soft lead underneath (Ogston, "The Effects of the Dum-Dum Bullet": 1425). This more powerful, improvised bullet had proven its effectiveness in the 1897–1898 Tirah campaign and came to be known as the Dum-Dum, after the Indian Army arsenal in the town of Dum Dum where the projectile was manufactured (Waldren: 3–4; Raugh: 63–64; Hutchinson; "1897–98 Tirah Expedition"). The bullet was taken out of service after the Second Boer War had been declared in October 1899. The British army service rifle was the .303 Lee-Enfield (Benton: 278). Neither the British nor the Boers were signatories to the 1899 Hague Conference (Waldren: 17–18).

Why did the British originally develop this more destructive bullet? Ogston explains that, in 1898, most European armies were employing small-bore weapons that fired "the mantled leaden bullet of very high velocity" ("The Wounds Produced": 814). Big-game hunters testing these bullets against large game found that they did not expand enough on impact to stop a lion or buffalo. To kill large game, hunters improvised either by slitting the mantle of the bullet to permit the lead core to expand on impact or by simply uncovering the apex of the projectile to expose the lead. Upon impact, instead of the tapered bullet passing cleanly through soft tissue, the Dum-Dum casing expanded with a rending effect, creating a very large wound.

Because the standard bullet had a limited effect in colonial warfare, British soldiers were hard-pressed if attacked by overwhelming numbers of indigenous forces. During the Chitral and Northwestern frontier wars against indigenous tribes, as a matter of survival rifle ammunition had to be manually altered to increase lethality ("The Wounds Produced": 814). In these engagements British soldiers abraded the mantle of the bullet at its tip and gained a marked advantage. This alteration, as Ogston mentions, was brought into production at the Indian arsenal: "the projectiles so supplied, with the core partly exposed, received the now familiar name of Dum-dum bullets. And the bullets so prepared were found to expand on impact, and thoroughly disable a wounded enemy" ("The Wounds Produced": 814). At least five variants of the altered projectile were known. Ogston lists (1) the penetrating bullet (a conventional, hard-mantled round, not subject to deformity on impact); (2) the setting-up bullet (a soft-leaden round that mushroomed at the tip, causing greater injury but presumably not fragmenting into metal shards); (3) the expanding bullet (a soft-leaden round with an air chamber at the apex that expanded on impact more than did the setting-up bullet); (4) the disintegrating bullet (soft-leaden or mantled bullets with leaden tips that break into pieces or disintegrate on impact at high velocity and great energy); and (5) the explosive bullet (detonated by a substance housed in the apex) ("The Wounds Produced": 814). Cheyne, it appears, had come across improvised ammunition which the Boers, in desperation against superior British manpower and munitions, had adopted for use in guerrilla warfare. In 1899, Ogston was explicit about the importance to British arms of the Dum-Dum and of Mark IV-C Woolwich bullets, both of which were preferred in colonial warfare if fixed positions were to repel native infantry or cavalry ("Continental Criticism of English Rifle Bullets": 752). Furthermore, to refute critics of expanding ammunition, Ogston noted that all high-velocity projectiles, whether fully mantled or expanded, had varying explosive effects (756–757).

In November 1901 Cheyne tried to account for his intense and frustrating experiences as a combat surgeon. It was paradoxical that modern surgery and aseptic measures could not appreciably improve survival rates for the wounded. In South African fighting, if one were wounded, the best outcome would be expected when struck by a high velocity, fully-jacketed bullet, through soft tissue, and to have been exposed to the dry climate before transportation. Wounds of this kind, if spared aggressive manipulation and antisepsis, were more likely to heal through the formation of a scab. Along with the maxim of keeping hands and dirty instruments away from a wound, on the South African battlefield survival was left to chance and to the environment.

Cheyne's 1901 paper "On the Treatment of Wounds in War" analyzed the problems he encountered in 1900 and enquired into why modern methods of surgery and of medical bacteriology had not met expectations in South Africa. Predicting continued failure in warfare unless improvements were made, he affirmed that "it is only by the study of these conditions that one can hope to lay down lines for progress in the future" (1591). He imagined the plight of an infantryman, wounded on the South African veldt ("On the Treatment of Wounds in War": 1591). First, the wounded soldier was in jeopardy since the open, unbroken landscape made it very difficult for medical personnel to reach him under fire. In the scope of the enemy marksman, the downed soldier had no choice but to wait for the action to cease, or until a comrade or doctor could arrive. Second, only rudimentary medical supplies were portable: a field-dressing consisting of a small piece of antiseptic gauze, a piece of wool, and a bandage. The dressings, according to Cheyne's inspection, were small and thin, nothing more than "a clean rag" that lost antiseptic potency once saturated with blood. Whether or not the medical attendant or team was under fire, it was virtually impossible to find a clean, flat surface upon which to arrange supplies as the soldier's clothes were cut away to expose the wound. The attendant's hands were dirty, the contents of the kit would inevitably gather dust, sand, and grease, and the turmoil was unnerving. Even if the soldier were lucky enough to have a Medical Officer on site, care was very limited. The doctor could neither carry antiseptic bottles to the firing line nor apply a clean bandage with dirty hands (1591). What Cheyne learned from his South African service was that current emergency care during battle was inadequate. Dirty hands and raggedy gauze notwithstanding, if a fully-jacketed bullet hit a soldier, and if it had not damaged bone, vital organs or arteries, and if it was a dry sunny day, the soldier was in luck—as long as no one touched his wound (1592). Nonintervention, ironically, was the preferred treatment, in many cases. Wounds resolved naturally as long as infection was controlled:

> The antiseptic dressing and the surgical treatment adopted had little or nothing to do with the result in these cases and the only way in which we can bring in modern surgery as of value is because it has taught us the meaning and value of healing under a scab and has shown us that unless the wound can be thoroughly aseptic the less such a scab is disturbed in the first instance the better ["On the Treatment of Wounds in War": 1593].

Logistical failures exacerbated battlefield conditions in South Africa for the British and, more so, for the Boers ("On the Treatment of Wounds in War": 1592). As noted, limited amounts of drinking water were brought in for the British camps, but contaminated river water had to be used for medical purposes. Since the water at Paardeberg was polluted with fecal and other microbes, it had to be boiled; but this necessity was hampered by the shortage of tinder and kettles. In addition, it was impossible to secure enough antiseptic; since the supply on hand, if used for lotions, depended on clean water, it could not be utilized efficiently. Getting the wounded away from the battlefield as quickly and as gently as possible was also difficult. Cheyne recalls, for example, that the three-day journey from the Paardeberg action to the nearest field hospital took two or three days. Transportation was slow, via ox-cart and over the "rough veld." This arrangement had a deleterious effect on anyone with a serious injury. Complications set in and deaths occurred while in transport or upon arrival at the battalion aid station (1592).

Cheyne revised his opinion of high-velocity ammunition. He learned that asepsis was not guaranteed for the common bullet wounds ("On the Treatment of Wounds in War": 1592). In a minority of cases, those struck by the mantled, high-velocity ammunition of the German Mauser developed infections even though these bullets rarely drove dirty clothing or debris into the wound. The probable causes were the absence of immediate care in the field and evacuative delays, for "in many cases the ride from the battlefield to the field hospital contributed to the death of a wounded soldier" (Benton 281). In cases where bones were broken, necrosis and other degenerative processes invariably followed, no matter the type or caliber of the projectile; for compound fracture patients, transport out of wagons onto stretchers and across bad ground often led to amputation (Benton: 281). Even if the transported patient reached the hospital early, Cheyne struggled to stabilize the patient and to manage acute septicemia, gangrene, and osteomyelitis, infections often resulting in amputations. In addition, small conical projectiles did not always traverse soft tissue cleanly. In the worst case, a conical, small-bore round could impact the body obliquely; or, if the rifle barrel was worn, a wobbly projectile could ricochet off bone, its splinters tearing through organs and exiting through gaping exit holes (1592).

To improve the existing system, Cheyne outlined a plan of his own ("On the Treatment of Wounds in War": 1593). His first point was an adaptation to circumstances: asepsis on the battlefield was not a realistic pursuit, but antisepsis, as a temporary measure, could ward off or inhibit infection at least for 48 hours. To promote the formation of scabs on smaller wounds which was the ideal outcome, an attendant or doctor should carry antiseptic powder, the pepper-box container he had improvised at Bloemfontein in the spring of 1900. The antiseptic powder would dry and protect the wound, adhering to tissues, instead of draining away or evaporating as did lotions. The temporary gauze placed over the wound could be saturated with zinc cyanide.

A most important innovation, in Cheyne's opinion, would be to move the field hospital closer to the Front so as to lessen transport time and to furnish "suitable supplies and medical comforts" ("On the Treatment of Wounds in War": 1594). Two serious drawbacks that Cheyne cited should have been addressed before the War: one was lack of intelligence about the geography; and the other, the logistics of medical supplies and evacuation (Baird: 28).

10

1914–1918

Listerism on the Battlefield

The scholarship on wound treatment in World War I is substantial. Researchers of British war-time medicine in this period, such as Cay-Rüdiger Prüll (1999), Mark Harrison (2010), R. L. Atenstaedt (2010), Ana Carden-Coyne (2014), and Jane Coutts (2015), along with periodical commentators, have surveyed its history widely and from differing perspectives.[1] Cheyne's contribution to the field, as I pointed out in chapter 9, has been given relatively limited attention, even though he was a Consulting Surgeon in the Second Anglo-Boer War. In World War I, he re-experienced the stress and frustration of combat medicine, was concerned about the inability to control deadly forms of bacterial sepsis, and, in total, investigated these problems in five papers on World War I.

1914–1916: Cheyne and No Man's Land

Cheyne held the rank of Royal Navy Commander, was surgeon at Chatham Naval Hospital, and had been appointed President of the Royal College of Surgeons of England. Clearly, he was well qualified and positioned to address the issues of modern warfare and of wound treatment in World War I, and he did so in a series of instructive papers, one of which was his "Remarks on the Treatment of Wounds in War" (16 November 1914), delivered at the opening session of the London Medical Society, and published in *The British Medical Journal* of 21 November (865). The picture he sketches was grim. Cheyne stated that all of the wounds that he had recently treated were infected, some very badly. Those working at army hospitals were struggling to manage sepsis, along with shrapnel, bullet, and burn injuries, all of which were complicated by tetanus and acute gangrene. Secondary infections, especially those from sporiferous, soil-borne organisms, did not affect naval personnel to the degree they did the foot soldier. That all larger wounds arriving at the base hospital were without exception infected was especially worrisome. Cases of compound fracture often resulted in amputation, protracted convalescence, chronic abscesses and pain, deformity, and lifelong disability (865).

To understand the causes of sepsis in traumatic wounds, Cheyne harkened back to Lister's earliest breakthroughs in the management of soiled compound fractures. Lister had been able to control infection and to achieve asepsis in many cases, and his early work provided insight into the present problem. Cheyne realized, as early as 1900, that the longer medical evacuation was delayed, the greater the chance of severe and irreversible infection. Whereas in civilian practice, a seriously-injured patient could be attended to in a relatively short period of time, both in naval and land warfare conditions delayed the transportation of the wounded, raising the mortality rate (Cheyne, "Remarks on the Treatment of Wounds": 865).

The environmental context of Cheyne's "Remarks," in the autumn of 1914, was the beginning of trench warfare. These insalubrious conditions gave rise to trench fever, a rickettsial disease transmitted by body lice, first reported among British forces in Flanders during the summer of 1915 (Atenstaedt, "Trench Fever"). By September 1915, mobile warfare had given way to fighting from fortified, earthwork positions. Constructing trench-lines along the high ground of Chemin des Dames, on the northern bank of the Aisne River, the Germans extended their defensive position for miles. By 1918, their trench-lines meandered along the entire length of the Western Front. The Chemin des Dames trenches were effective, stopping a British Expeditionary Force (B.E.F.) frontal assault ("Battles of the Western Front"). Stalemates, however, exacerbated health problems. Allied and German trenches that were cramped and dirty became breeding grounds for disease and vermin, and these conditions detracted from the overall fitness of the combatants.

The situation for a wounded soldier lying beyond his trench-line in No-Man's-Land was dire. Cheyne described his plight: in some instances, the patient would not receive medical attention for hours. During the Second Anglo-Boer War, in contrast, the wounded soldier had had a better chance of survival than did his World War I counterpart. A number of reasons accounted for this difference. In terms of scale, duration, intensity, and munitions, the typical South African engagement could not be compared to a World War I battle. In terms of artillery, for example, the entire Boer army, in 1899, had only 38 pieces (75-mm field guns and four 155-mm siege guns), whereas a *single* German infantry Division in World War I was equipped with 72 pieces (7.7 cm field guns and 10.5 cm light field howitzers) (*The Boer War*: 25; Zabecki:

112, respectively). In South Africa, medical personnel were more routinely recognized as noncombatants (although this was not always the case); distances
between combatants and the rear were not beyond the capacity of the transport
system in South Africa, although the rides were incommodious and dangerous
for seriously wounded patients; medical personnel in South Africa could gather
the wounded as the fighting was going on, the bumpy ride of the ox-cart
notwithstanding; and the munitions employed during the Anglo-Boer Wars
were not as destructive as those of World War I. Consequently, on the South
African veldt, unlike the rural Western European landscape, medical care and
antiseptic treatment could more readily be applied as a temporary medicament,
at the point of injury or while in transit.

In World War I, Cheyne explains, the Medical Corps could not venture
into a prolonged battle where artillery shells were falling and where, from strong-
points, Maxim MG08s (Maschinegewehr 08) were firing 400–450, 7.92 ×
57mm Mauser rounds per minute ("Remarks on the Treatment of Wounds":
865; D. Alex, "Maxim"). Even if the enemy were to respect the Red Cross on
an ambulance, the battlelines were extensive, desultory, and vehicle markings
indistinct. Since visibility was poor and communication sporadic, the safety
of medical personnel could not be guaranteed. The presence of medical wagons
and of noncombatants, moreover, could give away an entrenched unit's exact
position to artillery spotters. These conditions were unique to the history of
warfare: "The result," Cheyne writes, "is that the wounded often cannot be
attended to for a long time, sometimes, indeed, we hear of forty-eight hours
or longer having elapsed before they are collected and in the meantime the
wounds are becoming more and more infected" (865).

The distance between the point of injury and the field or temporary hospital, even under the best circumstances, as in a lull, was often very great. Enormous numbers of wounded in periods of serious fighting would have to be
transported many miles over muddy, shell-pocked terrain, the uncomfortable
ride worsening injuries. The gravity of the wounds received in modern warfare
meant that the field hospital could, at best, provide only basic care in serious
cases, "but in many cases it [was] not till the patient reaches the base that full
attention can be paid to the injuries caused by the missiles" ("Remarks on
Treatment of Wounds": 865–866).

Even in naval engagements, the wounded had to lay where they fell on
the ship until the battle was over or when a hiatus in the fighting occurred.
They were then transshipped from their vessel as soon as possible to a hospital
ship and eventually to a land-based facility. One advantage naval had over
infantry combatants was that, generally, treatment could begin faster. Naval
wounded, as noted above, were not subject to soil-based pathogens that became

embedded in multiple wounds, spawning tetanus or gangrene. But because sailors were also injured by batteries of powerful guns in confined areas, their wounds—second- and third-degree burns, exposure to fuel oil, immersion, along with lacerations and dismemberments by metallic shards and shrapnel— were in some respects worse than those of infantry, and they were not free of infection ("Remarks on Treatment of Wounds": 866). Although the army suffered the most casualties, the navy had problems unique to the service. The planning and elaboration of casualty care, for example, was a central concern for naval authorities because of the potential for mass casualties in the briefest of engagements (Herrick, "Casualty Care").

After 24 hours, the possibility of rendering an infantryman's wound aseptic was low, practically zero after 48 hours. After 48 hours, Cheyne advised *against* making any attempt at disinfection. Dismayed by these conditions, he regretted that Lister's foundational work had all but been forgotten, and, when recalled, some treated his system disdainfully. The high rate of serious infection Cheyne attributed, not to the inefficacy of antiseptics but, in many instances, to a lack of surgical precision, to outright carelessness, and to hectic conditions. Some surgeons who prided themselves on having the trappings and equipment of the profession had forgotten the tenet that bacteria were ubiquitous and that gloves and instruments, though sterilized at the outset of a procedure, became contaminated in the course of an operation; and that live bacteria were reintroduced into a wound, as a result ("Remarks on the Treatment of Wounds": 866–867). If suppuration were to occur, anti–Listerian operators tended to blame it on causes other than "faulty manipulations" with contaminated gloves and probes: thus, a surgeon who did not "disinfect his hands," and who relied exclusively on "the protection of boiled gloves" had not taken into account that the boiling could be insufficient to render them aseptic and that the gloves were put on with dirty hands (867).

Medical training, Cheyne complains, propagated a bias against chemical disinfection: "a good many men are being sent out from the schools every year who are not impressed with the bacteriological problem involved in the treatment of wounds, or who have come to believe that no such problem is involved," even to the point that they considered the use of antiseptics "an abomination" ("Remarks on the Treatment of Wounds": 867). This trend's worst effects were being felt in war when, confronted with battle-soiled wounds, the new surgeon either did not know how to proceed or expeditiously coated the skin with iodine and covered the wound with gauze. As a direct result, bacteria remained enclosed in an environment providing ideal conditions for propagation. Unfortunately, the extraordinary results Lister had achieved in the 1860s with compound fractures seemed to have been forgotten.

Cheyne's analysis of the problem correctly began with bacterial physiology and pathogenicity. The uncomplicated gunshot wound—the projectile cleanly-traversing soft tissue, missing vital organs and bone, and healing by first intention—had become a rarity; As he had learned in South Africa, an uncomplicated gunshot wound was a dangerous oxymoron. The Western European battlefield, in the autumn of 1914, had dangers more insidious than bullets and shrapnel. As noted, many wounded at the firing-line wound up with sepsis, and infectious diseases such as trench-foot were rampant: "Working and sleeping in polluted soil and standing water, surrounded by garbage and human and animal feces, inundated by lice and fleas, and lacking proper hygiene and recreations, soldiers fell victim to all forms of infectious disease" (Haller, *Battlefield Medicine*: 153). Skin, debris, and soil were often driven into irregular cavities in penetrating, large wounds. Because of delayed care, the severely injured soldier, subject to blood loss and shock, was prey to sporiferous pathogens that throve in moist, nutrient-rich tissues. Under these conditions, "myriads of organisms" sprouted and multiplied rapidly, filling the wound crevice in a matter of hours (Cheyne, "Remarks on the Treatment of Wounds": 867). Saprophytic pathogens subsisted on traumatized or devitalized tissues and, once established there, were difficult to eradicate. The only efficient way to prevent this from happening was to destroy them within 48 hours. Painting the skin, inundating the wound with caustic iodine, or filling the cavity with saturated gauze alone, could not prevent sepsis. Along with managing the trauma of the wound, the doctor had to understand that the patient's life and hope of convalescence depended equally on surgical repair and on infection control (867).

In a 1917 paper, Dr. Herbert Henry who was stationed at the laboratory of a base hospital in France described the ordeal of anaërobic infection so often contracted in the field. The soiling of a wound in a manured field or by clothing contaminated with fecal organisms created intense suffering. Because modern projectiles and shell fragments produced large, multiple wounds, conditions were especially favorable to the growth of anaërobes and to sepsis (H. Henry: 76). Henry outlines in four phases this lethal pathology: (1) the initial trauma, being a latent period, established a pre-clinical cavity of bacterial growth; (2) in the first active phase of anaërobic infection, saprophytic microbes, such as *Bacillus welchii*, began to thrive on fermentable carbohydrates which devitalized muscle provided. Acid and gas, metabolic by-products of the bacteria, accumulated; as gas pressure and edema increased, surrounding tissues died from lack of blood, and the next active phase of the process led to (3) anaërobic invasion. At this point, the bacteria digested the dead and dying tissues. Muscle became flaccid and, as it was consumed, turned from red to black. The breaking down of protein molecules in this way released toxic substances; and (4), once bacteria and toxins invaded the bloodstream, death ensued (H. Henry: 77).

From the second to the twelfth hour, natural healing was not absolutely ruled out (Cheyne, "Remarks on the Treatment of Wounds": 867). Wounds treated in this time-frame, if small or without comminuted bone, could be made aseptic. Even up to 24 hours, given the ideal conditions described above, infection might be "rooted out." But where wounds were soiled, a virtual certainty in rural, agricultural terrain, the danger of tetanus and gangrene loomed. Infected wounds are treatable inside 40 hours, observed Cheyne, after which "the disinfection will certainly fail, and the attempt will do more harm than good by injuring the tissues which are opposing the invasion of the bacteria and thus enabling the infection to spread more easily and rapidly" (867). The judicious employment of chemical antiseptics that Cheyne advocated had three caveats: (1) both actively-growing and spore-forming bacteria had to be destroyed; (2) the disinfection of wounds unsoiled with earth should be undertaken within the first 24 hours; (3) and the disinfection of wounds soiled with earth had to be undertaken within 48 hours (867).

Successful disinfection depended on a number of factors (Cheyne, "Remarks on the Treatment of Wounds": 868). Since spores had to be killed as quickly as possible, the inclination was to use highly-concentrated antiseptics. The problem was that many antiseptics, notably carbolic acid, were toxic if incorrectly or overly applied; moreover, they took time to destroy spores. Carbolic acid, a mainstay in the antibacterial arsenal, killed vulnerable bacteria in seconds, but it took 12 to 15 hours to penetrate and destroy spores. Many of Lister's compound-fracture cases had followed an aseptic course as long as he applied the acid in saturated lint, in paste, or in liquid form. The acid lost potency and took longer to work, however, if the spores were present in albumen or oily material. Thus, washing out a wound with 1–20 solution of carbolic acid-to-water would not cleanse the wound of live spores, although full-strength acid, at the risk of irritating tissue and compromising immunity, could do so over time, once it permeated spores. There were benefits to the use of carbolic acid beyond antisepsis. Even if the acid precipitated, and mixed with, albumen or oil, it remained bactericidal. Cheyne preferred its use because of its anesthetic effect and ability to penetrate the skin. In contrast, though iodine had antiseptic power on par with carbolic acid, it was very painful (868).

When Lister learned that carbolic acid was soluble and worked in oil or water, he later washed out compound fractures with 1-in-20 lotion, achieving aseptic conditions in most wounds—with the exception of those that were soiled. Through trial-and-error, he learned that the 1–20 lotion was unable to reach and destroy inaccessibly-embedded spores ("Remarks on the Treatment of Wounds": 868). Cheyne suspected that, in the midst of action, Lister's original plan to use liquefied carbolic acid as a germicide on the wound surface and in all cases of compound fracture was useful but not in deep bullet tracks. Conversely, disinfecting the interior of a wounded extremity, but leaving bacteria on the surface, was dangerous since cutaneous microbes were "free to multiply" and to re-infect the wound (868).

Cheyne, in line with von Reyher, also advocated debridement: the opening up of wounds to arrest bleeding, to remove devitalized tissue, and to reach inner recesses for thorough disinfection ("Remarks on the Treatment of War": 868). To disinfect the interior of a large wound, one had to cut away visibly soiled skin and fat, to retract the wound opening, and to apply sponges saturated with liquefied acid to the walls and recesses of the cavity (869). Cheyne described a similar procedure for trunk and neck wounds. Antiseptics, as opposed to aseptic dressings, had to be applied as soon as possible in the form of sterile gauze; the aseptic gauze, on the other hand, because it rapidly absorbed blood and serum was reduced to a contaminated rag. In the midst of action, serum-soluble antiseptics such as salicylic acid in wool could be used to pack the cavity temporarily (869).

Lister's 1870 plan was applicable to conditions in the field hospital because careful disinfection could only be achieved over time and without having to worry about enemy harassment. Whether some form of effective disinfection could be administered in the field or in trenches was a question Cheyne and colleagues tried to answer at the Chatham Naval Hospital. One possibility was to equip combatants with soluble, carbolic-acid suppositories (1.25 grains) that, once inserted in a wound, would melt at body temperature, underneath emergency dressing or an iodine solution; an alternative was to push saturated gauze into the cavity. Only in the field hospital or in sick bay could the patient undergo methodical disinfection ("Remarks on the Treatment of Wounds": 870). These measures and new products afforded temporary germicidal treatment for compound fractures, joint wounds, soil-contaminated injuries, and septic-related amputations. Vaccinations for tetanus, though no substitute for chemical antisepsis, also had prophylactic value (870). Emergency dressings simply bought time.

The most difficult challenge to battlefield medicine, as Cheyne observed firsthand, was the condition of infected wounds seen from 24 to 48 hours after the initial injury ("Remarks on the Treatment of Wounds": 870). In this period of time, antiseptics did not work effectively against known pathogens; and if such an attempt failed it could inadvertently injure "actively granulating tissue," impair natural immunity, and impede healing. Under these conditions, Cheyne recommended debridement and the dabbing of full-strength carbolic acid onto raw surfaces; but he proscribed "general disinfection" with any type or strength of antiseptic as "a very injudicious and harmful procedure." The general practice of surgeons seemed to have misconstrued Lister's instructions. It appears that his colleagues had routinely used contaminated instruments, neglecting to re-sterilize them during procedures. For septic wounds, Cheyne recommended irrigation with mild hydrogen peroxide and with solutions of normal or isotonic saline. The lingering bias against antiseptics, Cheyne surmised, was traceable to surgical overuse of these chemicals, to the point of impairing "the resisting power of the tissues." The ensuing bad outcomes had led some to blame Listerism rather than their own mishandling of the case. Cheyne reiterated the important point that antiseptics were intended for auxiliary use, not in the wound's interior, but on the surface to prevent pathogens from entering (870).

Just as misuse of antiseptic chemicals could be injurious, so, too, could the over-packing of draining wounds with gauze. Aseptic dressings soaked up discharges that, if left to decompose in the fabric, created a breeding ground for harmful bacteria. Dressings were best for small wounds, and the chemicals of choice would be cyanide gauze against the wound and salicylic wool as the outer dressing. Large masses of dressing should not be placed on large wounds, advised Cheyne; on this point, he departed from Lister's multilayered system. Water baths and saline irrigation, followed by boracic fomentations, were indicated for large bone and joint wounds. Cheyne subscribed to the use of antiseptics only at the very beginning of treatment; and only then were the most potent ones to be used (871).

In "The Hunterian Oration on *The Treatment of Wounds in War*," delivered on 15 February 1915 to the Royal College of Surgeons of England, Cheyne reconsidered the prevention and treatment of septic war wounds. He outlined three actions against secondary infection: (1) to strengthen natural immunity so the patient could resist "the parasitic invaders"; (2) to destroy bacteria that had already broken through natural defenses and established themselves in the body; and (3) to prevent or destroy these microbes, "at their point of entrance into the body," and before they colonized and spread. He discussed the possibility of using vaccines to enhance immunity prophylactically; but once an infection had become systemic, he doubted the value and the logic of injecting toxins and dead bacteria into a sick patient. Antitoxic sera for use against tetanus neutralized bacterial toxins as the immune system destroyed the bacteria; but positive results for tetanus vaccine, as noted, had been established only for prophylactic use, although research in this area was ongoing. Chemotherapy was the best means to treat bacterial infection; hence, in his view, it was the most effective treatment in the disinfection of war wounds.

Cheyne reviewed the origins of Listerism and how he had adapted it to his practice. In its original form, Lister's system had introduced impure liquid carbolic acid into wounds, had mixed it with blood to the consistency of paste, had left the mixture in place to clot, and in the interim had painted the clot's surface with the acid ("On a New Method of Treating Compound Fractures," *CP.* II: 1–36). The infections invariably accompanying compound fractures were thereby precluded; sloughing, suppuration, and inflammation, all signs of active infection and of the body's immune response to it, were prevented; and dreaded hospital infections were virtually eliminated. Because tetanus caused by the toxin of *Clostridium tetani* disappeared along with the microbes, this suggested, as early as 1867, that Lister had linked the presence of a microorganism to a specific disease.

According to Cheyne, Lister later modified the 1867 system ("Hunterian Oration"). Instead of using undiluted carbolic acid, he washed out compound-fracture wounds thoroughly with a strong lotion (1–20 acid-to-water); and, with a syringe and catheter, eradicated deeply-embedded colonies. Because war wounds incurred during the Franco-Prussian War were often soiled as well, Lister had to modify the second stage further. The third stage included washing out the wound with the 1–20 lotion and débriding soiled tissues, to which undiluted acid was sparingly applied. Cheyne successfully adapted this stage to his work but determined that the circumstances of the twentieth-century battlefield required even further modification: "I suggested a *fourth* plan—viz., in addition to clipping away the soiled parts, to apply the undiluted carbolic acid to the whole surface of the wound, opening it up if necessary" ("Hunterian Oration"). His rationale was to make certain that the entire surface of the wound had been rendered aseptic and to leave no recesses untreated. Washing the wound with bactericides temporarily rendered tissues inhospitable to bacterial growth. As long as the tissues were not injured by the acid, "their defensive action," enacted through antimicrobial substances in bodily fluids, phagocytosis, natural killer cells, and inflammation, remained active (Tortora and Derrickson: 438–440). Acknowledging the risks of using excessive quantities of the acid for large wounds, he believed that the danger of sepsis outweighed that of toxicity ("Hunterian Oration").

Cheyne disagreed with surgeons who considered the disinfection of gunshot wounds impossible ("Hunterian Oration"). He specifically referred to Louis A. La Garde's experimental claim, based on gunshot wounds in animal carcasses, that high-velocity bullets drove gunpowder into tissues as deep as 17 millimeters, lodging bacteria outside the reach of syringed antiseptics (Haller: 306). Colonel Louis A. La Garde (1849–1920), who served in the United States Army Medical Corps, wrote extensively about the treatment of war wounds and the management of sepsis. He argued for the merits of Almroth

Wright's hypertonic saline lavage as a way of increasing the flow of lymph "containing antibodies" and of leukocytes into a wound cavity. Just as blood cells oozed through capillary walls in a process called "diapedesis," salt solution allegedly drew antibacterial cells into the damaged area to the point of overflowing; an outflowing of stagnant lymph was simultaneously maintained through saline irrigation, reputedly as asepsis was maintained without resort to chemicals (La Garde: 148–151).

Neither La Garde's ballistics experiments nor his aseptic regimen appealed to Cheyne. With respect to ballistics, Cheyne cited his South African experiences as proof that high-velocity wounds, if uncomplicated, untampered with, and uncontaminated, for the most part would heal naturally and with no sign of infection (although this was not necessarily true of close-range gunshots) ("Hunterian Oration"). Of critical importance, in Cheyne's mind, "was the length of time which might elapse between the receipt of the injury and the patient's arrival at a suitable place where thorough disinfection could be carried out." To improve the patient's chances of avoiding sepsis, he offered two suggestions that were in development. At the Royal Naval Hospital at Chatham, he and co-workers were developing soluble bougies. These suppositories contained carbolic acid; once pushed into wounds, they would melt at body temperature, to seep into bodily recesses. The purpose of this treatment was not to achieve asepsis by eradicating all bacterial flora: rather, it was to "delay sepsis," until the patient could be safely retrieved from the site and transferred to the rear for extensive treatment.

Cheyne needed experimental data if he hoped to perfect his method and convince military superiors and medical colleagues as to its value ("Hunterian Oration"). At Chatham, he joined with colleagues Fleet Surgeon Bassett-Smith, R.N., and Cheyne's assistant and bacteriologist Arthur Edmunds to discover "whether it was possible to introduce an antiseptic into a wound soon after its infliction which would remain there, diffuse in the blood and tissues, and inhibit the growth of bacteria till such time as the wound could be thoroughly disinfected." In the tradition of battlefield first aid, he had in mind a durable compound that would temporarily inhibit the growth of pathogenic species, including sporulative organisms. His position, in November 1915, was that suppositories and iodine gauze were efficient for battlefield use and were an improvement over lotions that splashed unevenly over wound surfaces, that became contaminated or evaporated.

Cheyne and colleagues had to find an effective antiseptic compound and a dissolvable solid in which to contain it ("Hunterian Oration"). The cylindrical material or base, imbued with the antiseptic, had to melt at body temperature so as to allow the chemicals to diffuse widely, while simultaneously retaining potency for as long as possible. Since the base and the infused chemical inter-acted with each other, the problem was to find a balance between diffusion and potency. Cheyne and colleagues learned that pure lanoline (a wool grease) was the most diffusible base; and paraffin wax, the least. When carbolic acid, the primary antiseptic, was blended with lanoline, it proved too diffusive: the acid exuded from the base too quickly, with pungent odor and irritating effect. Adding wax to the lanoline hardened the suppository but lessened antiseptic diffusion by one-quarter; nevertheless, it was potent enough, and the pungency and irritation were eliminated. The greatest wound diffusion was obtained either with undiluted carbolic acid or with cresol, a coal-tar derivative manu-factured as Tricresol. Altogether, Cheyne's team tested the inhibitory power of 16 agents, each infused in a lanoline-wax base and in both agar and blood, using emulsions of *Staphylococcus pyogenes aureus* and of the sporiferous *Micrococcus prodigiosus* and *Bacillus subtilis*. Preliminary results suggested that 30 percent carbolic-acid paste or Tricresol inhibited *Bacillus pyocyaneus* for four days before pus confirmed reactivated infection. Although gangrene-producing organisms were checked, *Clostridium tetani* was not; but, in this instance, the prophylactic use of tetanus serum was useful.

In terms of tissue diffusion, carbolic acid, Tricresol, Lysol, corrosive sub-limate, paraform, oil of cinnamon, cyanide of mercury and zinc, salicylic acid, and balsam of Peru, all made the grade ("Hunterian Oration"). The tentative conclusion of the researchers was that either Tricresol or carbolic paste should be the first tried at 20 percent strength. Since, at the time, "no fresh naval wounded" were at the Chatham Hospital, they did not have the opportunity of testing it "on actual war wounds."

The object of the Chatham bacteriological experiments was "to find some means of averting sepsis in a wound for some hours, or at most a day or two, till thorough disinfection can be carried out" ("Hunterian Oration"). Incon-trovertibly, a combination of carbolic acid and Tricresol delayed infection four to five hours but only if paste or suppository were inserted in the wound within 15 minutes after the injury had been incurred. And only if the wound were subsequently cleaned out would rapid healing occur. They also observed that pus did not appear for ten days, even in dirty wounds with marginally-devitalized tissue. The danger of infection, between the point of the original trauma and arrival at the hospital, in Cheyne's view, could be lowered consid-erably through the use of portable bactericides.

Correspondence of 22 May 1915, published under the title "The Recom-mendations of the Naval Medical Committee on the Treatment of Wounds in War," recapitulated the aim and method of the experiments, described in the 15 November "Hunterian Oration," and republished in *The Lancet* of 27

February (419–430). Laboratory tests were performed to determine how numerous chemical agents affected pathogenic bacteria in blood clots, on agar, in animal wounds, and *in vitro*. The Chatham Committee's recommendations in this area were inconclusive since data had not been collected at the Front. Although Cheyne, as a Royal Navy Commander, requested but was denied permission to conduct infantry trials using temporary antiseptics, his colleague, Arthur Edmunds, who had been deployed to the Dardanelles was presented with the opportunity to test antiseptics near a war zone, and to compile the data his colleagues at Chatham needed. At this point in the 17 May correspondence, Cheyne reminded readers that the original aim of the research had been to find ways *to delay* the growth of harmful bacteria, "in the interval between the infliction of the wound and the arrival of the patient at the dressing station or the field hospital" (italics added; "The Recommendations": 912). Their assumption was that, once the patient was out of the war zone, his wound(s) could be surgically opened if necessary; the temporary antiseptic could then be removed from the cavity; and the wound could be thoroughly disinfected, drainage tubes could be inserted, antiseptic dressing administered, and treatment undertaken, "as in civilian practice." The implication was that the Listerian method in its latest incarnation could be applied with good effect but only in a secure and stable setting.

Cheyne then tried to set the record straight for critics who were misinterpreting or distorting his aim. Thus, he reiterated that the Chatham Committee had not expected to disinfect wounds with their battlefield pastes. Their tests had shown that the Tricresol and carbolic pastes could not destroy spores, at least not in the short run. Thorough disinfection, coupled with debridement, cleansing, and drainage, were to come at the field or base hospital or on board a hospital ship. The Chatham team found that a combination in equal parts of salicylic and boric acid (product acronym: *borsol*) was effective against tetanus bacilli, and against the micro-organisms responsible for gangrene. Cheyne's tripartite regimen designed for spore-forming microbes included anti-tetanic serum, borsol, and, as an adjunct, either Tricresol or carbolic acid. Despite its effectiveness, borsol had a drawback: administered only as a powder, it could not reach the inner recesses of a wound, whereas carbolic-acid paste was better suited for immediate use under fire; afterwards, it could be applied to blood clots and to recessive sites. Once in a hospital setting, surgeons could treat the cavity thoroughly with borsol, in the hope of obtaining asepsis. Lister's salicylic-carbolic acid cream in a glycerine base was also part of the Chatham pharmacopoeia (Cheyne, "The Recommendations": 912).

The fourth document in the series, "Observations on the Treatment of Wounds in War," was published in *The Lancet*, on 31 July 1915 (213–219). By this time, Cheyne had been appointed President of the Royal College of Surgeons of England and Surgeon-General of the Royal Navy. In "Observations," he reiterated the important finding at Chatham: conventional antiseptics failed to destroy the bacilli of tetanus and gangrene. But borsol powder, in conjunction with phenolic antiseptics, was "very efficacious against infections by these organisms." As a consequence, Cheyne asserted that "the chances of successful disinfection, if undertaken within eight to twelve hours after the infliction of the wound, are very considerably improved." His position restated, he then tried to unscramble the mess that was occurring at the Front, regarding the use of the temporary antiseptics that he and his co-workers had prescribed.

Having recommended the disinfection of wounds in the field as a transitory measure and calling for the use of an improved, initial field-dressing to implement this strategy, Cheyne was disturbed by military decisions confining him to laboratory research. In "The Hunterian Oration," he expressed the need to test his neo-Listerian plan under battlefield conditions and to oversee the implementation of the Chatham plan:

> For this purpose, a small section of the Front should be set apart, and the [Chatham] committee should be able to follow the cases to the base, so that they remained from first to last in their own hands and did not pass through the hands of other surgeons who might not understand what was wanted and might, with the best will in the world, completely spoil the investigation.

Cheyne's enthusiasm might have gotten the better of his judgment even though, as a Consulting Surgeon, he had experienced the medical exigencies of land warfare in South Africa. Although his latest plan was logical and scientifically justified, the construction of an experimental dressing station on the front-line could have been viewed as logistically impossible since maintaining vital signs under fire, rather than performing surgery, was the priority. Or perhaps Cheyne's plan calling for the exclusion of non–Chatham personnel insulted the Army Medical Corps? Responses to Cheyne from combat surgeons at the Front who had half-heartedly agreed to use the temporary antiseptics, it turned out, were either muddled or evasive. Despite having combat experience in South Africa, perhaps Cheyne had not appreciated the enormity of trench warfare and that the weaponry of the Great War dwarfed the damage veldt elephant guns or the Boer Mauser could inflict on living vertebrates, even with improvised ammunition? In the Great War, rifle bullets had become more powerful, the machine gun was used against massed infantry, and artillery delivered shrapnel and high-explosive shells that inflicted "horrifying wounds by fragments of all sizes on a scale never before witnessed" (Helling and Daon: 176). Beyond Cheyne's possible shortsightedness in this regard was another variable: professional competitiveness from several quarters—from those who rejected the Listerian tradition altogether, or from those who, like Almroth Wright, had other ideas about how to achieve asepsis.

One response to Cheyne's inquiry about a front-line trial, from someone whom he identified as Dr. A., should have been anticipated ("Hunterian Oration"). Dr. A. had taken the position that disinfection was impossible, so Cheyne hoped to change his mind and, in the process, to gather valuable data. Unfortunately, Dr. A. seemed to have misinterpreted Cheyne's intention: the former thought the field-kit paste was meant to be the equivalent of the hospital-administered antiseptic, rather than a temporary one prescribed only during transportation. In distributing the borsol and cresol paste to his colleagues, Dr. A. (Cheyne surmised) had conveyed an erroneous instruction to his colleagues. In later correspondence to Cheyne, Dr. A. is said to have rejected antiseptic treatment as ineffective, a claim flatly contradicting an earlier statement by the same doctor, reputedly about having achieved good results with carbolic acid and peroxide.

Opposition to the use of the Chatham pastes at the Front, though formidable, was not universal. Dr. B., for example, whom Cheyne had met and instructed at Chatham, had served at the Front and, upon the former's return to London, reported good results with the field antiseptics. Other reports were unenthusiastic. Dr. C., a Major in the Royal Army Medical Corps, said that debridement, wound enlargement, and the paste, though effective, were equal in benefit to "free drainage." To Cheyne, Dr. C.'s omission of statistics supporting this claim smacked of politics. Though Dr. C. claimed to have had no faith in antiseptic chemicals, he inadvertently admitted to having used methyl alcohol for cleaning wound surfaces and "aseptic gauze and wool" (which normally contains a chemical). These admissions revealed Dr. C.'s limited knowledge of Listerian antisepsis: methanol was a poor disinfectant while aseptic gauze clogged wound cavities, forming a nidus for bacterial infection. In cases where Dr. C. reputedly applied the Chatham regimen, he either did so incorrectly or disingenuously. On the one hand, according to Cheyne, overuse of antisepsis was counterproductive. External cleansing of wounds, introducing paste, and applying dressing plugged the cavity. When the festering plug was later removed, purulent discharge and gas were emitted. Results such as these ran counter to what Cheyne and his colleagues had achieved or expected in practice. He therefore attributed the negative claims either to a surgeon's failure to understand the temporary purpose of the paste or to their not having used it at all.

Cheyne was especially concerned about overtreatment. Small blood clots that were one-half inch apart, for example, since they retained antiseptic, did not require more ("Hunterian Oration"). With this and other concerns in mind, he had formally requested to visit the Front to oversee and prove the worth of the system, to instruct doctors, and to benefit soldiers. But his request, as noted above, was denied. An anti–Listerian, Dr. D. implied that the visit would be a waste of time since the disinfection of war wounds was fundamentally impossible. Cheyne, with tongue-in-cheek, knew that he was involved in a political rather than scientific debate: "There was, of course, no personal animus in the matter, and had I been an advocate of hypertonic salt solution I expect he would have welcomed me warmly!" This statement referred to Almroth Wright's competitive method of wound disinfection using salt solutions to achieve asepsis. We will turn to that debate shortly.

Cheyne's request for front-line duty had earned a hearing in France before an assembly of distinguished medical officers. It turned out to be a pro forma hearing, more a courtesy than a serious enquiry ("Hunterian Oration"). Dr. B's positive results were contrasted with those that were either negative or procedurally questionable. As it turned out, no careful decision on the Chatham plan was possible because of a dearth of evidence, ensured by the suspicious denial of Cheyne's request for front-line duty, and by the half-hearted work of sympathizers. The judges then proceeded to condemn pastes, powders, and the disinfection of wounds, to discredit Dr. B., and, in so doing, to settle "the whole question of antiseptics and disinfection to their entire satisfaction and comfort" ("Hunterian Oration"). As a post-scriptum to the narrative, we learn that the bacteriologist Dr. D., in a subsequent letter to Cheyne, would reveal that he had, in fact, correctly understood the role of the paste as a "temporary expedient"; nevertheless, though aware of its purpose, he "sat in judgement on and condemned the whole thing" ("Hunterian Oration").

Arthur Edmunds' medical work in the Dardanelles, as we noted earlier, afforded the Chatham committee with the opportunity to test antiseptic treatment on board hospital ships, under reasonably controlled conditions, and not under fire ("Hunterian Oration"). Edmunds reached the Dardanelles on 30 April 1915 and joined the hospital staff of the S.S. *Soudan* as Staff Surgeon. Casualties were soon brought aboard. In a 24 May letter to Cheyne, Edmunds described more than 17 cases he had treated for wounds and sepsis. Apparently nothing besides a field dressing had been applied to arrest microbial growth. Undaunted, Edmunds disinfected wounds as he routinely had done in civilian practice. Following Cheyne's directions, he dusted them with borsol powder and squeezed cresol into the recessed cavities. When patients were anesthetized, Edmunds and his surgical team scrubbed large areas of skin with soap and 1–20 carbolic lotion, cleaned out debris and devitalized tissue, washed recesses with 1–20 carbolic acid and 1–500 corrosive-sublimate-to-water lotion, dried wounds, added more borsol and cresol paste, inserted drainage tubes, and applied antiseptic dressings. To deal with shock and blood loss, they used sub-cutaneous saline infusion when anesthetic was administered. In short, he applied the revised Listerian method.

Edmunds was confronted with a variety of life-threatening injuries: skull wounds with exposed grey matter; gunshot perforations shattering bones; severe lacerations; bullet wounds to the face; and skeletal fractures ("Hunterian Oration"). Not all cases, as the following example shows, had a prognosis: "Extensive laceration of scalp. Right arm amputated on the beach before reaching hospital ship. Is now dying." But there were dramatic successes, such as one in which sepsis was cleared up with combined cresol and borsol in a soldier who had received a shell wound to the forehead and whose face and legs were "scorched." It is fair to say that Edmunds' experiences on the *Soudan* demonstrated the potential of the Chatham antisepsis plan, but substantial data had to be compiled for support. When Edmunds arrived in Malta on the hospital ship S. S. *Rewa*, there were on board 548 wounded who were headed home. At the end of his first voyage, Edmunds expressed some surprise that, given the severity of so many wounds and the prevailing sepsis, "several compound fractures and other wounds had already healed." The implication was that antisepsis had played a crucial role, but whether he eventually published his findings was not indicated in the text. His findings described in the extant literature remain of interest.

The *Rewa* returned to the Dardanelles on 10 June 1915 ("Hunterian Oration"). Letters to Cheyne, of 16 and 19 June, contain the results of Edmunds' second round of antiseptic surgery at the base hospital. The second voyage was marked by a regression in septic management. The wounded had been transferred directly to the S.S. *Soudan* from the battlefield in the first voyage, and most had infected wounds. On the second voyage and before a hospital ship arrived, dressing stations had been established onshore. What might sound like an improvement in emergency care, ironically, made things worse: this detour inadvertently delayed transport of the wounded to the *Soudan* where the Chatham plan was being practiced. On 10 June, Edmunds was heartened to learn that even though the wounds he saw were infected and the injuries severe, the judicious use of antiseptics had still produced good results: the powders and pastes worked well together. The serious antiseptic work began offshore.

One problem on the beach was that dressers had applied powder and paste haphazardly; hence, inadequately treated wounds, upon admittance to the *Soudan,* stank of gangrene, and *Bacillus tetani* was confirmed in many. No doubt aggravated by these regressive trends, Edmunds worked hard, once again opening wounds, excising gangrenous skin and dead muscle, slitting open subcutaneous pus cavities, cleaning out and covering wounds with borsol, and injecting tetanus vaccine. Twenty-four hours after following this routine, in one particular case, he reported that the odor was gone and the patient doing well, although the clinical picture was not fully disclosed. Edmunds acknowledged to Cheyne that his experiences with sepsis in the Dardanelles yielded incomplete results, as the care that patients received on the beach was out of his control. After reaching Malta on the last voyage, Edmunds boarded the *Rewa* and returned to the Dardanelles. His note of 4 July to Cheyne extolled the benefits of antisepsis, and he requested more supplies sent to the *Soudan*, but his reports, it seems, were not discussed in a professional forum.

Dr. A. M. Fauntleroy, a United States Navy surgeon, in August 1916, argued that the idea of battlefield disinfection at the point of initial injury was theoretically possible but unlikely since conditions on the Western Front appeared insurmountable, and since the first-aid kits currently issued were inadequate. He described the realities with which medical personnel, equipped with rudimentary kits, had to contend in the heat of battle. In trench warfare, the extent of a soldier's wounds could not be fully assessed unless he was thoroughly examined; and this was impracticable since his clothes could not be removed and first-aid dressings satisfactorily applied in that environment. In the midst of action, a soldier would ordinarily remain where he had fallen until a lull in the fighting occurred, at which point his comrades would try to evacuate him to safety hundreds of yards in the rear. In the interim, exposed to the environment, he would lose blood, and, by the time of rescue, his wounds would have been contaminated. For these reasons the ordinary first-aid packet, as distinguished from larger kits, was useless in the case of serious wounds (144).

Cheyne's South African tour had shaped his thinking on the use of the first-aid kit, with its portable cresol paste and borsol powder. Guerilla engagements, of low to moderate intensity in an arid climate, could not compare with the powerful munitions and weapons technology used by both sides in World War I. The ideas about battlefield antisepsis that Lister and Cheyne developed would have been better served had they been formulated in historical continuity with the Franco-Prussian War of 1870–1871, where high-velocity breachloaders and similar weapons were first used on a large scale (McCallum:124–125). Cheyne's idea of initial disinfection on World War I battlefields, however reasonable, could not be used because of the time it took to reach the patient and to inspect him thoroughly for hidden wounds. Too often because of the delay, the portable antiseptics then in use were, according to Dr. Fauntleroy, inadequate and ineffective.

1914–1916: The Wright-Cheyne Debate

Almroth Wright (1861–1947), an opponent of Listerism, believed that the antimicrobial properties of blood and tissues could be harnessed to keep war wounds aseptic without primary recourse to chemical antiseptics. Although Cheyne and Wright debated over the therapeutic benefits of their respective treatments, Wright's thinking was not diametrically opposed to Cheyne's, as far as aseptic surgery was concerned. A more nuanced understanding of their respective viewpoints will show that confusion and conservative obstructionism misrepresented their positions publicly. Cheyne agreed with Wright that natural antimicrobial powers were considerable. Although Cheyne heavily emphasized the role of antiseptics, he argued for a regimen that balanced natural healing with chemical intervention, and he warned that antiseptic misuse could interfere with healing. Whereas Wright either dismissed antisepsis completely or (in some contexts) allowed only for its auxiliary use, Cheyne believed chemical compounds were effective germicides that, if judiciously employed, could reinforce cellular defenses.

The rival treatment plans of Cheyne and Wright shared the same fate: each was misapplied during the War; and each, in the long run, was proven to be ineffective and harmful. Wright's treatment using hypertonic-saline lavage was misunderstood, misapplied, and shown to be physiologically ineffective; its employment actually impeded healing and worsened infections. Cheyne's temporary antiseptic for battlefield use was experimentally promising. But during the war his instructions were misconstrued, misapplied, and deprecated by anti–Listerians; today, in all of its forms, carbolic acid is recognized as a poisonous substance. Compounding the irony is the fact that Dakin's solution, the chemical disinfectant that had superseded both Cheyne's Tricresol and Wright's hypertonic-saline therapy, has recently been proven to be highly toxic to human cells.

Wright had been appointed, in 1913, as Director of the Bacteriological Department of the newly-established Medical Research Committee, but, when the war started, he went to France before accepting the civilian post. As a colonel in the Army Medical Service, he set up a lab in Boulogne to test ways of managing wound infections. Wright's contributions to battlefield medicine were significant: he developed an effective anti-typhoid vaccine; compiled statistics on the types of bacterial flora found in wounds; warned that overuse of antiseptics impaired natural antimicrobial activity; tracked and staged the development of infectious disease in war wounds, from the open-wound colonization of aerobic bacteria (*Streptococcus* and *Staphylococcus*) to anaërobic colonization (gangrene causing organisms) in oxygen-depleted, closed wounds

(28); and practiced debridement (the removal of devitalized tissues to deprive anaërobes of a food source), as well as the open-wound treatment (Hager: 28–29; Worboys, "Almroth Wright"). Opposed to the primary use of chemical antiseptics (except for external use), he devised a complicated system of saline irrigation, the logic of which was threefold: that innate, adaptive (cell- and antibody-mediated), and artificial immunity provided efficient defense; that salt in high concentration destroyed bacterial pathogens; and that chemical bactericides, if used in excess on damaged tissues, could interfere with natural defensive systems.

Although Cheyne and Wright became involved in a heated controversy over the effectiveness of antisepsis and over the employment of saline lavage, they shared a common concern for the soldiers who lived in and fought from trenches in the fields of France and Belgium. As a consulting physician to the B.E.F. and as a researcher at the No. 13 General Hospital in Boulogne, Wright observed in a 10 April 1915 article that in this war "practically every wound is heavily infected." His approach to wound treatment appropriately focused on the control of sepsis by promoting natural defenses and healing. The source of infection, in virtually all cases, was contamination: clothing, skin, and soil-borne pathogens, driven into wounds with projectiles, shrapnel, and natural debris, were implanted "far beyond the reach of any prophylactic applications of antiseptics" (Wright [10 April 1915]: 625). The wounds Wright routinely had examined were found to contain a multitude of bacterial species, among which were fecal *Streptococci*, the *Bacillus aerogenes capsulatus* discovered by Welch (a cause of gas gangrene), and more rarely *Bacillus tetani*.

During the earliest stage of an injury, *Streptococci* and the fecal microbes grew rapidly in "imprisoned discharges," making this a critical period for the patient. Dr. William W. Keen, in 1917, describes how rural topography was conducive to bacterial infection: "The soil of Belgium and France ... [had] been roamed over by horses, swine, and other animals ... manured thousands of times [the soil] is deeply and thoroughly impregnated with fecal bacteria in addition to the ordinary pyogenic bacteria" (30). Wright, too, was well aware of the problems that diverse flora posed for the injured soldier. The ideal medical approach to treatment, in his view, remained "healing by first intention"; but this ideal was difficult to realize, as nearly all wounded soldiers upon arrival at base hospital had severe infections; as a result, many lives were imperiled ("An Address on Wound Infections" [10 April 1915]: 625).

In response to the urgent situation, Wright weighed the effectiveness of three "therapeutic measures": (1) antiseptics; (2) physiological methods (opening and draining the wound to bring the antibacterial properties in the blood to bear on the pathogens); and (3) vaccine therapy. In order of priority, he listed the physiological method, vaccination, and antisepsis (an "ancillary" third) (625).

Although unenthusiastic about antisepsis in a 30 October 1915 article, to some degree Wright espoused a synthetic and open-minded approach to wound treatment. The development of an effective mode of treatment, he wrote, required "strenuous study of the infecting microbes, the conditions in the wound, and the therapeutic agents which we employ, and the defensive operations of the organism" ("A Lecture on Wound Infections" [30 October 1915]: 629). His ostensible aim was to learn which proportional combination of methods would best meet the needs of wounded servicemen.

As published accounts of experiments show, Wright was a systematic and improvisational thinker. One only has to consider the hydraulic apparatuses that he had invented to irrigate and drain wound-surfaces, while cleansing them with therapeutic solutions. He constructed ten models and variants, using funnels, beakers, rubber hoses, and siphon bandages, to distribute saline solution via a graduated drip along the walls of wounds; each apparatus was even modified to treat a specific type of wound ("Memorandum on the Employment of Bandages": 564–567). The drawings of these intricate models, by his colleague, Lieutenant H. H. Tanner, are elegant and precise.

Wright, like Cheyne, understood that war-wound treatment was a multistage process, beginning on the field, en route to the battalion aid station and ending (hopefully) in the hospital. After the incurrence of injury, the wounded soldier endured a series of pathological developments; the first was "the incubation period" of his affliction as microbes had just been projected or had migrated into tissues. In World War I, it became clear that, in view of the environmental factors, even the high-velocity projectile could drive microbes into the body. Wright was convinced, however, that despite this strong possibility, the body still had the capacity to respond to this kind of injury without much help—that is, if outflowing blood were permitted to scour the bullet track and tissue walls and, once clotting had occurred, if antimicrobial substances and phagocytic cells could be drawn to the wound, to destroy adherent pathogens ("A Lecture on Wound Infections" [30 October 1915]: 629). The process of wound control, Wright recognized, was contingent on several factors that required close observation and adjustments. During the imprisoned-discharge phase, for example, blood clots and migrating leukocytes gradually disintegrated in the wound cavity, releasing the protein-digesting enzyme, trypsin. At this critical point in the immunity response, an infection could resurge unless the wound were flushed out, and unless fresh lymph-bearing antimicrobial cells, proteins, and antibodies were drawn to the site. To prevent the microbes from recolonizing in a purulent cavity, the physician had, therefore, to ensure both the migration of fresh microbe-fighting proteins and cells and the unin-hibited drainage of old lymph and blood ("A Lecture on Wound Infections" [30 October 1915]: 630). Wright was aware that, without careful monitoring and the readiness to intervene with wound opening and drainage, an uncom-plicated gunshot wounds could rapidly become life-threatening.

Many wounds in this conflict were complicated. If a rifle bullet commin-uted a bone, both projectile and bone fragmented implosively, tearing internal structures, hemorrhaging blood vessels, herniating muscles, and exiting through a large, ragged hole. Multiple wounds from machine-gun fire, mortars, and artillery shells were difficult to manage. The implanted microbes in these wounds could be sown over a wide internal area, into vital organs, and through-out folds and recesses of flesh. With the explosive effect, blood supply to bone and soft tissue was impaired or interrupted, viscera burst or gelatinized, and dead tissue became a food source for invasive and digestive microbes. As tissue desiccated, and if saprophytic spores were present, gangrene could occur (Wright, "A Lecture on Wound Infections" [30 October 1915]: 630). An "imprisoned infection" could then develop since antimicrobial elements and phagocytic cells, as they invariably deteriorated, would remain in a stagnant, polluted cavity. Varieties of bacteria could then develop coextensively. Without the influx of fresh lymph, serious pyogenic and saprophytic infections could set up (630).

In the 13 November 1915 installment of the series, Wright outlined an immunological alternative to chemotherapy ("A Lecture on Wound Infections" [13 November 1915]: 717–721). His system was based on the assumption that hypertonic salt solution has a higher osmotic pressure that lymph and, there-fore, could draw this clear fluid, rich in bactericidal blood cells and proteins, to the area of the wound. To understand Wright's theory, several basic defini-tions are helpful: (1) the process through which solvent particles (ions and molecules) move from high to low areas of concentration is called, *diffusion*; (2) and the process by which these ions and molecules diffuse *through* semi-permeable cell membrane is called, *osmosis* (Tortora and Derrickson: G-9, G-19). Since salt concentrations were central to Wright's aseptic method, two more terms require definition: (3) an *isotonic* or *physiological* solution is one in which the salt content is equal to that of blood and cells; and (4) a *hypertonic solution* is one in which the salt content is higher than that of blood and cells (Campbell, Reece, et al.: G-18, G-20).

Wright's treatment, therefore, hinged on the idea that, if hypertonic saline (with its high saline content) were to contact blood and cells, it would osmot-ically draw water out of them; antimicrobial elements in the blood and lymph, he assumed, would then diffuse into the hyper-salinated wound. Theoretically, the chemical migration (in Wright's scheme) would eventually level off, at the point when equilibrium between the hypertonic saline solution and the antimicrobial lymph had been reached. If Wright's idea of equilibrium signified a midpoint between 0.85 percent isotonic and 5.0 percent hypertonic, then the hypertonicity of the wound would range between 2.0 and 3.0 percent, a level experimentally proven to slow bacterial growth. This dynamic system, catalyzed by the hypertonic solution, supposedly promoted natural healing and asepsis, as long as the osmotic influx of antimicrobials was maintained, and as long as wound-cavity stagnation was avoided.

A number of physiological caveats were raised. Some argued that hypertonic salt, in drawing water from cells, would actually cause them to dehydrate, collapse, and then deteriorate, prior to or while destroying bacteria. If that happened, as was the case with antiseptic overuse, the treatment actually would interfere with rather than promote healing. The crux of the argument was whether Wright's influx-efflux system could be continuously maintained. His basic idea, one should recognize, had been in the mainstream of immunology. The idea of attracting antimicrobials to a wound is called *chemotaxis*, a principle owing its early development to the botanists, Theodor W. Engelmann (1883) and Wilhelm Pfeffer (1906), both of Leipzig (Tortora and Derrickson: 438–439; Adler: 708–716, respectively).

Through a series of experiments, Wright determined that the correct concentration of saline solution with which to cleanse a wound was 0.85 percent (isotonic or physiological). Because isotonic salt was equal in saline concentration to that of lymph, it could be safely used to wash out a wound. Two important thresholds had been established: at 2.0 percent (hypertonic) saline concentration, bacterial growth slowed; and, at 5.0 percent, it ceased. Experiments, detailed in the 17 April paper of the series, convinced him that achieving the correct saline concentration was absolutely essential: thus, salinity had to be increased from 0.85 to 2.0 percent if microbial growth was to be slowed. Although, as the concentration level approached 5.0 percent and microbes began to die off, the danger was that hypertonic salt would kill leukocytes, as well. So the challenge was to stay somewhere in the 2.0–5.0 percent range, at a balancing point where bacteria were suppressed and leukocytes could destroy them (Wright, "An Address on Wound Infections" [17 April 1915]: 668). If the balance was off and if blood products died, their disintegration and release of proteins and enzymes would provide nutrition to existing bacteria, allowing them to rejuvenate without opposition. To prevent stagnation and bacterial re-growth, Wright advised that, as soon as asepsis was attained, the wound should immediately be closed; at that point, bacteria in the wound were presumed to be dead and the incursion of epidermal bacteria would have been prevented ("A Lecture on Wound Infections" [13 November 1915]: 717–718). Wright's entire system rested precariously on two assumptions: (1) hypertonicity drew antimicrobial blood products to a wound site; (2) and, within the 2.0–5.0 percent hypertonicity range, a specific concentration level existed that, if held, would stunt microbial growth, attract antimicrobials; and preserve both inflowing leukocytes and healthy tissue.

The protective powers of an organism, Wright theorized, could also be brought to bear against bacteria through the use of vaccines. Immediately upon being wounded, and as the patient was entering the incubation period, he advised that an inoculation of mixed and attenuated bacterial pathogens should be administered. A second vaccination would be intended to neutralize pyogenic bacteria ("A Lecture on Wound Infections" [13 November 1915]: 718). He called for a change in perspective, with which Cheyne would have agreed: to combat bacterial infection, one had to think "in terms of microbes," of protective substances, of auto-inoculations and vaccines, and of laboratory techniques (721). The antiseptic strategy of Lister, in his opinion, had "completely broken down" (721). In the 17 April article, however, Wright was less adamant about proscribing the use of antiseptics, in conjunction with physiological or isotonic saline ("An Address on Wound Infection" [17 April 1915]: 629–635).

By 1916, the British medical establishment had to contend with intellectual warfare in its ranks. A medical polemic had broken out between Cheyne and the antiseptic camp, on the one hand, and Wright and the aseptic camp, on the other; in some contexts, it is important to remember, the latter assigned chemical disinfectants to an auxiliary position. Cheyne's opposition to Wright's system began with *The Lancet* article "Sir Almroth Wright's Lectures on the Treatment of Wounds in War" ([8 May 1915]: 961–962). Leading medical journals took sides. A 28 October 1916 editorial in *The Journal of the American Medical Association*, "The Treatment of War Wounds," misrepresented Cheyne's evidential presentations as "vigorous"; his practice of Listerism, unyielding. The ad hoc committee, alluded to in the JAMA editorial, had been working at the Chatham Naval Hospital. The object of those experiments, which Wright and his colleagues rejected, had been to devise a suitable antiseptic compound and delivery system for temporary use on battlefield wounds. Early reports of the antiseptic system's failure on the battlefield, according to Cheyne, had been imputed, not to the ineffectiveness of the process, but to the failure of those who applied it incorrectly. The JAMA editors who were skeptical about Lister's method did not give Cheyne a fair hearing:

> Practically, Sir Watson Cheyne seems to have constituted himself the uncompromising champion of the so-called [L]isterian method. According to Sir Watson Cheyne the unfavorable results, which were given by his method in the early part of the war, were due to

defective technic. He states that the improved technic-devised by himself and other members of the commission ad hoc did give in some hands excellent results, and would have given them more generally but for the fact that justice was not done in the official trials, owing to the prejudice entertained by the surgeons. He intimated that discipline alone prevents the thousands of surgeons from expressing their dissatisfaction [Editorial, "The Treatment of War Wounds" [28 October 1916]: 1304].

Cheyne criticized Wright's method in a series of 1915 papers, and Wright's lengthiest rejoinder was "The Question as to How Septic War Wounds Should Be Treated," which featured a convoluted discourse on polemical ethics ([16 September 1916]: 503–516).

The two systems—the antiseptic and the aseptic—were not diametrically opposed to one another. Both Cheyne and Wright sought a balance between the two methods of controlling post-operative infection, and their points of view need to be clarified.

We gain a clearer understanding of Cheyne's thinking on antiseptics during the war years and on scientific investigation generally if we begin by consulting a pre-war publication. In "The Bradshaw Lecture on *The Treatment of Wounds*" (4 December 1908), delivered before the Royal College of the Surgeons of England, Cheyne tried to account for the then-current unpopularity of Lister's work. To explain adversarial views and the rush to judgment, Cheyne believed that those who opposed Listerism had forgotten the logic of scientific inquiry. Hypotheses were to be tested experimentally and either abandoned or reformulated in the light of substantiated, competitive views. If a medical scientist sought to explain biological phenomena and to use this knowledge to support public health at home and abroad, then he or she had to acknowledge membership in a professional community; to guard against entrapment in one's own system; and to recognize either the partial or complete truth of another's work, even if a competitive discovery superseded or invalidated one's own. In medicine, notes Cheyne, "oscillation[s] in the views which are current from time to time" were common. These swings of "the pendulum of medical opinion in one or other direction" had tended to be "very extreme and far beyond what is justified by the facts of the case." Thus, to maintain impartiality and a clear focus, one should not

discard too hastily the results of previous research and experience, and on the other hand not to overlook what seems to be sound in the new work. Nowhere is oscillation of opinion more marked at the present time than in the views founded on experimental and chemical pathology, and more especially in connection with the new science of bacteriology and the relation of bacteria to the living body. And when the wish was expressed to me [that] I should take up the subject of the treatment of wounds at the present time, I was the less reluctant to do so, seeing that as regards that matter we are just now in one of those extreme oscillations ["The Bradford Lecture on *The Treatment of Wounds*"].

Much of what Cheyne had said about the antiseptic system in 1908 adumbrated the sentiment Wright would express in the 1915 papers, notably that the prevailing role of the immune system in treating war injuries should be emphasized. Cheyne observed, in 1908, that the chemical destruction of bacteria should not risk damaging healthy tissues. He was interested in striking a balance: to use antiseptics up to, but not beyond, the point of infringement upon natural healing, and where "excessive effusion of serum … necessitated the constant use of drainage-tubes." Cheyne understood the rebound effect: excessive use of antiseptics, using the wrong ones or one in too-high a concentration, could hamper "the resisting powers of the tissues," giving microbes an opportunity to colonize and cause further damage. Ironically, the views of Cheyne and Wright, from 1908 to 1915, were similar to each other with respect to natural antimicrobial capacities.

More can be learned about Cheyne's idea of cellular immunity if we review Lister's antecedent understanding of inflammation. We now know that inflammation, as a sign of tissue damage and of the body's response to trauma, is a component of internal defense. It stimulates vasodilation chemically, makes blood vessels more permeable, and induces chemotaxis, the immigration of various kinds of antimicrobial proteins and phagocytes to an injury (Tortora and Derrickson: 438–439). It took time and careful work in the laboratory and clinic for Lister to realize this. In 1857, he had not yet associated inflammation with pathogenic bacteria. But, in a 26 March 1870 pamphlet, "Remarks on a Case of Compound Dislocation," he revised his thinking in part: inflammation could be the consequence either of a noxious agent or of tissue reparation (CP. II: 147, 150 respectively). By 1870, he had also come to realize that inflammation was an aspect of the healing process, inducing granulation, and temporarily replacing devitalized tissue; and he profited from the abscess research of Alexander Ogston: that suppuration, neither a disease process nor a concomitant, actually evidenced antimicrobial activity (CP. II: 150). In this manner, Lister correlated inflammation with the effects of chemical antiseptics that helped a wound to develop a granulating sore (CP. II: 151). Cheyne's line of argument, paralleling Wright's in 1915, extended from the Listerian tenet that "living tissues have the power in some way or other of disposing of bacteria." Both Cheyne and Wright agreed that the excessive application of chemical antiseptics could nullify natural antimicrobial power, kill harmless bacteria, and promote the growth of pathogens.

Cheyne targeted two elements in Wright's systems. One was the latter's reliance on a hydraulic system to maintain asepsis. Cheyne had reservations about this since water tended to swell and devitalize cells essential to healing. This physiological fact, in Cheyne's opinion, appeared to undermine Wright's advocacy of saline asepsis and demotion of chemical antisepsis. Yet, on a related point, Wright and the followers of Lister were in agreement. Cheyne had pointed out in the "Bradshaw Lecture" that successful wound treatment depended not only on the exclusion of pathogenic organisms during and after an operation, but also on the avoidance of wound irritation ("so as not to interfere with healing") and on the ability of the tissues to prevent "the growth of any bacteria which may have entered." It is important to recognize that Lister had also approached wound treatment from an immunological rather than antiseptic direction. The aim of Lister's work in the 1880s, according to Cheyne, was to find "efficient antiseptics, which could be as little irritating as possible to the wound" ("Bradshaw Lecture").

In short, the Lister-Cheyne model for wound management balanced internal with external defensive mechanisms. Wright, it seems, either was unaware of, or ignored, the fact that he and his perceived adversary shared common ground. For Lister and Cheyne, antiseptics were not dispensable auxiliaries; if used appropriately, they were beneficial in wound management and supported immunity. Cheyne attempted to account for the prevailing, conceptual error. His 1885 metaphor of a pendulum swinging to the counter-antiseptic extreme was apt. Because physicians had not studied Lister's system carefully enough, they proceeded to overuse carbolic acid to the point of impairing antimicrobial activity. The results of this imbalance were that wounds became worse, making it appear that the antiseptics were either ineffective or themselves responsible for the patient's disorder. Wright seemed to have made a similar prejudgment about antisepsis as the cause rather than as the solution to wound infection. Surgeons who disparaged antisepsis had failed to realize that treatment had to be maintained throughout. Ironically, one of the reasons that Wright's method would lose traction, as he would admit in a 1916 apologia, was that its practitioners had trouble implementing it correctly ("The Question" [16 September 1916]: 503–16).

Overall, disaffection with the Listerian system stemmed from five sources: (1) the excessive use of chemicals, with unsatisfactory results ascribed to the system and not to the operator; (2) the popular assertion that cleanliness could support asepsis; (3) the precipitous conclusion that antiseptics were completely ineffective against bacterial spores when the agents were not given the time necessary to destroy them; (4) the difficulty involved in eradicating bacteria from the skin led some to believe that it could not be done; and (5) (what probably was Wright's chief motivation) an exaggerated belief in the immunological power of the body.

Cheyne's argument in regard to the management of battlefield wounds took into consideration a number of contingencies, one of which had largely to do with evacuation time. In 1917, an injured soldier had to be moved under the cover of darkness after being kept for at least a day in a makeshift dugout (Bowlby and Wallace: 30). Behind the firing line on the Somme, stretcher bearers had to traverse miles of mud for hours carrying a man to the rear because ambulances could not travel through muddy fields. In the worst case, downed soldiers in No-Man's-Land had to wait for days, exposed to the elements before they were rescued. Under the best conditions, an ambulatory, slightly-injured patient could make his way to the Casualty Clearing Station in one hour. If a great battle were going on, on the other hand, stretcher-bearers would be overwhelmed with casualties and exhausted by each trip (30). Cheyne was fully aware of the difficulties. Thus, his plan was limited and practical: chemical antiseptics in powder and paste media, as von Esmarch had suggested decades before, constituted a temporary measure, a way of inhibiting bacterial growth in the interim between injury and evacuation. Wound care had to proceed, quickly and systematically, in view of the life-threatening conditions.

We now turn to Wright's complete plan. In the 13 November 1915 article, he assigned antisepsis to a third level in his regimen, vaccination and hypertonic salt solution having been given priority. His approach was, therefore, predominantly immunological. Against microbes that vaccines and chemotaxis had not destroyed, he advised the adjunctive use of antiseptics (he did not specify which ones): "if we neglect any antiseptic precaution, [we shall] be running the risk of superadding to the already subsisting infection another infection; and in hospital we shall be running the risk of transferring infections from patient to patient." It appears that Wright was referring to post-surgical pathogens as distinguished from those acquired on the field of battle from contaminated soil, clothing, and other debris ([13 November 1915]: 721). The crucial point is that he conceded the importance of antisepsis in the overall plan, in order to contain the spread of the disease.

Lister had become a kind of straw man in Wright's critique of antisepsis and promulgation of hypertonic saline. For example, in the monograph *Wound Infections* (1915), Wright flatly stated that antiseptics were useless against bullet wounds, a proposition contrary to current military medicine: "The principle that microbes can be held off from wounds by an antecedent employment of antiseptics has no application to projectile wounds" (52). This comment had no bearing on Cheyne's published assessments of antiseptic use in war. As we have seen, his version of Listerism subscribed, not to the extensive use of antiseptics on the battlefield or in the hospital, but rather to their immediate application from the field kit in the form of borsol powder and cresol paste; the rapid application of antimicrobials was vital until the patient reached the Clearing Station. Wright, on the other hand, went so far as to affirm that antiseptics were inactive against severe wounds, the chemicals reputedly having "no penetrative power" to reach deeply-embedded pathogens (*Wound Infections and Some New Methods*: 52, 57). But Cheyne had not argued that they were to be used in this way. In fact, he warned against deluging a wound that was seriously infected or against impacting it with antiseptic gauze. He had been explicit about these points in the professional literature, but front-line practitioners (as noted above) either misunderstood, or intentionally ignored, his directions about when and how to use disinfectants.

Wright confessed that the use of diluted antiseptic lotions in recovery and during convalescence could preclude hospital infections (*Wound Infections*: 62). Antiseptics should never be used in high concentration. Those that inhibited leukocytes' movement, that paralyzed phagocytosis, that led to the corruption of lymph, or that inadvertently stimulated microbial growth, should be proscribed (63). Ironically, this comment was consonant with Listerian doctrine and with Cheyne's practice of it. Paradoxically, the overuse of antiseptics, the failure to provide for free drainage, and reinfection from poor hygiene and dirty instruments, were what gave Listerism the bad reputation that Cheyne sought to rehabilitate.

In the 8 May 1915 *Lancet* article "Sir Almroth Wright's Lecture on the Treatment of Wounds in War," Cheyne indicated precisely where he either agreed or disagreed with Wright's method. He began with the statement that his article was not polemically intended. In fact, as was borne out by the facts, he agreed with "a good many" of Wright's statements, "especially with those on the use of antiseptics in septic wounds" (961). Wright had objected to the practice of syringing out and flushing these wounds with antiseptics. Cheyne concurred because this practice washed away superficial pus but did not diminish sepsis to any appreciable degree. Instead of using antiseptics to saturate an open wound, Cheyne actually preferred the use of milder, isotonic salt solution. The most compelling reason for avoiding the overuse of antiseptics against bacterial infection was that strong chemical lavage would interfere with the natural bactericidal properties of the tissues. Instead of the antiseptic wash, Cheyne simply preferred the common practice of free drainage and of mild isotonic irrigation, with what Wright called "physiologic salt."

Cheyne had a different opinion of Wright's idea of using hypertonic salt solution ostensibly to increase the flow of lymphatic cells to the injury. He was diplomatic: "Curiously enough, it is stated that [the induction of lymph with hypertonic salt] acts by osmosis, but that must surely be a slip of the pen." Hypertonic salt—pouring salt into the wound—was an irritant, not an antimicrobial catalyst: "Colloids such as lymph dialyse very slowly or not at all.... Any increased flow of lymph which may arise from the introduction of a hypertonic salt solution into a wound will be the result of irritation, and the question then arises whether salt is the best irritant, or whether, indeed, it is wise to use any irritant at all" ("Sir Almroth Wright's Lecture": 961).

Cheyne's refutation of Wright was built on the understanding that the most potent bactericidal substances "reside in the tissues themselves to a much greater degree than in the blood and lymph ... the application of an irritant may seriously interfere with the natural processes" ("Sir Almroth Wright's Lecture": 961). By 1915, Cheyne had not yet abandoned the cellular-humoral dichotomy for a unified construct in which antimicrobial activity of any kind was, by definition, cellular in origin and mediation (Tortora and Derrickson: 442). That excessive salinity irritated tissues was, however, common knowledge.

Consistent with the aim set forth in the opening paragraph of the 1915 article on Wright's treatment, Cheyne tried to correct misrepresentations of his system in the context of the War. Wright abrogated the use of antiseptics when patients reached him in base hospitals, "already in a highly septic state and in a stage in which antiseptics as ordinarily employed are of no avail." With this point, Cheyne agreed. But he was most concerned with "a preceding period, which [Wright] does not come across, in which the septic process has not yet taken root in the wound, and in which it may be eradicated by the proper use of antiseptics" ("Sir Almroth Wright's Lecture": 961). Cheyne's research, as we have seen, was chiefly concerned with the earliest stage in a wounded soldier's crisis, the moment when he was most vulnerable, and when antiseptics presented "the only possible chances of success" (961). He observed, in general, that his *and* Wright's work were "in no way antagonistic," as they were both devoted to preserving life and limb. For Cheyne, antiseptic auxiliaries were called for at the very beginning, not at the end, of treatment. The use of antiseptics on the battlefield, of the right kind and in the correct measure, was essential if septic germs were to be inhibited during the period it took to transport a soldier to the hospital—a period of hours, even of days. This early phase in wound treatment could determine the outcome (961).

Echoing the sentiment of Arthur Edmunds in the Dardanelle letters, with respect to compound fractures and sepsis, Cheyne noted a difference in degree rather than in kind between civilian and combat cases ("Sir Almroth Wright's Lecture": 961). The same bacteriological flora got into civilian injuries soiled with contaminated earth as they did in battlefield wounds; and they could be just as serious, whether incurred on a British farm or in rural France and West Flanders. Of course, Cheyne recognized that civilian, as opposed to military, cases were in number and severity very different: "there are seldom the long channels in the tissues, difficult of access, in the ... [civilian practice] which occur in the ... [military], but there is the same blood clot and the same torn and bulging muscles, rendering the channel irregular, in the one as in the other" (961). Cheyne might have been thinking in terms of veldt wounds and of Ogston's point about high-velocity modern projectiles, shedding "the septic material which they carry in with them, as shown by the frequency with which these bodies become encapsulated" (961). Cheyne's comparison between civilian and combat wounds did not hold up very well in World War I, considering the frequency of multiple wounds, complicated by burns, adverse environmental conditions, and the unavoidable delay in treatment. John Hager provides a graphic picture: "The British wounded in Flanders often fell into stagnant water, pooled at the bottom of bomb craters, and they would lie for hours, sometimes days, before evacuation. In Flanders there was no such thing as a clean wound" (26).

Some investigators subscribed to Wright's saline regimen. H. H. Tanner, noted for his mechanical drawings in "The Treatment of Wounds by Saline Solution" (1916) vindicated Wright's method, charging that his opponents had confounded the two distinct, but complementary, uses of saline solution: one was to rinse debris and dead cells out of the wound; and the other was to facilitate the migration of lymph as a natural and continuous bactericide. Thus, Tanner based his pro–Wright stance on the indisputable fact that strong saline solutions (2.0–5.0 percent) inhibited and even killed microbes; and he claimed that normal or isotonic saline solution (0.85–0.90 percent) was a helpful lavage (70–81). For Cheyne, on the other hand, isotonic salt solution was, at best, a useful rinse, while the hypertonic was counter-productive.

Wright, like Cheyne, had to defend himself from those who misused his system, confused its complicated steps, or had not monitored wound conditions closely. In response to these failures, in the 3 June 1916 "Memorandum on the Treatment of Wounds" he included an instructive section (the 16th of 17), stressing that his system demanded protracted and meticulous care so as neither to endanger patients nor to misrepresent his system ("Memorandum": 797). In Section 16 of the article, he outlined in four subsets how failure could be precluded: (1) packing a wound with hypertonic salt while blood was still oozing should be avoided since salt inhibited clotting and granulation, the very processes upon which Wright's system depended for natural healing. The adjustment here was to postpone salting the wound until bleeding had stopped, at which time hypertonic solution at 5.0 percent strength could be applied and gradually diluted with no danger of dissolving clots; (2) if the wound were covered with "a firmly adherent glutinous coating," this indicated that strong salt had destroyed leukocytes and that the dead cells had formed a membranous covering. To avoid what was a serious obstacle to proper drainage and irrigation, essential if sepsis was to be avoided, the operator was instructed to clean away surface build-up and pus inside the wound. On rare occasions, the protracted application of concentrated salt solutions could also cause healthy granulations to swell. The strong salt, through osmosis, induced fluid infiltration, weakening healing tissue. If this arose, the salt application had to be discontinued and hot fomentations begun, to activate circulation (thereby removing salt) and to promote absorption; (3) if granulations looked coral red and bled easily, salt solution had to be reduced. If drainage of the tissues was still needed, the solution should be gradually reduced from 5.0 to 2.50 percent (hypertonic), and further to 0.85 percent (isotonic), if the lymph-drawing effect was still required; (4) routinely, when the surface of the wound was free of induration and sloughs (necrotic tissue), "physiological" or isotonic salt solution (0.85 percent) should be used exclusively in place of the stronger hypertonic solution. The rationale here was twofold: continued use of strong salt solution could break down healthy clots and granulations, undoing the natural progress of healing; and diluted salt was sufficient to clear away and inhibit residual pyogenic bacteria.

In Section 17, Wright reiterated that once the hypertonic salt had begun to act, its overuse could become destructive ("Memorandum": 797). Employing isotonic or physiological salt, on the other hand, promoted drainage and phagocytosis; and the frequent cleansing and redressing of the wound prevented it from becoming "lymph-bound"—a condition in which stagnating lymph and dead leukocytes disintegrated, providing nutrition in support of bacterial re-growth. Another aspect of the delicate balance that Wright sought (at a salinity point between 2.5 and 5.0 percent) involved the life-cycle of WBCs. On the one hand, once they are dead, WBCs release the digestive enzyme, trypsin, which kills bacteria; to a degree, this works to the advantage of the patient. On the other hand, if fresh lymph were unable to reach the wound, and if drainage were impeded, then the accumulating trypsin from the desiccating blood cells and the released chemicals would destroy residual leukocytes; consequently, the advantage would shift to the unopposed pathogens in and on the wound.

Wright reiterated in Section 17 that, by adjusting salinity levels, natural defenses could be harnessed in the control of sepsis, obviating the use of chemical agents (except as auxiliaries) ("Memorandum" [3 June 1916]: 795). Hypertonic salt solution (5.0 percent), he reaffirmed, would exterminate microbes "in the depth[s] of tissues"; and in the wound cavity (794). But he had become uncomfortably aware, by the summer of 1916, that his system, too, might have been inapplicable in the mayhem of war, especially at the crowded base hospitals. Miscalculations or inattention could easily subvert a process that required constant monitoring and delicate readjustments. Like Lister's 1870 method, Wright's was complicated, required strict attention, and, in the opinions of some, was ill-suited to the conditions of a war where logistics were poor, the casualty count high, and the hospitals understaffed.

In the May 1915 paper, Cheyne directly questioned the physiological basis of Wright's system on the ground that lymph which was thick in consistency did not diffuse well through semipermeable membranes. Hypertonic salt solution (Cheyne's range: 3.0–7.0 percent) had the ability to hydrate thickened secretions if applied directly (although, if overdone, the opposite effect could result); but, in Cheyne's judgment, because of its viscosity, hypertonic saline would be unable to draw lymph efficiently to the wound in order to harness its natural antibacterial effects. If Cheyne was correct, then Wright's chemotaxis method was intrinsically flawed. Later researchers raised similar doubts. In 1917, Dr. W. Parry Morgan, formerly Wright's assistant in the Bacteriological Department of the Medical Research Committee, published a physiological rebuttal of hypertonicity. According to Morgan, Wright had misunderstood the biochemistry of osmosis: hypertonic salt *did not* "draw lymph out of a wound" (italics added; Morgan: 685): "Salt cannot ... draw lymph to itself even by osmosis, because [as Cheyne had noted above] albuminous substances can neither pass through membranes impermeable to salt," nor could hypertonic solutions ensure a continual lymph flow (685). Furthermore, "once a flow is established, the current of lymph outwards will tend to counteract the diffusion of salt inwards. Thus, as soon as the surface becomes pervious, the presence of hypertonic solutions cannot be expected still further to increase the flow ... it does not necessarily happen that the osmotic effect of hypertonic solutions will always produce conditions favourable to flow" (685). At some point, according to Morgan, once the outflow stagnated, a cascade of bad effects followed: as antibacterial cells and proteins disintegrated, bacterial growth invariably resumed.

Morgan's attack on Wright, which was harsher than anything Cheyne had written, accused him of having misunderstood chemotaxis, the process whereby chemical stimulus attracted phagocytes to microbes. Morgan's conclusion was blunt: hypertonic saline did not stimulate leukocytic migration to any greater degree than did blood serum, the salt concentration of which was isotonic (0.85 percent); genuine chemotaxis, on the other hand, incited migrations over greater distances and over a "particular axis" (686). Ironically, Morgan's experiments demonstrated that, in practice, Wright's hypertonic saline, when applied to wounds, actually rendered natural defenses against *Streptococcus* and *Staphylococcus* ineffective (686). Incidentally, Morgan observed that antiseptic solutions, though useful for washing out wounds, had their limitations, too. On this point, he would seem to have agreed with Cheyne that, if an antiseptic agent could reduce the quantity of pathogens, "even temporarily, without at the same time unduly interfering with physiological processes," then this was an important gain (Morgan: 687–688). Morgan closed with the pro forma acknowledgement of his past association with Colonel Wright, despite having arrived at conclusions "divergent from many of those at which he ha[d] arrived" (688).

Consulting surgeons, Anthony Bowlby and Wallace Cuthbert, in 1917, determined that hypertonic salt solutions (their range: 3.0–7.0 percent) were ineffective. For Bowlby and Cuthbert, the verdict was unequivocally negative: the solution slowed healing, rendering granulations pale, sodden, flabby, and overgrown. Furthermore, Wright's saline solution, unlike antiseptics then in use, was implicated in secondary hemorrhages. The salt pack (salt tablets in gauze), another of Wright's inventions, had been supplanted by a multiplicity of new and presumably more effective chemical agents (e.g., Eusol [Edinburgh University Solution] and Dakin's fluid) (Bowlby and Cuthbert: 33–34; "Eusol," OED. **II**; Supplement: 3958; J. Lorrain Smith, et al.). A great number of agents were being tested on wounds in addition to salts. One of these compounds, combining calcium hypochlorite (bleaching powder), boric acid, and sodium hypochlorite (Dakin's solution), was reported to have had excellent results, whereas antiseptic powders, notably Rutherford Morrison's Bismuth-Iodoform-liquid paraffin, had undesirable effects. At base hospitals, mercurial salts, carbolic acid, hydrogen peroxide, brilliant green (a dye), and flavine, were among the chemicals tested (Carden-Coyne: 123–124).

In a 15 August 1920 review of new surgical methods that had been used during World War I, Dr. E. M. Larson elucidated what Cheyne had diplomatically called Wright's "slip of the pen." Larson who understood Wright's method well believed the use of hypertonic salt to be self-defeating, and his assessment is as follows. Wright had correctly assumed that, as pus and cellular debris accumulated in a wound, a stagnant condition arose, leading to the destruction of phagocytes, the vitality of which was essential if sepsis was to be prevented. Because of the stagnation, phagocytes died and disintegrated (along with other WBCs); and, as they disintegrated, released digestive ferments into the wound. These ferments broke down albumen in the wound serum, releasing peptones (water-soluble by-products of hydrolyzed proteins). Bacteria having survived the initial wave of antimicrobial cells, and once provided with a renewed food source, could rapidly multiply, and the infection would then worsen. Wright knew that if serum and lymph coagulated, the cellular detritus would become deposited on the wound wall, theoretically preventing the influx of fresh blood serum, a substance conveying bactericidal and phagocytic cells and proteins. A wound clogged with stagnating and putrefying blood products, Wright had called "lymph-bound." It was precisely to prevent this from occurring that he had turned to hypertonic salt solution to cleanse the inner surfaces of wounds and (in theory) to stimulate the migration of antimicrobial lymph and blood products. The paradox Larson underscored stemmed from Wright's error in biochemistry: to rely mistakenly on hypertonic salt to replenish the lymph. Larson, like Morgan and others, stated that Wright's hypertonic solution (3.0–7.0 percent), actually *destroyed* WBCs and *inhibited*, rather than enhanced, bactericidal activity, quite the opposite of what was intended.

Surgeon General George H. Makins pointed out, in 1917, that a new antiseptic approach was needed and was being explored:

> In the earlier stages of the campaign numerous antiseptic solutions were employed, also the hypertonic saline solution, but of late, in the great majority of cases, solutions of which the active constituent is chlorine have found most favour and have proved the most satisfactory in practice. Eusol, and with gradually increasing frequency, the Dakin-Daufresne solution of hypochlorite of sodium, are those now most resorted to [Makins: 59].

For the biochemist Henry D. Dakin, the strategies of Cheyne and Wright had fallen short of expectations. Cheyne's cresol, a methyl-phenolic compound (commercially called Tricresol), had been routinely used to sterilize hands and instruments. By 1918, however, the cresol paste that Cheyne had recommended as a temporary regimen "for the early treatment of infected war wounds" was said to have had unfavorable results and was henceforth discontinued (Dakin and Kellogg: 45). The only favorable result Cheyne had in mind, we must remember, was simply to delay serious infection for up to 48 hours, nothing more. To call his temporary paste-and-powder regimen a failure, within the parameters of its prescribed use, implied that it had had no noticeable effect on infected wounds, and that this alleged failure had been statistically documented. Cheyne, if we recall, had been denied the opportunity to test his hypothesis and had justifiably complained how front-line physicians, perhaps favoring Wright's aseptic theory, were being obstructive, incompetent in the use of the regimen, or simply confused about its temporary purpose. Who or what, then, is the source of Dakin's information, and does it reflect a serious attempt to use Cheyne's emergency antiseptics according to directions? Cheyne's supplemental salicylic-boric acid powder, called borsol, was also found to be ineffective (according to Dakin) (46). Cheyne's misgivings and frustration are understandable. Both the paste and the powder were explicitly intended for emergency use under fire and while awaiting evacuation, not for the field hospital, before surgery, or in the long term (46).

New antiseptic regimens, as Makins notes above, were introduced in World War I. Most promising was Henry Drysdale Dakin's 0.45–0.50 percent concentration of sodium hypochlorite, developed and applied under his direction, with the assistance of practitioners such as Carrel, Daufresne, Dehelly, and others. Eventually, the Dakin-Carrel antiseptic system supplanted both Wright's hypertonic salt treatment and Cheyne's cresol (13; Carrel and Dehelly). Its application with a system of rubber hoses ensured "uniform distribution" of the antiseptic, supplemented by a chlorine paste formulation: chloramine-T. This compound was designed to maintain wound sterility after the hypochlorite had been vigorously applied (15).

In 2013, U.S. Army researchers determined that Dakin's solution which had been used as a topical antimicrobial in the military, and which was commercially available at strengths ranging from 0.5 to 0.25 percent, was actually toxic to human cells; specifically, *in vitro* tests indicated that it destroyed macrophages, blood products central to natural immunity (Cardile, et al.). Thus, Dakin's solution, employed during World War I at a caustic strength of 0.45–0.50 percent, had the potential for harm; and, in that respect, it was a chemical counterpart to Wright's hypertonic salt. Cheyne's cresol compound, though rigorously tested at Chatham and found to be an effective disinfectant, is today also considered a poison, acute exposure to which causes weakness, confusion, central nervous system depression, respiratory failure, liver and kidney damage, dermatitis, genetic mutations, and is possibly carcinogenic ("Occupational Safety": 1–6; Vincoli: 703–705).

On the one hand, the terrible paradox for the wounded soldier in World War I was the choice between two kinds of toxicity: one of bacterial and the other of industrial origin. On the other hand, if one were to compare the mortality statistics of different wars, one might find that, wherever antiseptics were routinely used in patient care, fewer died of infection.

11

1886–1899
Bacteriology and Medicine

Cheyne's research as a medical bacteriologist began in the 1870s when he gathered wound discharges from patients and tried, with limited success, to identify bacterial pathogens. Although his earliest efforts had mixed results, for reasons discussed in chapters 3 and 4, he subscribed unwaveringly to the conviction that bacteriology was indispensable to medical practice, as the antiseptic system had dramatically proven. The significance of bacteriology to surgery is reflected in Frederick Treves' two-volume collection, *A System of Surgery* (1895–1896). Contributing four articles to volume **1**, Cheyne wrote extensively on inflammation, on suppuration, on ulceration, and on gangrene, 100 pages in all (Worboys, *Spreading Germs*: 181; *System of Surgery*, volume **1**: 53–153). From 1886 to 1899, having acquired modern equipment and aptitude in its use, he was better able to correlate pathogen to pathology ("Bacteriology" [July1886]: 66). He had learned from Koch that, if a bacterial pathogen were isolated, cultured, and reintroduced into animal models with a confirmatory result, every aspect of patient care would immeasurably benefit. From 1884 to 1913, Cheyne's writings emphasized that bacteriology and all branches of medicine were interdependent.

Like Pasteur two decades earlier, Cheyne realized in the mid–1880s that the medical specializations of his times complemented each other. In 1884, when the allied disciplines of bacteriology and of public health had begun to diverge from each other, as a result of bacteriology's maturation and need for a correspondingly exclusive forum and literary genre, Cheyne recognized that too-wide a breach between the two branches diminished the effectiveness of both. Because of the biological connection between bacteriology and public health, as exemplified by the common interest in pathogens borne by air, food, or water, he argued that a Medical Officer had to know the bacteriological literature and, in fact, "must himself be a bacteriologist" ("Bacteriology" [July 1886]: 66). Cheyne reminded hygienists of the common ground bacteriology and environmental medicine shared. If a Medical Officer of Health was to be truly qualified, he had to be adept in the use of basic laboratory methods in order to identify microbial contaminants in food (e.g., tuberculosis in milk) and water (e.g., cholera in drinking water), and, as an epidemiologist, to ascertain rapidly "the exact source and commencement" of an outbreak (66). In the case of cholera, these were very important skills since clinical and post-mortem examinations to determine the cause of a suspected choleraic death were not diagnostically reliable (66–67). Without bacteriological data, an investigator, in 1886, could only confirm the environmental presence of the cholera bacilli after, not prior to, an outbreak: "if the first case be recognized, measures can be taken which may prevent the spread of the disease" (67). Koch's methods of cultivation and staining, Cheyne affirmed, showed that bacteriology was the foundation of clinical medicine, of public health strategies, and of epidemiology.

Competence in bacteriology, Cheyne thought, was an obligation for surgeons if they hoped to utilize antiseptics correctly and to associate a particular germ with a disease; and this applied to treatment either for a primary bacterial disease or for a post-operative/post-accident infection. Lister taught that antiseptic surgery depended on keeping harmful organisms out of open wounds, a precept Cheyne never forgot. The exclusion of micro-organisms and the prevention of their development could only be achieved if the surgeon had practical knowledge of laboratory techniques ("Bacteriology" [July 1886]: 67). Working with micro-organisms in the Listerian tradition provided knowledge essential to surgical medicine (67). Knowing the biology of tubercle bacillus, for example, revealed its susceptibilities. Cultivating the germ provided scientists with the means to simulate human disease in animal models and to develop preventative measures and antimicrobial regimens (67). Seeing a specific disorder from multiple perspectives, moreover, improved patient care: if the pathological effects of an organism were known and predictable, treatment could be devised accordingly. The multi-perspectival approach, Cheyne emphasized, had to be made part of the British medical school curriculum. The curricula of leading German universities that had already incorporated Listerism provided instructional models.

Knowledge of bacterial pathogens was a direct aid to the surgeon preparing for and performing an operation. It sharpened the operator's focus. Cheyne wrote, in July 1899, that "a thorough practical acquaintance with bacteriological work is of the first importance to the surgeon, for in that case the manipulations necessary to keep bacteria out of his wounds become automatic, and he is thus able to concentrate his whole attention on the operative details without having to fear that he may be omitting some detail essential for securing the asepticity of the wound" (Preface, *The Treatment of Wounds*: viii). The availability of antiseptics allowed the surgeon to work at a less harried pace and to undertake complicated or protracted procedures, eschewed in the past for fear of postoperative infection. The integration of bacteriology into surgical practice gradually transformed medicine.

The multidisciplinary concept enunciated in Cheyne's 1886 "Bacteriology" paper informs the New Sydenham publication, *Recent Researches* (1886). This collection of 33 papers by continental researchers fostered an Anglo-European renascence in medical science. Cheyne's editorial purpose was to make available to English-speakers current bacteriological discoveries and active lines of inquiry (Preface: vii-viii). He consistently asserted in the publications of this period how essential the Henle-Koch Postulates were to medicine. The use of these criteria, three of which follow, afforded proof of pathogenicity: during the early stages of a bacterial disease if the same species of micro-organism was constantly present in diseased parts; if the organisms were separated and cultivable from the morbid parts; and, once reintroduced into a susceptible animal if they caused pathology similar to that which occurred in humans. If these and associated criteria were satisfied, then strong (though not absolute) evidence for causality had been obtained (Preface: viii; Koch, "Etiology of Tuberculosis": 116–118). Although the criteria could not be applied to animals insusceptible to a particular disease, in the absence of definitive proof and even if the conditions of infection were unknown, a study of the diseased parts of the anatomy (and analogies to nonhuman vertebrates) helped to associate a micro-organism with a particular malady.

A perceptive editor, Cheyne selected papers for the 1886 collection that dealt primarily with infectious diseases, and he widened his scope to include important research on global health issues (Preface: ix). Thus, ongoing experiments on syphilis, gonorrhea, relapsing fever, and malaria, although yielding contradictory results and beset at times by methodological errors, were not excluded (Preface: xx-xii). One important area of investigation warranting further study was bacterial toxicity, implicated in erysipelas, diphtheria, and cholera (xii). How these organisms damaged tissue, for example, was not fully known. On the basis of clinical presentations and of laboratory experience, and according to the scientific literature, he reasonably assumed that in these conditions "there is every reason to believe that the disease is essentially a local one, and that the general constitutional disturbance is not due to the passage of the living and multiplying ... [microbe] into the blood, but [to] the absorption of the poisonous chemical products of its growth" (xii-xiii). Other than

Ludwig Brieger's research on toxins and ptomaïnes, discoveries in this area had been "few and imperfect" (xiii). Cheyne perceived the intrinsic linkage of bacteriology (in diphtheria and tetanus) to both toxicology and immunology, as the antitoxin discoveries of von Behring and Kitasato in 1890 would dramatically show.

Carl Friedländer's (1847–1887) paper "The Micrococci of Pneumonia," which Cheyne had translated for the 1886 collection, exemplified the Pasteurian idea of the biochemical nature of bacterial life. Friedländer's newly-discovered *Bacillus pneumonia* is a pathogenic ferment. Its metabolism of carbohydrates into formic and acetic acid and ethyl alcohol demonstrated the interrelatedness of biochemistry and medicine (Preface: xiv). Cheyne made the prescient claim that "the advances made in the future will probably depend as much on the chemistry of bacteria as on any other department of this [i.e. medical] science" (xiv). This declaration bolstered the idea that biology, medicine, and sanitary science (with its emphasis on chemistry) were kindred disciplines.

Cheyne, like Lister, had to persuade those who rejected or who minimized the idea of bacterial pathogenicity that the study of these micro-organisms was essential to medical practice. In the 1889 republication of *Lectures* on *Suppuration and Septic Disease*, he declared his dual aim: to summarize what was known; and to give "uniformity to the knowledge which has been gained ... during the last few years" (Preface: v). Reiterating the 1886 message that bacteriology and medicine were interdependent fields, he was dismayed by artificial barriers erected between practitioners in these disciplines. A counterproductive trend was that the physician neither possessed nor hoped to acquire "a practical knowledge of bacteriology" (Preface: v). The bifurcation of bacteriology and clinical practice, he thought, was fundamentally a matter more of emphasis than of difference: the bacteriologist overestimating the importance of microbial life; the clinician underestimating the importance of "causal organisms" (Preface: v). Cheyne arbitrated between the two departments: "As one who has studied the subject from both points of view, though more from the clinical than the bacteriological side, I have naturally directed my efforts to the attempt to reconcile the apparently conflicting evidence," and in the *Lectures* "to estimate ... the relative importance of the various factors which come into play in the production of suppuration and septic diseases" (Preface: v-vi). Along with commentary on bacteriology, etiology, and clinical medicine, the *Lectures* also brought recent studies on the immunological role of blood products within the widening scope of medical science.

The idea of multidisciplinary research and practice informs Cheyne's periodical history of bacteriology, as shown by the outlines in *Suppuration* (1889)

and "On the Progress and Results of Pathological Work" (1897). He enumerated bacteriological Landmarks or nodes along the ca. 1860-to-1897 timeline
and characterized the period as "the great revolution in *Surgery* of modern
times" (italics added). As we have seen, from Cheyne's perspective, microbiology was absolutely essential to clinical and surgical medicine. The consecutive milestones, in the 1889 article, were Lister's antiseptic surgery; Ogston's
1880-to-1882 surgical and laboratory work on abscess pathology; and Rosenbach's 1884 isolation of pure cultures of both Billroth's *Streptococcus* and *Staphylococcus*. Conspicuously absent from the 1888 list were Pasteur and Koch. In
1897, as I pointed out in the Introduction, Cheyne named as representative
figures Lister, Koch, and Metchnikoff and allotted Landmark status to their
respective works. The modified paradigm, combining the rosters of 1889 and
of 1897, lists six representative figures and five turning points or stages in early-
modern bacteriology, c. 1860–1900: Pasteur and Lister (I), Koch (II), Ogston
(III), Rosenbach and others (IV), and Metchnikoff (V). Rosenbach who nom-
inally represents Landmark IV on the strength of the 1884 paper "On the Micro-
Organisms in Infective Diseases of Wounds," assessed the characteristics of
diverse pyogenic organisms and suggested their connection to suppuration.
Because his investigations had been confirmed by colleagues, and because other
pyogenic organisms were discovered in this period, Cheyne included the complementary works of Fehleisen, of Passet, and of Flügge in the discussion. Like
Lister and Pasteur, Cheyne maintained that medical science owed its continuity
to intuition, to inventiveness, and to cooperation between allied fields.

The Address "On the Progress and Results of Pathological Work" was
delivered at the Annual Meeting of the British Medical Association in Montreal,
September 1897. Although this lecture was concerned primarily with disease,
Cheyne surveyed and emphasized the importance of bacteriology (586). He
recalled that Edinburgh University training in the 1870s had given little attention to medical microbiology; nevertheless, from 1867 to 1876, Lister had been
elucidating its relevance in 15 antisepsis papers. The prolific work of Lister and
others notwithstanding, from the late 1870s to 1897, Cheyne characterized the
academic study of micro-organisms as an "absolute blank" (586). As Lister's
House Surgeon, Cheyne recollected objections set against the bacterial theory
of infection, ranging from denial that microbes existed to their being irrelevant
to disease (586). These "objections," Cheyne admits, "[had] led [him] to take
up bacteriology, for it is of great importance to ascertain whether or not, as a
result of the antiseptic treatment, organisms were absent from the discharges
from the wounds" (587).

In the seven-volume *Manual of Surgical Treatment* (1899–1913), which
Cheyne co-authored with Frederic Francis Burghard, the capabilities of the

multidisciplinary concept are revealed in the description of pyemia. In the *Manual*, he described the interactions of bacteriology, pathology, and surgery in the diagnosis and management of pyemia (**1**: 213–216). A serious bacterial infection, pyemia is characterized by fever, blood toxicity, and the formation of abscesses. Cheyne suspected its cause to be an ordinary pus-producing coccus, *Streptococcus pyogenes aureus*, the etiology of which had been described in the 1880s (*Suppuration*: 41–42). These micro-organisms enter wounds, invade the bloodstream, and attach to blood clots or to other solid material (41–42). With this knowledge, an investigator could determine the concentration of micro-organisms in the body, the virulence factor, and the susceptibility of the patient (42). In volume **1** of the *Manual* (1899), Cheyne reiterated how bacteriology, pathology, and surgery could work together in the diagnosis and treatment of pyemia. From a surgical perspective, he outlined how *Streptococcus pyogenes aureus*, once gaining a foothold, causes rigor, fluctuating fever, jaundice, organ and joint abscesses, blood clots and emboli, kidney damage, suppuration, and death in eight to ten days. Post-mortems provided additional information on pyemic disease. Investigators learned that this disorder could be present if a thrombosed vein was near the wound. Microscopic analysis showed that the organisms grew in blood clots. When the germ-laden clot broke up, the particles circulated in the bloodstream, reaching the kidneys, spleen, and synovial tissues, where they lodged and where the microbes recolonized; the chain-forming *Streptococcus pyogenes aureus* could actually clog veins, causing clot formation, creating the potential for further metastasis as the clot disintegrated (**1**: 213–216). Knowing how the germ behaved and of its harmful potential, the surgeon would be certain to clean and disinfect wounds; to incise and ligate thrombosed veins; to drain and irrigate abscesses; and to rehabilitate the patient through the administration of medications, daily wound care, and proper diet (**1**: 213–216). Antiseptic surgery for pyemia, as well as for any other bacterial disorder, encompassed all of these activities.

Bacteriology, in its nascent stage, 1860–1900, developed from a recondite domain to a branch of interdisciplinary medicine. From 1880 to 1938, as Rosemary Wall has shown, physicians became accustomed to using bacteriological laboratories in their practices; technology and education, she points out, helped to make laboratories for routine examinations commonplace; in addition, both the legal profession and the general public became aware of infectious diseases in the workplace; and diagnostic procedures in hospitals were regularly performed (5–13, 177–183). Cheyne contributed to medical research and education, establishing the bacteriological assay as routine in hospital care; acquiring tissue samples from operations at King's College Hospital, including the results of these investigations into case histories; funding the construction of a tissue archive and pathology center; developing courses in bacteriology at King's; and, in 1888, assisting Edgar Crookshank in the delivery of inaugural bacteriology lectures at the Royal Veterinary College (Edmunds: 1013; Worboys, *Spreading Germs*: 69; Coutts, "Illustrating Microorganisms": 6). As a founder of medical bacteriology in Britain, Cheyne chronicled, and participated in, its institutionalization (Coutts, "Illustrating Microorganisms": 4, 9). From 1878 to 1927, he produced 60 articles on bacteriology, wrote and co-authored 13 books on bacteriology and surgery, was clinician and surgeon, an instructor at King's College Hospital, a prolific *litterateur* (author, editor, and translator), an international liaison, and a naval Commander to whom the welfare of military personnel was his greatest concern.

Appendix I

Cheyne and His Achievements:
A Timeline of His Work
in Medicine and Bacteriology[1]

by PATRICK A. DEPAOLO

Winter session 1870: Courses in botany and chemistry; attains First Class Honors and University medal.

May 1871: Begins medical studies; registers for natural history; earns 98 percent, first medal in chemistry.

October 1872–1873: Studies physiology and surgery, achieving 96 percent in the junior division; third prize in junior surgery; and first medal in practical physiology.

Summer 1873: Lister names Cheyne dresser.

1875: Graduates with M.B. and C.M.

June 1875: Visits medical institutions in Leipzig, Halle, and Berlin; attends lectures in Vienna.

Spring 1876: Visits Strasbourg; laboratory work in bacteriology with F. D. von Recklinghausen.

October 1876: Becomes Lister's House Surgeon in Edinburgh.

1876–1879: Begins bacteriological experimentation.

1877: Wins the Syme Scholarship for thesis on relation of microbes to wounds treated by Lister.

Spring 1877: Lister invites Cheyne and select students to join his new practice in King's College, London.

1878–May 1879: Presents experimental findings to London Pathological Society. "On the Relation of Organisms to Antiseptic Dressings" published in May 1879.

29 November 1879: Publishes "Statistical Report of all Operations Performed on Healthy Joints."

1880: At request of the New Sydenham Society, translates Koch's 1878 *Investigations into the Etiology of Traumatic Infective Diseases*.

1880: Receives Jacksonian Prize for dissertation, "The Principles, Practice, History and Results of Antiseptic Surgery."

1880 or 1881: Obtains initial licensing to conduct animal experimentation.

August 1881: Attacks Ogston's ideas on micrococcal pathogenicity at International Medical Congress.

1882: Publishes *Antiseptic Surgery, its Principles, Practice, History and Results*, incorporating updated material from 1877 Syme Fellowship in Edinburgh, Boyston Prize essay for Harvard University, and the Jacksonian prize-winning essay, translated into German and published in 1883.

June 1882: Displays tuberculosis bacillus at Norfolk and Norwich Hospitals before the Medico-Chirurgical Society.

12 July 1882: Goes to France and Germany for the Association for the Advancement of Medicine by Research to test claims for the cause of tuberculosis.

21 July 1882: Visits H. Toussaint's laboratory in Toulouse, then the laboratories of Schüller and Robert Koch.

1 February 1883: Presents report "On the Relations of Micro-Organisms to Tuberculosis" to the AAMR.

3 March 1883: Delivers "Tubercle: Its Etiology and Modern History" at the meeting of The British Medical Association.

24 January 1884: Responds to Percy Kidd's argument against the connection between tuberculosis and the bacilli.

17 May 1884: Sets up working laboratories to display the interaction of public health and medicine for the International Health Exhibitions in South Kensington.

1884: Co-authors *Public Health Laboratory Work*, with W. H. Corfield and C. E. Cassal.

20 September–4 October 1884: Publishes three seriatim papers, "Report on Micrococci in Relation to Wounds, Abscesses, and Septic Processes."

24 January 1885: Publishes "The Bacillus of Tubercle."

1885: Publishes *Manual of the Antiseptic Treatment of Wounds for Student and Practitioners*.

24 March 1885: Publishes "Report to the Royal Medical and Chirurgical Society."

11 April 1885: Publishes "Correspondence: The Cholera Bacillus of Koch."

25 April–23 May 1885: Publishes in four installments "Report on the Cholera-Bacillus."

10 October 1885: Publishes "On 'Foulbrood.'"

1886: Edits and publishes for The New Sydenham Society *Recent Researches on Microorganisms in Relation to Suppuration and Septic Diseases*.

1886: Publishes "Bacteriology: Relations of Bacteria to the Soil on which They Grow" in *The American Journal of the Medical Sciences*.

July 1886: Publishes "Bacteriology" in *The American Journal of the Medical Sciences*.

31 July 1886: Publishes "Report on a Study of Certain of the Conditions of Infection."

Mid–1880s: Serves as Assistant Surgeon to King's College Hospital and Senior Surgeon at Paddington Green Children's Hospital.

1886: Visits Koch in Berlin to experiment on the causes of infection and disease.

1886: Contributes to Gant's *The Science and Practice of Surgery.*

5 March 1887: Publishes "On Early Tracheotomy in Diphtheria."

July 1887: Publishes "Bacteriology: II. Study of Bacteria by Means of Cultivation," in *The American Journal of the Medical Sciences.*

4 February–10 March 1888: Publishes in three installments, "Lectures on Suppuration and Septic Diseases."

1888: Receives the Astley Cooper Prize for "Tuberculous Diseases of Bones and Joints"; becomes Hunterian Professor of Comparative Anatomy and Physiology at the Royal College of Surgeons of England.

August 1888: Co-authors with Frank R. Cheshire "The Pathogenic History and History under Cultivation of a New Bacillus (*B. alvei*), the Cause of a Disease of the Hive Bee hitherto Known as Foul Brood."

1889: Publishes *Suppuration and Septic Diseases: Three Lectures Delivered at the Royal College of Surgeons of England,* in February 1888.

1890: Translates and edits Flugge's *Micro-organisms, with Special Reference to the Etiology of the Infective Diseases.*

19 August 1890: Elected Honorary Consulting Surgeon at Bexley Cottage Hospital, London.

29 November–20 December 1890: Publishes in four installments "Extracts from 'Three Lectures on Tubercular Diseases of Bones and Joints.'"

1891: Publishes "On the Value of Tuberculin in the Treatment of Surgical Tubercular Diseases."

4–25 April 1891: Publishes in four installments "Lectures on the Pathology of Tuberculous Diseases of Bones and Joints."

9 May 1891: Publishes "The Value of Tuberculin in Surgery."

2 July 1892: Publishes "Abstract of 'Lectures On the Treatment of Surgical Tuberculous Diseases.'"

1893: Becomes Professor of Surgery at King's College and delivers three Lettsomian Lectures on Surgery to the Medical Society of London.

2 December 1893: Elected Honorary Consulting Surgeon at the North London Hospital for Consumption in Hampstead.

February–March 1894: Delivers three Lettsomian lectures on cancer surgery to the Medical Society of London.

1895: Publishes *The Treatment of Wounds, Ulcers and Abscesses* and *Tuberculous Disease of Bones and Joints: Its Pathology, Symptoms and Treatment.*

March 1895: Assists in editing of the first Annual Report of King's College Hospital.

July–August 1895: At the 63rd Annual meeting of the British Medical Association, delivers the paper "Operations for Malignant Disease of the Pharynx and Naso-Pharynx with Cases."

1895–1896: Contributes to Edward Treves' *A System of Surgery* chapters on inflammation, suppuration, ulceration and gangrene (Volume **1**); and a chapter on breast diseases (Volume **2**).

1896: Publishes *Objects and Limits of Operations for Cancer, being the Lettsomian Lectures for 1896.*

27 June 1896: Becomes Examiner in Surgery at Cambridge.

August 1897: Publishes "The Disinfection of Hands and Instruments."

31 August–3 September 1897: Becomes Acting President of the Section of Pathology and Bacteriology at the British Medical Association's Annual Meeting in Montreal.

4 September 1897: Published "On the Progress and Results of Pathological Work."

27 November 1897: Accepted as an honorary Surgeon to the Scottish Hospital in London.

6 August 1898: Lectures on hydrocephalus at the 66th annual meeting of the British Medical Association.

1899: Contributes to the multi-volume *Encyclopedia Medica* and a section on Antiseptic Surgery to the *Dictionary of Practical Surgery*.

10 April 1899–1902: Becomes President of the Pathological Society of London.

1899–1907: Co-authors with F. F. Burghard the seven-volume *Manual of Surgical Treatment*.

1900: Publishes "On the Treatment of Tuberculous Diseases in Their Surgical Aspect, Being the Harveian Lecture for 1899" and becomes consulting Surgeon during the Second Anglo-Boer War.

12 May 1900: Publishes "The War in South Africa. The Wounded from the Actions between Modder and Driefontein."

22 June 1901: Publishes "The Organisation of Medical Aid in a Great War."

30 November 1901: Publishes "On the Treatment of Wounds in War."

13 December 1902: Publishes "Listerism and the Development of Operative Surgery."

1908: Publishes "The Bradshaw Lecture on *The Treatment of Wounds*."

1 January 1910: "Remarks on the Treatment of Wounds in Connexion with the Recent Results Obtained at St. George's Hospital."

June 1910: Becomes Surgeon Surgeon in Ordinary to King George V.

24 June 1910: Elected President of the Edinburgh Royal Infirmary Residents' Club.

1912: Publishes "A Discussion on the Treatment of Tuberculous Joint Disease in Children."

1914–1916: While President of the Royal College of Surgeons of England, serves in the rank of Commander and as Surgeon at Chatham Naval Hospital.

November 1914: Publishes "Remarks on the Treatment of Wounds in Wars."

February 1915: Publishes "The Hunterian Oration on *The Treatment of Wounds in War*."

8 May 1915: Publishes "Sir Almroth Wright's Lecture on the Treatment of Wounds in War."

22 May 1915: Publishes "Correspondence: The Recommendations of the Naval Committee on the Treatments of Wounds in War."

31 July 1915: Publishes "Observations on the Treatment of Wounds of War."

1915: Publishes "The Treatment of Wounds in War" in *The British Journal of Surgery* and "The Treatment of Wounds in War" in *The Lancet*.

17 March 1920: Delivers "Testimony: "Dogs' Protection Bill.""

1925: Publishes *Lister and His Achievement*.

1927: Publishes *Three Orations: The Lister Centenary* and "Reminiscences of 'The Chief'" in *Joseph, Baron Lister: Centenary Volume*.

Appendix II
Lister's 1875 Visit to Germany

While visiting Munich in 1875 to monitor antiseptic outcomes, Lister learned of inconsistencies which he discussed in the 4 August 1875 paper "An Address on the Effect of the Antiseptic Treatment upon the General Salubrity of Surgical Hospitals" (*CP*. **II**: 247–255). In this text, he recounted his visit to the large *Allgemeines Kranken-haus* where hospital diseases, such as pyemia, erysipelas, and a form of gangrene, had been rampant; "hospital gangrene," from 1872 to 1874, had occurred in 80 percent of the patients who had open traumatic wounds or who had undergone surgery; and those post-operative infections were responsible for prolonged hospital stays and for many fatalities (*CP*. **II**: 248; West: 76–77).

Early in 1875, these conditions improved markedly once the surgeon, Johann N. R. von Nussbaum (1829–1890), applied Listerian antisepsis correctly, eradicating "hospital gangrene," pyemia, and erysipelas, and emptying the once overpopulated convalescent wards (*CP*. **II**: 248–249). Von Nussbaum attested to this dramatic change in the 1875 pamphlet, *The Munich Surgical Clinic in 1875: A Memoir for his Students*, and he attributed the success to the revolutionary benefits of Lister's system (*CP*. **II**: 249n.). As von Nussbaum explained, the hard-working Munich staff had previously tried everything to control post-operative infections, including irrigation with chlorine water and even carbolic-acid solutions, but nothing seemed to work. The problem, however, was to use the antiseptic in strict accordance with Lister's latest directions. The results, to which von Nussbaum testifies in the pamphlet, *Die Chirurgische Clinik zu München im Jahr 1875*, improved very rapidly "in the course of a single week" when, "with great energy and industry, [they] applied to all our patients the newest antiseptic method, now in many respects improved by Lister, and did all operations according to his directions … not a single other case of hospital gangrene occurred" (*CP*. **II**: 249n). Von Nussbaum profited specifically from the 1875 *Lancet* paper "On Recent Improvements in the Details of Antiseptic Surgery," in which Lister emended the carbolic-acid dosage that he had presented in the 10 August 1871 British Medical Association, "Address in Surgery" (*CP*. **II**: 172–205; and 206–246). Von Nussbaum's 1875 reference was to the initial ineffectiveness of carbolic-acid strength prescribed in 1871 ("On Recent

Improvements," *CP*. **II**: 206). Lister, in 1871, had been experimenting to maximize its bactericidal properties while minimizing irritative and toxic side-effects. Sepsis in the Munich Surgical Clinic was the consequence of a diluted solution of carbolic acid. Since the acid-water proportion of 1-to-100 had proven to be too weak, Lister prescribed the stronger 1-to-40 concentration in surgical wounds and the even stronger 1-to-20, for the disinfection of skin, instruments, sponges, and in emergency cases involving pre-operative, accidental wounds (*CP*. **II**: 206). Three applications of the acid—in watery solutions, in fixed oil, and in unctuous resin mixture—had specific therapeutic benefits (*CP*. **II**: 212). The stronger regimen worked and convinced von Nussbaum as to its merit.

Lister's Munich visit was followed by stops in Leipzig and in Halle. In Leipzig, he learned that the surgeon, Karl Thiersch (1822–1895), had been using the antiseptic system successfully in his 300-bed hospital. Thiersch innovated as well, choosing salicylic over carbolic acid for the external dressings, although Lister had doubts about the former's efficacy; but Thiersch still relied on carbolic lotion and spray (*CP*. **II**: 249). In Halle, the famed surgeon, Richard von Volkmann (1830–1889), replicated Lister's system and had been teaching the method to his countrymen (Willy, et al.). Lister was impressed by von Volkmann's initiative. Although the latter had appropriated the method without training at the Edinburgh Royal Infirmary, he had read translations of Lister's papers to learn about the antiseptic system.

The Halle wards, like those in Munich, witnessed improvement, despite the decrepitude of the building and its small, overcrowded rooms (*CP*. **II**: 250). Von Volkmann wrote that, since using the antiseptic system (late November 1872), no compound-fracture case had ended in mortality. Since July 1873, no pyemia had been reported, although 60 amputations had taken place in this period (*CP*. **II**: 250). Von Volkmann even challenged the accuracy of Thiersch's results, as they pertained to the use of carbolic acid in the control of infection (*CP*. **II**: 251). Other testimonies, from hospitals in Berlin, Madgeburg, and Bonn, supported claims favorable to Lister's treatment as it had been articulated in 1875 (*CP*. **II**: 252–253).

Chapter Notes

Introduction

1. In a 1932 Obituary Notice, Bulloch called Cheyne "a pioneer in bacteriological research," a major disseminator of continental science in Britain, and a tireless experimenter. Walter Martin's 1932 Obituary Notice mentioned Cheyne's quantitative approach to pathogenesis, his calculation of the numerical relation between *Staphylococci* and infection, which had become a standard reference in bacteriological textbooks; in 1879, for example, Cheyne had estimated that one minim (0.016 milliliters) of cucumber infusion contains at least two million micro-organisms (336–337). In 1932, Cheyne's colleague, Arthur Edmunds, recalled how the former had compiled a surgical tissue registry, referred to tissue samples in case histories, established a pathology institute and clinic at King's College Hospital in London, and, in 1886, collaborated with Edgar M. Crookshank to deliver bacteriology courses at King's (1013). W. R. Bett, in 1952, observed that Cheyne contributed significantly to tuberculosis etiology, pointing out that the milk of tuberculous cows posed a public-health risk to consumers (366). Leonard G. Wilson, in 1987, credited Cheyne for recognizing that micrococci exist in many forms, for distinguishing their pathogenic properties from those of rod-shaped organisms, and for proving experimentally that bacteria are not present in the bodies of healthy animals (405–407). Among other facts, Claire E. J. Herrick (2006) cited Cheyne's translations of continental authors; in 1883, his support of Koch's hypothesis of the tubercle bacillus as the cause of tuberculosis, and his

trip to Berlin, in 1886, to work in Koch's laboratory.

Michael Worboys, in *Spreading Germs: Disease Theories and Medical Practice in Britain, 1865–1900* (2000), contributed significantly to our understanding of Cheyne's bacteriology, 1879 to 1890. Although only a portion of this comprehensive study is devoted to Cheyne's activities, important points are covered, primary texts cited, and theoretical background set. Worboys discusses Cheyne's laboratory investigations and debates involving micrococci, tubercle bacilli, *Vibrio choleræ*, and other organisms (170–181); his role as gatekeeper for continental researches (175–181); how he revised Listerism (182–183); and his definition of aseptic surgery (188–189). He concludes that, in the early 1880s, both Cheyne and Ogston emerged as "leading international researchers on germs"; but, by 1890, both are said to have abandoned bacteriology for clinical medicine, as British bacteriology had become the province of physicians and pathologists (170). Caroline C. Watson and co-authors (2013) discussed Cheyne's treatment of congenital hydrocephalus.

Jane Coutts, in the 2015 paper "Illustrating Micro-Organisms," incisively covers Cheyne's early laboratory work, his role as observer, chronicler, assessor, and collaborator with international colleagues (1–10). This paper was followed by her *Microbes and the Fetlar Man: The Life of Sir William Watson Cheyne* (Edinburgh: Humming Earth Publishers, 2015). The first full-length biography of Cheyne, it was released in August 2015, while the present study was in its final stages. The author used

archival and published sources to survey
Cheyne's family background and early years;
the beginnings of his scientific and surgical
career at Edinburgh; his study tour of European universities and relationship with Lister;
daily life in Lister's wards from the perspective
of the patient Margaret Mathewson; the early
stages of his scientific and medical career at
King's College Hospital, London; his indebtedness to Koch and other prominent figures;
his introduction of continental research
methodology to Britain and public-health instruction; his work on the etiology and epidemiology of infectious diseases, such as
cholera, tuberculosis, typhoid, and diphtheria;
his practical use, defense, and development,
of antiseptic regimens in civilian life and during war. In terms of purpose, genre, style, and
method, my study differs from, and hopefully
will complement, Dr. Coutts'portrayal of
Cheyne.

Chapter 1

1. Commentaries on the Pasteur-Lister relationship, from 1912 to 2002, have appeared
intermittently in the scholarly literature. These
brief commentaries are embedded in broader
discussions of spontaneous generation and of
the germ theories of disease. Biographies,
monographs, and essay collections containing
limited discussions and summaries are chrono-
logically listed here: Hart, "An Address on
Pasteur and Lister," *The British Medical Journal*
(1902); **2** (2189): 1838–1840; F. C. Clark, "A
Brief History of Antiseptic Surgery," *Medical
Library and Historical Journal.* September
1907; **5**(3): 145–172, esp. 156–172; Metchnikoff, *The Founders of Modern Medicine,*
(1912; New York: Walden, 1939), 106; Paget,
Pasteur and After Pasteur, edited by John D.
Comrie (London: Adam and Charles Black,
1914), 34–35; Vallery-Radot, *The Life of Pasteur,* translated by R. L. Devonshire (New
York: Doubleday, Page, & Company,1915),
186–187, 237–240, 430–431, 448–449;
Godlee, *Lord Lister* (London: Macmillan,
1918), 162–164, 171–177, 435–439, 533–534;
Cheyne, *Lister and His Achievement* (London:
Longmans, Green and Company, 1925), 6–
10, 44–45; Creed, "Pasteur and Lister," *The
British Medical Journal* (12 July 1936); **2**:
3942; Bulloch, *The History of Bacteriology*

(1938; New York: Dover, 1979), 141, 183;
Cameron, *Joseph Lister: The Friend of Man*
(London: William Heinemann Medical Books,
1948), 102–105, 118, 120, 134, 173–174;
Guthrie, *Lord Lister: His Life and Doctrine*
(Edinburgh: E. & S. Livingstone, 1949), 54–
55, 83–85, 87, 100, 110, 112; Dubos, *Louis Pas-
teur: Free Lance of Science* (New York: Da
Capo, 1950), 244–245; and *Pasteur and Mod-
ern Science* (Garden City, NY: Anchor Books,
Doubleday & Company, 1960), 99–101; Fisher,
Joseph Lister: 1827–1912 (New York: Stein and
Day, Publishers, 1977), 131–134, 145, 149–
150, 153, 156–157, 271; Lawrence and Dixey,
"Practicing on Principle: Joseph Lister and
the Germ Theories of Disease," in *Medical
Theory, Surgical Practice,* edited by Christopher
Lawrence, Wellcome Institute Series in the
History of Medicine (London: Routledge,
1992), 153–215; Bynum, *Science and the Practice of Medicine in the Nineteenth Century,* Cambridge History of Science Series (New York
and Cambridge: Cambridge University Press,
1994), 113, 144, 216, 219–220, 238, 240; Geison, *The Private Science of Louis Pasteur* (Princeton: Princeton University Press, 1995), 33,
36, 164, 262; Debré, *Louis Pasteur,* translated
by Elborg Forster, 2nd edition (Baltimore:
The Johns Hopkins University Press, 1998),
275–276, 402; Jerry L. Gaw, *A Time to Heal:
The Diffusion of Listerism in Victorian Britain,*
Transactions of the American Philosophical
Society, Vol. **89**.1 (Philadelphia: American
Philosophical Society, 1999), 31–36, 63–65,
71–72, 74–84; Worboys, *Spreading Germs:
Diseases, Theories, and Medical Practices in
Britain, 1865–1900, Cambridge History of
Medicine,* edited by Charles Rosenberg (Cam-
bridge: Cambridge University Press, 2000),
96–97, 150; and James E. Strick, *Sparks of
Life: Darwinism and the Victorian Debate over
Spontaneous Generation,* 2nd edition (Cam-
bridge, Massachusetts: Harvard University
Press, 2002), 22, 36, 133–134, 146, 159, 200,
204, 209.

2. The Listerian texts, published by the
University Press of the Pacific and cited in this
study were reprinted from the original and
rare 1909 Oxford University Press edition of
Lister's *Collected Papers.* The editorial committee, in 1909, consisted of Lister's closest
associates and his nephew: Drs. Hector C.
Cameron, William Watson Cheyne, Rickman
J. Godlee (nephew), C. J. Martin, and Dawson

Williams. Their intent in making the collection was to honor the eighty-two-year-old surgeon by gathering together "all the papers and addresses" into a memorial and archive of his scientific work, which Lister appropriately called his magnum opus (Godlee: 582–4; see also, "Preface" and "Introduction to *CP*. **I**: v–vi and xi–xliv, resp.). The original papers had appeared in *The Lancet*, *The British Medical Journal*, *The Edinburgh Medical Journal*, the *Quarterly Journal of Microscopical Science*, the *Proceedings of the Royal Society of London*, the *Transactions of the Royal Society of Edinburgh*, and other journals.

3. John Farley and G. L. Geison, "Science, Politics and Spontaneous Generation in Nineteenth Century France: The Pasteur-Pouchet Debate," 161–198; J. Farley, "The Spontaneous Generation Controversy," 285–319; Lawrence and Dixey, "Practicing on Principle: Joseph Lister and the Germ Theories of Disease," 153–215; James E. Strick, "New Details Add to Our Understanding of Spontaneous Generation Controversies" 192–198; and "Darwinism and the Origin of Life: The Role of H. C. Bastian in the British Spontaneous Generation Debates, 1868–1873," 51–92; and Worboys, *Spreading Germs*, 86–90.

Chapter 2

1. *Lister and His Achievement* (1925), 3–4, 13–18, 23–32, 42, 61, 69, 74, 100; "Lister, the Investigator and Surgeon" (16 May 1925), 923–926; and *Three Orations: The Lister Centenary* (1927); and Cheyne's "Reminiscences of 'The Chief,'" in *Joseph, Baron Lister: Centenary Volume, 1827–1927*, edited by A. Logan Turner, pp. 122–128.

2. The typical, later nineteenth-century laboratory had an extensive supply-inventory. Edgar M. Crookshank, in his *Text-book of Bacteriology*, describes the amount of glass and hardware needed for a moderately-equipped bacteriological laboratory. Not only was a good oil-immersion microscope (of Leitz, Zeiss, Hartnack, Powell, or Lealand manufacture) and a condenser (such as Abbé's) indispensable, but the lenses had to be up to specifications (1/12 and 1/18 oil-immersion or 1/12 and 1/15 for Powell and Lealand); in addition, one could not do without a Zeiss micrometer eyepiece to measure bacterial size.

Accessories were sundry and many: a bell-glass microscope cover; one square-foot blackened plate glass and one white porcelain slab, respectively; the glass and hardware including bottles with funnels; small dishes, watch glasses for staining, glass slides, various cover-glasses, platinum needles, fine-point glass rods used for work with acids, copper lifters, brass tongs, a turn-table for slide preparation, and so on (Crookshank, *Text-book* [1887]: 6).

Chapter 4

1. "Sir Alexander Ogston," *The British Medical Journal* (16 February 1929); **1**(3554): 325–327; I. A. Porter, "Alexander Ogston; Bacteriologist," *The British Medical Journal* (7 August 1954); **2**(4883): 355; A. Lyell, "Alexander Ogston (1844–1929)—Staphylococci," *Scottish Medical Journal* (October 1977); **22**(4): 277–278; and "Alexander Ogston, Micrococci, and Joseph Lister," *Journal of the American Academy of Dermatology* (February 1989); **20** (2, Part 1): 302–310; James Baird, "Sir Alexander Ogston and the Royal Army Medical Corps," *The Staphylococci: Proceedings of the Alexander Ogston Centenary Conference* (Aberdeen, 1981), 22–32; reprinted in *The Medical War: British Military Medicine in the First World War*, edited by Mark Harrison (Oxford: Oxford University Press, 2010), 22–29; A. Adam, "Alexander Ogston and the Army Medical Services' Formation of the Royal Army Medical Corps, July 1898," *Scottish Medical Journal* (October 1998); **43**(5): 156–157; Alexander George Ogston, "Ogston, Sir Alexander (1844–1929)." *Oxford Dictionary of National Biography* (Oxford: Oxford University Press, 2004–2014); S. W. Newsome, "Ogston's Coccus," *The Journal of Hospital Infections* (December 2008); **70**(4): 369–372.

Chapter 7

1. According to *The Oxford English Dictionary*, the first use of the term *vivisection* is attributed to the physician and founder of the British Museum, Sir Hans Sloane (1660–1753). In 1707, Sloane used the term in his *Natural History of Jamaica* (volume **1**:2). A

Latinate neologism (*vivus* [alive] + *section* [section]), *vivisection* is defined as: "the action or cutting or dissecting of some part of a living organism; specifically, the action or practice of performing dissection, or other painful experiment, upon living animals as a method of physiological or pathological study" (*OED*. **II**: 3648).

Antithetical interpretations of this concept became prominent in the Victorian Age. An anonymous contributor to the 30 January 1875 edition of *The Medical Times and Gazette* understood the term literally: "Vivisection, as the word was originally used, occurs nowhere in Britain, as far as our knowledge extends. The word was applied to the dissection of living animals for anatomical purposes.... Physiological experiments have no relation to these; they are performed for purposes of instruction or direct investigation, both of which are held by the vast majority of the profession to be perfectly justifiable … not for a single moment would anything like cruelty to animals in the original sense be tolerated among them" ("Vivisection Again!"). The denotative understanding differentiates between unnecessary experimentation and that which was humane, legal, and undertaken in the interests of medical science and public health. For dedicated opponents of animal experimentation, such as Mrs. Fairchild Allen, editor of the journal, *Anti-Vivisection: Official Organ of the Illinois Anti-Vivisection Society*, the standard definition of vivisection was broadly defined to include torturing and killing animals. In March 1897, she wrote that, "Vivisection is the cutting up of Live Animals—also poisoning, burning, smothering, freezing, breaking the bones, irritating the bare nerves with electricity, dissecting out the stomach and other organs" (1).

Each definition had a measure of truth. A proponent of animal models in science would argue that living nonhuman vertebrates were essential to the pursuit of biomedical knowledge and to the control of human disease; and many agreed that investigators must abide by the law, and, wherever possible, avoid or mitigate animal suffering.

2. Koch's Postulates, though providing a useful historical reference point, were not intended as rigid criteria to establish causality (Alfred Evans, "Causation and Disease": 192). Biological phenomena, Koch implied, were just too complex to be absolutely defined through a train of presuppositions or premises that incrementally modify a hypothesis. To demonstrate this, Evans augmented the Henle-Koch criteria with six etiological variables, for which one would have to account when trying to prove bacteria-disease causation: (1) a particular clinical state could be produced by different pathogens; (2) causative agents might vary in effect, according to geography, age group, or host susceptibility; (3) some diseases were produced only if two or more agents acted together; (4) clinical responses to a pathogen might vary according to setting; (5) pathogens, whether individual or combined, could produce a range of clinical responses; and (6) the responses of a host either to an infectious or to a noninfectious agent could also vary widely, depending on behavioral patterns, genetic constitution, immunological status and other factors (191).

Chapter 9

1. The use of Lister's antiseptic method in war, from 1870 to 1872, had received only cursory attention in the historical literature, until the publication of Coutts' biography. Coutts provides historical background, along with a substantive discussion of Cheyne's experiences and lessons learned during the Second Anglo-Boer War (Godlee: 359–361; Fisher: 182–183; Haller: 305–306; Gaw: 53, 99; Coutts, *Microbes*: 240–295).

2. See: Jane Coutts' summary of Cheyne's experiences during Field Marshal Lord Frederick Sleigh Roberts" (1832–1914) march from Modder River to Bloemfontein (*Microbes*: 245–273).

Chapter 10

1. William C. Hannigan, "Neurological Surgery during the Great War: The Influence of Colonel Cushing," *Neurosurgery* (September 1988); **23**(3): 283–410; J. D. C. Bennett, "Medical Advance Consequent to the Great War, 1914–1918," *Journal of the Royal Society of Medicine* (November 1990); **83**: 738–742; John S. Haller, Jr., "Treatment of Infected Wounds during the Great War, 1914 to 1918," *Southern Medical Journal* (March 1992);

85(3): 303–315; Thomas S. Helling and Emmanuel Daon, "In Flanders Fields: The Great War, Antoine Depage, and the Resurgence of Debridement," *Annals of Surgery* (1998); 228(2): 173–181; Cay-Rüdiger Prüll, "Pathology at War, 1914–1918: Germany and Britain in Comparison," in *Medicine and Modern Warfare*, edited by Roger Cooter, Mark Harrison, and Steve Sturdy, Papers presented at the Conference": *Medicine and the Management of Modern Warfare,"* London, July 1995, Wellcome Institute Series in the History of Medicine (Amsterdam and Atlanta, Georgia: Rodopi, 1999), 131–161; Claire E. J. Herrick, "Casualty Care during the First World War: The Experience of the Royal Navy," *War in History* (April 2000); 7(2): 154–179; Clinton K. Murray, Mary K. Hinkle, and Heather C. Yun, "History of Infections Associated with Combat-Related Injuries," *The Journal of Trauma: Injury, Infection, and Critical Care* (2008); 64: S221–231; Jack Edward McCallum, *Military Medicine: From Ancient Times to the 21st Century* (Santa Barbara, California: ABC-CLIO, 2008); M. M. Manring, Alan Hawk, Jason H. Calhoun, Romney C. Andersen, "Treatment of War Wounds: A Historical Review," *Clinical Orthopaedics and Related Research"* (2009); 467: 2168–2191; Robert Atenstaedt, "The Development of Bacteriology, Sanitation Science and Allied Research in the British Army 1850–1918: Equipping the RAMC for War," *Journal of the Royal Army Medical Corps* (2010); 156: 154–158; Mark Harrison, *The Medical War: British Military Medicine in the First World War* (New York and Oxford: Oxford University Press, 2010); W. G. P. Eardley, K. V. Brown, T. J. Bonner, A. D. Green, and J. C. Clasper, "Infection in Conflict Wounded," *Philosophical Transactions of the Royal Society: B:Biological Sciences* (2011); 366: 204–218; Christine G. Krüger, "German Suffering in the Franco-German War, 1870/71," *German History*, German History Society (Oxford: Oxford University Press, 2011); 29 (3): 404–422; Ana Carden-Coyne, *The Politics of War: Military Patients and Medical Power in the First World War* (New York and Oxford: Oxford University Press, 2014); Jane Coutts, *Microbes and the Fetlar Man* (Edinburgh: Humming Earth), pp. 238– 239, 290, 331, 341, 348, 350, 357–358, 361, 368–369, 371.

Appendix I

1. This Chronology is indebted to the following sources: Sir William Watson Cheyne, *Lister and His Achievement* (London: Longmans, Green and Company); *Three Orations: The Lister Centenary* (London: John Bale, Sons and Danielsson, 1927), pp. 83–92. https://www.watson-cheyne.com; and "Reminiscence of 'The Chief'" [Joseph, Baron Lister], *Joseph, Baron Lister: Centenary Volume, 1827–1927* (Edinburgh: Oliver and Boyd, 1927), 122–128; "Cheyne, William Watson (1852–1932)," Obituary Notice: *Biographical Memoirs of Fellows of the Royal Society*, (December, 1932); 1(1): 26–30; "Death of Eminent Surgeon: Sir William Watson Cheyne, former Associate of Lister." *The Glasgow Herald* (21 April 1932); 96:11. https://news.google.com; Arthur Edmunds, "The Late Sir Watson Cheyne," Obituary Notice: *The Lancet* (7 May 1932), 219 (5691): 1013; William Bulloch, "Obituary Notice: Sir William Watson Cheyne, 1852- 1932." *Fellows of the Royal Society* (December, 1932). http://rsbm.royalsocietypublishing.org; Walton Martin, "Sir William Watson Cheyne, 14 December 1852– 19 April 1932," *Bulletin of the New York Academy of Medicine* (May, 1932); 8(5): 336–337. http://link.springer.com; W. R. Bett, "Sir William Watson Cheyne, Bart., F. R. S. (1852– 1932)," *Annals of the Royal College of Surgeons of England* (December 1952); 11(6): 346– 366; Claire Herrick, "Cheyne, Sir (William) Watson, first baronet (1852–1932), Bacteriologist and Surgeon," *The Oxford Dictionary of National Biography* (Oxford University Press: online edition, May, 2006). http://www.oxforddnb.com; Caroline C. Watson, et al. "William Watson Cheyne (1852–1932): A Life in Medicine and His Innovative Surgical Treatment of Congenital Hydrocephalous." *Child's Nervous System* (2013); 29(11): 1961– 1965. http://link.springer.com; and Jane Coutts, *Microbes and the Fetlar Man: The Life of Sir William Watson Cheyne* (Edinburgh: Humming Earth Publishers, 2015).

Bibliography

Abbé, Ernst. "Beiträge zür Theorie des Mikroskops und der Mikroskopischen Wahrnehmung." *Archiv für Mikroskopische Anatomie*. 1873; Volumen **IX**: 413–468.

"Abiogenesis." *Compact Oxford English Dictionary*. Volume **I**, A-O. Oxford: At the Clarendon Press, 1971. P. 5.

"Absinthe." *American Journal of Pharmacy*. 1868; Volume **40**: 356–360. *The Wormwood Society: America's Premier Absinthe Association & Network*. http://www.wormwoodsociety.org / accessed 23 February 2015.

"Absinthe." *Scientific American*. 3 April 1869; **20**(14). *The Wormwood Society*. http://www.worwoodsociety.org / accessed 23 February 2015.

Adam, A. "Alexander Ogston and the Army Medical Services['] Formation of the Royal Army Medical Corps, July 1898." *The Scottish Medical Journal*. October, 1998; **43**(5): 156–157. http://www.ncbi.nlm.nih.gov / accessed 8 October 2014.

Adler, A. "Chemotaxis in Bacteria." *Science*. 12 August 1966; **153**(3737): 708–716.

Alex, Dan. "Maxim MG08 (Maschinegewehr 08) Machine Gun (1908)." 8 April 2014. www.militaryfactory.com / accessed 22 March 2016.

Allen, Mrs. Fairchild, editor. *Anti-Vivisection. Official Organ of the Illinois Anti-Vivisection Society*. Aurora, Illinois: Illinois-Anti-Vivisection Society 1893; March, 1897; Volume **IV**: 1–24. https://www.archive.org / accessed 14 April 2015.

Ambrose, Charles T. "The Osler Slide, a Demonstration of Phagocytes from 1876: Reports of Phagocytosis before Metchnikoff's 1880 Paper." *Cellular Immunology*. March 2006; **240**(1): 1–4. http://www.sciencedirect.com / accessed 29 December 2014.

Amery, L. S., editor. *The Times History of the War in South Africa, 1899–1902*. 7 volumes. Preface by L. S. Amery. London: Sampson, Low, Marston & Company, 1909. Vols. **VI**: 537; **VII**: 18–22.

"Amici, Giovanni Battista [Biography]." *Molecular Expressions: Science, Optics, & You/ Pioneers in Optics*. http://micro.magnet.fsu.edu / accessed 22 September 2015.

Anft, Berthold. "Friedlieb Ferdinand Runge: A Forgotten Chemist of the Nineteenth Century." Translated by R. E. Oesper. *Journal of Chemical Education*. 1955; **32**: 566–574.

The Animal's Defender and Zoophilist. 2 May 1904; **24–25**(1): 129–130. https://www.google.books.com / accessed 14 September 2014.

Archibald, Edward. "The Mind and Character of Lister." *The Canadian Medical Association Journal*. November, 1936; **35**(5): 475–496. http://www.ncbi.nlm.nih.gov/ accessed 6 January 2016.

Arnold, David. *Colonizing the Body: State Medicine and Epidemic Disease in Nineteenth-Century India*. Berkeley: University of California Press, 1993.

Atalić, Bruno. "1885 Cholera Controversy: Klein versus Koch." *Medical Humanities*. 2010; **36**: 43–47. http://mh.bmj.com / accessed 7 August 2014.

_____, et al. "Emanuel Edward Klein, a Diligent and Industrious Plodder or the Father of

British Microbiology." *Medicinski Glasnik*. August, 2010; **7**(2): 111–115. http://www. ncbi.nlm.nih.gov / accessed 21 August 2015.

Atenstaedt, Robert L. "The Development of Bacteriology, Sanitation Science and Allied Research in the British Army, 1850–1918: Equipping the RAMC for War." *Journal of the Royal Army Medical Corps*. 2010; **156**: 154–158. http://jramc.bmj.com / accessed 23 May 2014.

_____. *The Medical Response to Trench Diseases in World War One*. Newcastle-upon- Tyne: Cambridge Scholars Publishing, 2011.

_____. "Trench Fever: The British Medical Response to the Great War." *Journal of the Royal Society of Medicine*. November, 2006; **99**(11): 564–568. http://www.ncbi.nlm.nih.gov / accessed 2 December 2014.

"Bacteriology." *The Compact Oxford English Dictionary*. Complete Text Reproduced Micro-graphically. 2 volumes. Oxford: At the Clarendon Press. Volume **I**. A-O. P. 155.

Baird, James. "Sir Alexander Ogston and the Royal Army Medical Corps." *The Staphylococci, Proceedings of the Alexander Ogston Centenary Conference*. Aberdeen: Repub Erasmus University, 1981, pp. 22–32. http://www.google.com / accessed 7 October 2014.

Baird-Parker, A. C. "Staphylococci and Their Classification." Section I. *An Analysis of the Staphylococcus: Inherent Properties of the Organism. Annals of the New York Academy of Sciences*. Volume **128**: *The Staphylococci: Ecologic Perspectives*. July 1965: 4–25.

Ball, Michael Valentine. *Essentials of Bacteriology: Being a Concise and Systematic Introduction to the Study of Micro-Organisms*. Saunders' Question-Compends, No. 2. October, 1891. 5th edition. Revised by Karl M. Vogel. Philadelphia and London: W. B. Saunders Company, 1907.

Barnett, Brendan. "Pasteur, Pouchet, and Heterogenesis." www.pasteurbrewing.com / accessed 8 November 2014.

Bastian, Henry Charlton. "The Bearing of Experimental Evidence upon the Germ-Theory of Disease." *The British Medical Journal*. 12 January 1878; **1**(889): 49–52; http://www.ncbi. nlm.nih.gov / accessed 4 November 2014.

_____. *The Beginnings of Life*. London: Macmillan, 1872.

_____. "Epidemic and Specific Contagious Diseases: Considerations as to Their Nature and Mode of Origin." *The British Medical Journal*. 7 October 1871; **2**(562): 400–409. http:// www.ncbi.nlm.nih.org / accessed 5 November 2014.

_____. *Evolution and the Origin of Life*. London: Macmillan, 1874.

_____. *The Evolution of Life*. London: Methuen & Company, 1907.

_____. "Facts and Reasonings Concerning the Heterogeneous Evolution of Living Things. I-III" *Nature*. 30 June 7 and 14 July 1870: 170–177, 193–201, and 219–228. Volume **I** of *Evolution and the Spontaneous Generation Debate*. 6 volumes. Introduction and Editorial Selection by James E. Strick. Series Preface by John M. Lynch. Sterling, Virginia, and Bristol, UK: Thoemmes Press, 2001.

_____. *The Mode of Origin of Lowest Organisms*. London: Macmillan, 1871.

_____. *The Nature and Origin of Living Matter*. London: Fisher Unwin, 1905.

_____. "On the Conditions Favouring Fermentation and the Appearance of Bacilli, Micro-cocci, and Torulae in Previously Boiled Fluid" (Read 21 June 1877). *The Journal of the Linnean Society*. 1877. Volume **XIV**: 1–83. Reprinted in *The Origin of Life, Being an Account of Experiments with Certain Superheated Saline Solutions in Hermetically Sealed Vessels*. New York and London: G. P. Putnam's Sons, 1911.

_____. "On the Great Importance from the Point of View of Medical Science of the Proof That Bacteria and Their Allies Are Capable of Arising *De Novo*." *The Lancet*. 1903. **162** (4183): 1220–1224. Reprinted in *The Lancet*. Volume II for 1903. Edited by Thomas H. Wakley and Thomas Wakley, Jr. London: *The Lancet*, 1903.

_____. *The Origin of Life: Being an Account of Experiments with Certain Superheated Saline Solutions in Hermetically Sealed Vessels*. New York: G. P. Putnam, 1911.

_____. *Studies in Heterogenesis*. London: Methuen & Company, 1904.

"Battles of the Western Front, 1914–1918." *The Great War, 1914–1918*. http://www.greatwar. com.uk / accessed 12 October 2014.

Behring, Emil von. "Studies on the Mechanism of Immunity to Diphtheria in Animals" (1890). In *Milestones in Microbiology: 1546–1940*. Preface, Historical Introduction, and Translations by Thomas D. Brock. Washington, D.C." American Society for Microbiology Press [ASM], 1999, pp. 141–144.

_____, and Shibasaburo Kitasato. "The Mechanism of Immunity in Animals to Diphtheria and Tetanus" (1890). *Milestones in Microbiology: 1546 to 1940*, pp. 138–140.

Bennett, J. D. C. "Medical Advances Consequent to the Great War 1914–1918." *Journal of the Royal Society of Medicine*. November 1990. Volume **83**: 738–742.

Benton, Edward H. "British Surgery in the South African War: The Work of Major Frederick Porter." *Medical History*. 1977; 21: 275–290. http://www.ncbi.nlm.nih.gov / accessed 20 March 2016.

Bergey, David Hendricks. *Bergey's Manual of Determinative Bacteriology*. Edited by John G. Holt. 9th edition. Philadelphia and Maryland: Lippincott, Williams, & Wilkins, 1994.

Bergey's Manual of Systematics of Archaea and Bacteria [BMSAB]. Hoboken, New Jersey: John Wiley & Sons, 2014.

Bett, W. R. "Obituary Notice: Sir William Watson Cheyne, Bart, F.R.S. (1852–1932)." *Annals of the Royal College of Surgeons of England*. 1952; **11**(6): 364–366. http://www.ncbi.nlm. nih.gov / accessed 18 April 2014.

Billroth, Theodor. "Traumatic and Inflammatory Complications, which may Accidentally Befall Wounds." *On Surgical Pathology and Therapeutics: A Handbook for Students in Practitioners* (Lecture XXIV). Translated from the 8th edition. Volumes **I–II** (with Index). Volume **76** of The New Sydenham Society. London: The New Sydenham Society, 1878. https://www.books.google.com / accessed 30 June 2014.

_____. *Untersuchengen über die Vegetationisformen von Coccobacteria Septica*. Berlin: Georg Reimer, 1874.

"The Biological Laboratory at the International Health Exhibition." *The British Medical Journal*. 5 July 1885; **2**(1227): 27. http://www.ncbi.nlm.nih.gov / accessed 28 July 2015.

Blevins, Steve M., and Michael S. Bronze. "Robert Koch and the 'Golden Age' of Bacteriology." *International Journal of Infectious Diseases*. September, 2010; **14**(9): e744-e751. http:// www.ijidonline.com / accessed 4 July 2014.

Blondel-Megrilis, M. "Auguste Laurent and Alcaloids." *Revue d'histoire de pharmacie*. 2001; **49**(331): 303–314.

The Boer War, Selected Translations Pertaining to. April 1, 1905. Washington, D.C.: Government Printing Office, 1905. https://www.books.google.com / accessed 14 March 2016.

Bonnin, J. G., and W. R. LeFanu. "Joseph Lister 1827–1912: A Bibliographical Biography." *The Journal of Bone and Joint Surgery*. February, 1967; **49B** (1): 4–23. http://boneandjoint. org.uk / accessed 28 February 2014.

Bono, Michael Joseph. "Micoplasmal pneumonia." *Medscape*. Edited by Robert E. O'Connor. www.medscape.com / accessed 17 June 2013.

Bordet, Jules. "Leucocytes and the Active Property of Serum from Vaccinated Animals." Translated by Thomas D. Brock. *Milestones in Microbiology*, pp. 144–148.

Bowlby, Anthony, and Cuthbert Wallace. "The Development of British Surgery at the Front." *British Medicine in the War, 1914–1917, Being Essays on Problems of Medicine, Surgery and Pathology, arising among the British Armed Forces engaged in the War and the manner of their Solution*. Collected out of *The British Medical Journal*, April-October, 1917. London: The British Medical Journal, 1917, pp. 30–46.

Boyce, Rubert W. "Methods Employed in Bacteriological Research, with Special Reference to the Examination of Air, Water and Food." Part VI. of *Public Health Laboratory Work*.

By Henry Richard Kenwood. Philadelphia: P. Blakiston, Son & Company, 1893, pp. 425–482. https://www.archive.org / accessed 31 March 2015.

"Boyce, Rubert W. (1863–1911)." Obituary. *The British Medical Journal*. 1 July 1911; **2**(2635): 53–54. http://www.ncbi.nlm.nih.gov / accessed 31 March 2015.

Breed, Robert S., and George J. Hucker. "The Status of the Genus Micrococcus Cohn, 1872." *International Bulletin of Bacteriological Nomenclature and Taxonomy*. 15 July 1957; **7**(3): 113–116.

British Medicine in the War, 1914–1917. London: The British Medical Association, 1917. https:// www.archive.org / accessed 6 October 2015.

Brock, Thomas D. "Comment": "On the organized bodies which exist in the atmosphere; examination of the doctrine of spontaneous generation." By Louis Pasteur. *Milestones in Microbiology*, pp. 42–43.

_____. "Comment": "Report on the lactic acid fermentation." By Louis Pasteur. *Milestones in Microbiology*. P. 30.

_____. *Robert Koch: A Life in Medicine and Bacteriology*. 1988. Washington, D.C.: American Society for Microbiology [ASM], 1999.

_____, ed. Preface, Historical Introduction, Translations, and Comments by Thomas D. Brock. *Milestones in Microbiology: 1546 to 1940*.

Brunton, T. Lauder. "Physiology." In *A Handbook for the Physiological Laboratory*. Edited by John Burdon Sanderson. 1873. Philadelphia: P. Blakiston, Son & Company, 1884, pp. 421–557. https://www.archive.org / accessed 16 August 2015.

Buchanan, R. E. *General Systematics Bacteriology History: Nomenclature, Groups of Bacteria*. Baltimore: Williams & Wilkins Company, 1925. https://www.archive.org / accessed 4 July 2014.

Buchner, Hans Ernst August. "Sind die Alexine einfache oder complexe Körper?" *Berliner Klinische Wochenschrift*. 1901; **3**(38): 855.

_____. "Üeber die Nahere Natur der Bacterientötenden Substanz in Blutserum." *Zentralblatt für Bakteriologie, Mikrobiologie und Hygiene*. 1889; **6**: 561–572.

Bulloch, William. *The History of Bacteriology*. 2nd edition. 1938. New York: Dover, 1979.

_____. "Obituary Notice: Sir William Watson Cheyne, 1852–1932." *Fellows of the Royal Society*. December, 1932. http://rsbm.royalsocietypublishing.org / accessed 18 November 2014.

Bynum, William F. *Science and the Practice of Medicine in the Nineteenth Century*. Cambridge History of Science Series. Edited by George Basalla and Owen Hannaway. New York and Cambridge: Cambridge University Press, 1994.

Cagniard-Latour, Charles. "Memoir on Alcoholic Fermentation." *Milestones in Microbiology: 1546 to 1940*, pp. 20–24.

Caird, Francis M. "An Address on the Principles of Wound Treatment as Established by Lister and the Subsequent Modifications in Treatment." *The British Medical Journal*. 21 November 1914; **2**(2812): 872–874. http://www.ncbi.nlm.nih.gov / accessed 25 September 2014.

"Camera Lucida." *Webster's Seventh New Collegiate Dictionary*. Springfield, Massachusetts: G. & C. Merriam Company, Publishers, 1971. P. 120.

Cameron, Hector Charles. *Joseph Lister: The Friend of Man*. London: William Heinemann Medical Books, 1948.

_____. "Lord Lister, 1827–1912: An Oration. Delivered in the University of Glasgow on Commemoration Day 23rd June, 1914." Glasgow: James MacLehose and Sons, 1914. https:// www.archive.org / accessed 8 September 2015.

Cameron, Hector Clare. "Lord Lister and the Evolution of Wound Treatment during the Last Forty Years." *The British Medical Journal*. 6 April 1907; **1**(2414): 789–803. http://www.ncbi.nlm.nih.gov / accessed 21 October 2014.

Campbell, Neil A., Jane B. Reece, et al. *Biology*. 8th edition. San Francisco, Boston, New York: Pearson/Benjamin Cummings, 2008.

"Canada balsam." *Webster's Seventh New Collegiate Dictionary*. P. 121.

Carden-Coyne, Ana. *The Politics of War: Military Patients and Medical Power in the First World War*. New York and Oxford: Oxford University Press, 2014.

Cardile, Anthony P., et al. "Dakin's Solution Alters Macrophage Viability and Function In Vitro." Post-Abstract Session/*Studies of the Interface of Host-Microbial Interaction. IDWeek*. 2–6 October 2013. San Francisco, California. www.Idsa.confex.com /accessed 17 July 2015.

Carling, P. C., and A. G. N. Moore. "Lister's Granuligera." *Journal of Hospital Infection*. 2009; **73**(1): 82–84. http://www.cabdirect.org / accessed 6 July 2014.

Carr, Kevin. "What Happens to an Animal Cell in a Hypotonic Solution?" *Synonym*. http://classroom.synonym.com / accessed 1 December 2014.

Carrel, Alexis, and Georges Dehelly. *The Treatment of Infected Wounds*. Translated by Herbert Child. Introduction by Sir Anthony Bowlby. New York: Paul B. Hoeber, 1917; London: Forgotten Books, 2012.

_____, and A. Hartmann. "Cicatrization of Wounds: VII. Sterilization of Wounds with Chloramine-T." *Journal of Experimental Medicine*. 1 July 1917; **26**(1): 95–118. http://www.ncbi.nlm.nih.gov / accessed 25 September 2014.

"Carrel, Alexis (1873–1941): Biography." *American Association for Thoracic Surgery*. http://www.aats.org / accessed 24 September 2014.

Carter, K. Codell. Preface, Introduction, and Translations. *Essays of Robert Koch*. Contributions in Medical Studies, Number 20. New York, Westport, Connecticut, and London, 1987.

_____. *The Rise of Causal Concepts of Disease: Case Histories*. The History of Medicine in Context. Hants, U.K. and Burlington, Vermont: Ashgate, 2003.

"Cassal, F. I. C., Mr. Charles Edward." [Death Notice]. *The British Medical Journal*. 31 December 1921; **2**(3183): 1136. http://www.ncbi.nlm.nih.gov/ accessed 7 March 2016.

Cathcart, Charles W. "Clockwork Control of the Carrel-Dakin Treatment." *The British Medical Journal*. 21 November 1925; **2**(3386): 933–934. http://www.ncbi.nlm.nih.gov / accessed 25 September 2014.

Cavaillon, Jean-Marc. "The Historical Milestones in the Understanding of Leukocytic Biology Initiated by Elie Metchnikoff." *Journal of Leukocyte Biology*. Published Online 31 May 2011; **90**(3): 413–424. http://www.jleukbio.org / accessed 26 December 2014.

Chaudurí, Keya, and S. N. Chatterjee. *Cholera Toxins*. New York: Springer Publishing Company, 2009.

Cheatle, Lenthal G. "The War in South Africa: A First Field Dressing." *The British Medical Journal*. 8 September 1900; **2**(2071): 668. http://www.ncbi.nlm.nih.gov / accessed 12 March 2016.

Chesebasi, A., K. Oelhafen, B. J. Shayota, Z. Klaassen, R. S. Tubbs, and M. Loukas. "A Historical Perspective: Bernhard von Langenbeck, German Surgeon (1810–1887)." *Clinical Anatomy*. October, 2014; 7: 972–975. http://www.ncbi.nlm.nih.gov / accessed 10 March 2015.

Cheyne, William Watson. "Abstract of an 'Address on the Value of Tuberculin in the Treatment of Surgical Tuberculosis.'" *The British Medical Journal*. 2 May 1891; **1**(1583): 951–961. http://www.ncbi.nlm.nih.gov / accessed 2 March 2015.

_____. "Abstract of 'Lectures on the Treatment of Surgical Tuberculous Diseases.'" *The British Medical Journal*. 2 July 1892; **2**(1644): 11–16. http://www.ncbi.nlm.nih.gov / accessed 16 August 2014.

_____. "Abstract of the Report on the Relation of Microorganism to Tuberculosis: Presented to the Association for the Advancement of Medicine by Research on Feb.[ruary] 1st, 1883." *The Lancet*. 17 March 1883; **121**(3107): 444–445.

_____. *Antiseptic Surgery: Its Principles, Practice, History, and Results*. London: Smith, Elder, & Company, 1882. https://www.archive.org / accessed 16 July 2014.

_____. "The Bacillus of Tubercle." *The British Medical Journal*. 24 January 1885; **1**(1256): 169–171. http://www.ncbi.nlm.nih.gov / accessed 6 May 2014.

_____. "Bacteriology." *The American Journal of the Medical Sciences* [*The International Journal of the Medical Sciences*]. Edited by Isaac Minis Hays and Malcolm Morris. July, 1886. New Series; Volume **92** (183): 66–93. Philadelphia: Lea Brothers & Company, 1886. https:// www.archive.org / accessed 20 March 2015.

_____. "Bacteriology: Relation of Bacteria to the Living Animal Body." *The American Journal of the Medical Sciences* [*The International Journal of the Medical Sciences*]. Edited by Isaac Minis Hays and Malcolm Morris. January, 1887. New Series—Volume **93** (185): 101–121. Philadelphia: Lea Brothers & Company, 1887. https://www.archive.org / accessed 26 March 2015.

_____. "Bacteriology: Relations of Bacteria to the Soil on which They Grow." *The American Journal of the Medical Sciences* [*The International Journal of the Medical Sciences*]. Edited by Isaac Minis Hays and Malcolm Morris. Volume **92** (184): 346- 363—New Series. Philadelphia: Lea Brothers & Company, 1886. https://www.archive.org / accessed 30 January 2015.

_____. "Bacteriology: II. Study of Bacteria by Means of Cultivation." *The American Journal of the Medical Sciences* [*International Journal of the Medical Sciences*]. Edited by Isaac Minis Hays and Malcolm Morris. July, 1887. Volume **94** (187): 69–107. https://www.archive. org / accessed 26 March 2015.

_____. "The Bradshaw Lecture on *The Treatment of Wounds*." London: J. Bale Sons & Danielsson, 1908. https://www.archive.org; http://www.watson-cheyne.com / accessed 17 April 2014.

_____. Case Reports #84 & #87 (1900). In *Report on the Surgical Cases Noted in the South African War, 1899–1902*. By William Flack Stevenson. London: His Majesty's Stationery Office; Harrison and Sons, 1905. P. 92. https://www.books.google.com / accessed 12 March 2016.

_____. "Correspondence: The Cholera-Bacillus of Koch." *The British Medical Journal*. 11 April 1885; **1**(1267): 756–757. http://www.ncbi.nlm.nih.gov / accessed 3 July 2014.

_____. "Correspondence: The Recommendations of the Naval Committee on the Treatment of Wounds in War." *The British Medical Journal*. 22 May 1915; **1** (2838): 912. http:// www.ncbi.nlm.nih.gov / accessed 10 September 2014.

_____. "A Discussion on the Treatment of Tuberculous Joint Disease in Children." *Proceedings of the Royal Society of Medicine*. 1912; **5** (Sect. Study Dis. Children): 109–111. http:// www.ncbi.nlm.nih.gov / accessed 2 March 2015.

_____. "The Disinfection of Hands and Instruments." *The British Medical Journal*. 7 August 1897; **2**(1910): 373–374. http://www.ncbi.nlm.nih.gov / accessed 11 May 2014.

_____. "Extracts from 'Three Lectures on Tubercular Diseases of Bones and Joints.'" *The British Medical Journal*. Lecture I. 29 November 1890; **2**(1561): 1227–1229. http://www. ncbi.nlm.nih.gov / accessed 17 September 2014.

_____. "Extracts from 'Three Lectures on Tubercular Diseases of Bones and Joints.'" *The British Medical Journal*. Lecture I. 6 December 1890; **2**(1562): 1283–1286. http://www. ncbi.nlm.nih.gov / accessed 17 September 2014.

_____. "Extracts from 'Three Lectures on Tubercular Diseases of Bones and Joints.'" *The British Medical Journal*. Lecture II. 13 December 1890; **2**(1563): 1348–1353. http:// www.ncbi.nlm.nih.gov / accessed 17 September 2014.

_____. "Extracts from 'Three Lectures on Tubercular Diseases of Bones and Joints.'" *The British Medical Journal*. Lecture III. 20 December 1890; **2**(1564): 1418–1422. http:// www.ncbi.nlm.nih.gov / accessed 17 September 2014.

_____. "Gangrene." *A System of Surgery*. Volume **1**: Pp. 128–153.

_____. "The Hunterian Oration on *The Treatment of Wounds in War*." *The Lancet*. 27 February 1915; **185** (4774): 419–430. http://www.watson-cheyne.com / accessed 14 October 2014.

_____. "Inflammation." *A System of Surgery*. Edited by Frederick Treves. 2 volumes. Philadelphia: Lea Brothers & Company, 1895–1896. Volume **1**: Pp. 53–78.

_____. "Lectures on the Pathology of Tuberculous Diseases of Bones and Joints." *The British Medical Journal*. 4 April 1891; **1**(1579): 739–743. http://www.ncbi.nlm.nih.gov / accessed 1 September 2014.

_____. "Lectures on the Pathology of Tuberculous Diseases of Bones and Joints." *The British Medical Journal*. 11 April 1891; **1**(1580): 790–795. http://www.ncbi.nlm.nih.gov / accessed 1 September 2014.

_____. "Lectures on the Pathology of Tuberculous Diseases of Bones and Joints." *The British Medical Journal*. 18 April 1891; **1**(1581): 840–844. http://www.ncbi.nlm.nih.gov / accessed 10 September 2014.

_____. "Lectures on the Pathology of Tuberculous Diseases of Bones and Joints." *The British Medical Journal*. 25 April 1891; **1**(1582): 896–901. http://www.ncbi.nlm.nih.gov / accessed 1 September 2014.

_____. "Lectures on Suppuration and Septic Diseases." Lecture I. *The British Medical Journal*. 4 February 1888; **1**(1414): 404–409. http://www.ncbi.nlm.nih.gov / accessed 25 January 2015.

_____. "Lectures on Suppuration and Septic Diseases." Lecture II. *The British Medical Journal*. 3 March 1888; **1**(1418): 452–458. http://www.ncbi.nlm.nih.gov / accessed 25 January 2015.

_____. "Lectures on Suppuration and Septic Diseases." Lecture III. *The British Medical Journal*. 10 March 1888; **1**(1419): 524–530. http://www.ncbi.nlm.nih.gov / accessed 25 January 2015.

_____. *Lister and His Achievement*. London: Longmans, Green, and Company, 1925.

_____. "Lister, the Investigator and Surgeon." *The British Medical Journal*. 16 May 1925; **1**(3359): 923–926. http://www.ncbi.nlm.nih.gov / accessed 7 May 2014.

_____. "Listerism and the Development of Operative Surgery." *The British Medical Journal*. 13 December 1902; **2**(2189): 1851–1852. http://www.ncbi.nlm.nih.gov / accessed 16 August 2014.

_____. *Manual of the Antiseptic Treatment of Wounds for Students and Practitioners*. New York: J. H. Vail & Company, 1885.

_____. "Micro-organisms in Disease." *Report of Societies: Royal Medical and Chirurgical Society* (23 May 1882). *The British Medical Journal*. 27 May 1882; **1**(1117): 777–783, esp. 778. http://ncbi.nlm.nih.gov / accessed 8 March 2016.

_____. "Observations on the Treatment of Wounds of War." *The Lancet*. 31 July 1915; **186** (4796): 213–29. London: John Bale, Sons and Danielsson, 1915. http://www.watson-cheyne.com / accessed 14 October 2014.

_____. "On Early Tracheotomy in Diphtheria." Reports of the Collective Committee of the British Medical Association. *The British Medical Journal*. 5 March 1887; **1**(1366): 504–506. http://www.ncbi.nlm.nih.gov / accessed 8 March 2016.

_____. "On 'Foulbrood.'" *The British Medical Journal*. 10 October 1885; **2**(1293): 697. http://www.ncbi.nlm.nih.gov / accessed 13 March 2015.

_____. "On the Progress and Results of Pathological Work." *The British Medical Journal*. 4 September 1897; **2**(1914): 586–589. http://www.ncbi.nlm.nih.gov / accessed 1 September 2014.

_____. "On the Relation of Organisms to Antiseptic Dressings." *Transactions of the Pathological Society of London. Report of the Proceedings for the Session, 1878–1879*. 1879; Volume **30**: 557–582. London: J. E. Adlard, 1879. https://www.archive.org / accessed 28 April 2014.

_____. "On the Treatment of Tuberculous Diseases in Their Surgical Aspect, Being the Harveian Lecture for 1899." London: John Bale, Sons & Danielson, 1900. Lecture I: 3–44.

_____. "On the Treatment of Wounds in War." *The British Medical Journal*. 30 November 1901; **2**(2135): 1591–1594. http://www.ncbi.nlm.nih.gov / accessed 10 September 2014.

_____. "On the Value of Tuberculin in the Treatment of Surgical Tubercular Diseases." *Medico-*

Chirurgical Transactions. 1891; **74**: 235–340. http://www.ncbi.nlm.nih.gov / accessed 2 March 2015.

_____. "The Organisation of Medical Aid in a Great War." *The British Medical Journal.* 22 June 1901; **1**(2112): 1558–1560. http://www.ncbi.nlm.nih.gov / accessed 6 May 2014.

_____. "Public Health Laboratory Work: Part I.—Biological Laboratory, with Catalogue of the Exhibits in the Laboratory." In *Public Health Laboratory Work.* By W. Watson Cheyne, W. H. Corfield, and Charles E. Cassal. Published for the Executive Council of the International Health Exhibition [London, 1884], and for the Council of the Society of the Arts. London: William Clowes and Sons, 1884, pp. 3–37. https://www.archive.org / accessed 9 April 2015.

_____. "Remarks on the Treatment of Wounds in Connexion with the Recent Results Obtained at St. George's Hospital." *The Lancet.* 1 January 1910. London: John Bale, Sons and Danielsson, 1910. http://www.watson-cheyne.com / accessed 14 March 2015.

_____. "Remarks on the Treatment of Wounds in Wars." *The British Medical Journal.* 21 November 1914; **2**(2812): 865–871. http://www.ncbi.nlm.nih.gov / accessed 10 September 2014.

_____. "Reminiscences of 'The Chief.'" *Joseph, Baron Lister: Centenary Volume, 1827–1927.* Edited for the Lister Centenary Committee of the British Medical Association. Edited and Foreword by A. Logan Turner. "Biographical Sketch (1827–1912)," by George Thomas Beatson. Edinburgh: Oliver and Boyd, 1927, pp. 122–128.

_____. "Report on the Cholera-Bacillus." *The British Medical Journal.* 25 April 1885; **1**(1269): 821–823. http://www.ncbi.nlm.nih.gov / accessed 8 August 2014.

_____. "Report on the Cholera-Bacillus." *The British Medical Journal.* 2 May 1885; **1**(1270): 877–879. http://www.ncbi.nlm.nih.gov / accessed 8 August 2014.

_____. "Report on the Cholera-Bacillus." *The British Medical Journal.* 9 May 1885; **1**(1271): 931–936. http://www.ncbi.nlm.nih.gov / accessed 8 August 2014.

_____. "Report on the Cholera-Bacillus." *The British Medical Journal.* 16 May 1885; **1**(1272): 975–977. http://www.ncbi.nlm.nih.gov / accessed 8 August 2014.

_____. "Report on the Cholera-Bacillus." *The British Medical Journal.* 23 May 1885; **1**(1273): 1027–1031. http://www.ncbi.nlm.nih.gov / accessed 3 July 2014.

_____. "Report on Micrococci in Relation to Wounds, Abscesses, and Septic Processes." *The British Medical Journal.* 20 September 1884; **2**(1238): 553–556. http://www.bmj.com / accessed 3 July 2014.

_____. "Report on Micrococci in Relation to Wounds, Abscesses, and Septic Processes." *The British Medical Journal.* 27 September 1884; **2**(1239): 599–603. http://www.ncbi.nlm.nih.gov / accessed 3 July 2014.

_____. "Report on Micrococci in Relation to Wounds, Abscesses, and Septic Processes." *The British Medical Journal.* 4 October 1884; **2**(1240): 645–648. http://www.ncbi.nlm.nih.gov / accessed 3 July 2014.

_____. "Report on a Study of Certain of the Conditions of Infection." *The British Medical Journal.* 31 July 1886; **2**(1335): 197–207. http://www.ncbi.nlm.nih.gov / accessed 28 February 2015.

_____. "Report to the Association for the Advancement of Medicine by Research on the Relation of Micro-Organisms to Tuberculosis." 1 February 1883. http://www.watson-cheyne.com / accessed 5 March 2014.

_____. "Report to the Royal Medical and Chirurgical Society" (Tuesday, March 24th, 1885). *The British Medical Journal.* 28 March 1885; **1**(1265): 654–656. http://www.ncbi.nlm.nih.gov / accessed 7 August 2014.

_____. "Sir Almroth Wright's Lecture on the Treatment of Wounds in War." *The Lancet.* 8 May 1915; **185**(4784): 961–962.

_____. "Statistical Report of All Operations Performed on Healthy Joints in Hospital Practice, by Mr. Lister, from September 1871 to the Present Time, Together with Such Accidental

Wounds of Joints as Occurred During the Same Period." *The British Medical Journal*. 29 November 1879; **2**(987): 859–864. http://www.ncbi.nlm.nih.gov / accessed 8 March 2016.

_____. "Suppuration." *A System of Surgery*. Volume **1**: Pp. 79–104.

_____. *Suppuration and Septic Diseases: Three Lectures Delivered at the Royal College of Surgeons of England in February 1888*. Edinburgh & London: Young J. Pentland, 1889.

_____. *Three Orations: The Lister Centenary*. London: John Bale, Sons and Danielsson, 1927. 83–92. http://www.watson-cheyne.com / accessed 14 October 2014.

_____. "The Treatment of Wounds in War." *The Lancet*. 1915; **186**(4803): 671–672.

_____. "The Treatment of Wounds in War." *The British Journal of Surgery*. 1915; **3**(11): 427–450.

_____. *The Treatment of Wounds, Ulcers, & Abscesses*. Philadelphia: Lea Brothers & Company, 1895.

_____. *Tuberculous Disease of Bones and Joints; Its Pathology, Symptoms, and Treatment*. Edinburgh and London: Young J. Pentland; Philadelphia: J. B. Lippincott Company, 1895. https://www.books.google.com / accessed 24 May 2014.

_____. "Ulceration." *A System of Surgery*. Volume **1**: Pp. 105–127.

_____. "The Value of Tuberculin in Surgery." *The British Medical Journal*. 9 May 1891; **1**(1584): 1043. http://www.ncbi.nlm.nih.gov / accessed on 2 March 2015.

_____. "The War in South Africa: The March from Modder River to Bloemfontein." *The British Medical Journal*. 5 May 1900; 1(2053): 1093–1096. http://www.ncbi.nlm.nih.gov / accessed 10 March 2016.

_____. "The War in South Africa. The Wounded from the Actions between Modder and Driefontein." *The British Medical Journal*. 12 May 1900; **1**(2054): 1193–1198. http://www.ncbi.nlm.nih.gov / accessed 6 May 2014.

_____, and F. F. Burghard. *Manual of Surgical Treatment*. 7 vols. Philadelphia and New York: Lea Brothers & Company, 1899–1913. Volume **1**, pp. 205–223. http://www.hathitrust.org / accessed 22 July 2014.

_____, and Frank R. Cheshire. "The Pathogenic History and History under Cultivation of a New Bacillus (*B. alvei*), the Cause of a Disease of the Hive Bee hitherto Known as Foul Brood." *Journal of the Royal Microscopical Society*. Transactions of the Society. August 1888; **XI**: 581–601.

_____, et al. "Testimony: 'Dogs' Protection Bill.'" Transcribed by M. David. *Parliamentary Intelligence*. House of Commons. Wednesday, 17 March 1920. *The Lancet*. 27 March 1920; **195**(5039): 739–741.

_____, editor. *Recent Researches on Micro-organisms in Relation to Suppuration and Septic Diseases*. Preface, Translated, and Edited by W. Watson Cheyne. Volume **115**. London: The New Sydenham Society, 1886. https://www.books.google.com / accessed 22 July 2014.

"Cheyne [W. Watson] on Tubercle-Bacilli." *Medical Times and Gazette: A Journal of Medical Science, Literature, Criticism, and News*. Volume **1**. 31 March 1883. London: J & A. Churchill, 1883. https://www.books.google.com / accessed 31 August 2014.

"Cheyne, Mr. Watson." *The Zoophilist*. 2 April 1883; Volume II, New Series **4**: 52. https://www.books.google.com / accessed 19 February 2015.

"Cheyne, Sir William Watson." *Fetlar Interpretive Centre*. http://www.watson-cheyne.com / accessed 1 December 2014.

Childs, Christopher. "Bacteriological Examinations of Air, Water, Soil, Food, &c." In *Public Health Laboratory Work*. By Henry Richard Kenwood. 2nd edition. London: H. K. Lewis, 1896, pp. 425–444. https://www.archive.org / accessed 9 April 2015.

"Chinosol [potassium oxychinolin sulphonate]." *A Handbook of Antiseptics*. By Henry Drysdale Dakin and Edward Kellogg Dunham. New York: The Macmillan Company, 1918. P.75. https://www.google.books.com / accessed 16 May 2014.

"Cholera." *World Health Organization*. Media Centre: Fact Sheet No. 107. February, 2014. http://www.who.int / accessed 4 August 2014.

"Cholera in Egypt" [Special Correspondent]. *The British Medical Journal*. 14 December 1899; **2**(1824): 1505–1506. http://www.ncbi.nlm.nih.gov / accessed 19 November 2014.

Clark, Franklin C. "A Brief History of Antiseptic Surgery." *Medical Library and Historical Journal*. September, 1907; **5**(3): 145–172. http://www.ncbi.nlm.nih / accessed 2 July 2015.

Clark, P. F. "Joseph Lister: His Life and Work." *The Scientific Monthly*. December, 1920; **11**(6): 518–539.

Clarke, H. T. "Henry Drysdale Dakin, 1880–1952." *The Journal of Biological Chemistry*. 1952; **198**:491–494. http://www.jbc.org / accessed 24 September 2014.

Clarke, John. "Antiseptic Surgery—Lister's Spray." In *Anti-Vivisection Evidences: a Collection of Authentic Statements by Competent Witnesses as to the Immorality, Cruelty, and Futility of Experiments on Living Animals*. Edited by Benjamin Bryan. London: Society for the Protection of Animals from Vivisection, 1895, pp. 40–42. https://www.archive.org / accessed 10 August 2015.

_____. "Returns, Value of the Official." In *Anti-Vivisection Evidences*, pp. 217–218. https:// www.archive.org / accessed 10 August 2015.

Cohn, Ferdinand. *Bacteria: The Smallest Living Organisms*. 1872. Translated by Charles S. Dolley (1881). Introduction by Morris C. Leikind. Baltimore: The Johns Hopkins University Press, 1939.

_____. "Studies on Bacteria" (1875). *Milestones in Microbiology: 1546–1940*, pp. 210–215.

_____. "Studies on the Biology of the Bacilli" (1876). *Milestones in Microbiology: 1546–1940*, pp. 49–56.

Cohnheim, Julius Friedrich. *Lectures on General Pathology: A Handbook for Practitioners and Students*. Translated, Preface, and Memoir by Alexander B. McKee. 2nd edition. London: The New Sydenham Society, 1889. http://babel.hathitrust.org / accessed 30 April 2015.

_____. *Üeber Entzündung und Eiterung (On Inflammation and Suppuration)* (1867). *Klassiker der Medizin*. Herausgegeben von Karl Sudhoff. Eingeleitet von Rudolf Benecke. Leipzig: Verlag von Johann Ambrosius Barth, 1914. http://babel.hathitrust.org / accessed 30 April 2015.

Coleridge, Stephen. "The Cruelty of Antivivisection." *The British Medical Journal*. 15 June 1901; **2**(2111): 1511–1513. http://www.ncbi.nlm.nih.gov / accessed 31 August 2014.

_____. "King's College Hospital, A New Departure." *The Zoophilist and Animals' Defender*. 2 May 1904; **24**(2): 129–130. https://www.books.google.com / accessed 31 August 2014.

Coley, William B. "The Treatment of Inoperable Sarcoma with the Mixed Toxins of Erysipelas and *Bacillus Prodigiosus*: Immediate and Final results in One Hundred and Forty Cases." *The Journal of the American Medical Association*. 27 August 1898; **31**(9): 456–465.

_____. "The Treatment of Malignant Tumors by Repeated Inoculations of Erysipelas: With a Report of Ten Original Cases." *The American Journal of the Medical Sciences*. May, 1893; **105**(5): 487–541.

Collingwood, R. G. *The Idea of History*. 1946; New York, London, and Oxford: Oxford University Press, 1975.

Cooter, Roger, Mark Harrison, and Steve Sturdy. *Papers presented at the Conference "Medicine and the Management of Modern Warfare," London, July 1995*. Wellcome Institute Series in the History of Medicine. Amsterdam and Atlanta, Georgia: Rodopi, 1999.

Corfield, W. H. *The Etiology of Typhoid Fever and Its Prevention, Being the Milroy Lectures*, Delivered at the Royal College of Physicians in 1902. London: H. K. Lewis, 1902.

Coutts, Jane. "Illustrating Microorganisms: Sir William Watson Cheyne (1852–1932) and Bacteriology." *Journal of Medical Biography*. Online Publication: 19 February 2015. http:// www.jmb.sagepub.com / accessed 27 March 2015.

_____. *Microbes and the Fetlar Man: The Life of Sir William Watson Cheyne*. Edinburgh: Humming Earth Publishers, 2015.

Cowan, S. T. "Staphylococcus Aureus versus Staphylococcus Pyogenes." *International Bulletin of Bacteriological Nomenclature and Taxonomy*. 15 July 1956. **6**(3): 99–100.

Cox, F. E. G. "History of Human Parasitology." *Clinical Microbiology Reviews*. October 2002; **15**(4): 595–612. http://www.ncbi.nlm.nih.gov / accessed 15 March 2006.

Creed, R. S. "Pasteur and Lister." *British Medical Journal*. 12 July 1936; **2**(3942): 200. http://www.ncbi.nlm.nih.gov / accessed 17 August 2015.

"Cresol, All Isomers, Occupational Safety and Health Guidance for." *The Centers for Disease Control. United States Department of Health & Human Services*. 1988: 1–6. www.cdc.gov / accessed 20 August 2015.

Crookshank, Edgar M. *Manual of Bacteriology*. 2nd edition. London: H. K. Lewis, 1887. https://www.biodiversitylibrary.org / accessed 8 April 2015.

_____. "Remarks on the Antiseptic Treatment of the Wounded on the Battle-Field." *The Lancet*. 8 March 1884; **123**(3158): 422. https://www.books.google.com / accessed 3 March 2014.

_____. *A Text-Book of Bacteriology*. 2nd edition. Revised. London: H. K. Lewis, 1887. Reprint. London: Forgotten Books, 2013.

_____. *A Text-Book of Bacteriology*. 4th edition. London: H. K. Lewis, 1896. https://www.archive.org / accessed 19 April 2015.

"Cruelty to Animals/Anti-Vivisection Act 1876." Great Britain Parliament. *Animal Rights History*. www.animalrightshistory.org / accessed 8 July 2012.

"The Cruelty of Antivivisectionism." *The British Medical Journal*. 15 June 1901; **2**(2111): 1511–1513. http://www.ncbi.nlm.nih.gov / accessed 31 August 2015.

"Cruelty Prevention Act, 1879." The Sessional Papers Printed by the Order of the House of Lords. (*42 & 43 Victoriae*). Volume **3** of 4 volumes. Arm-Gre: Public Bills, 1878–1879. Oxford: Oxford University Press, 1879, pp. 392–398.

Cruse, Julius M., and Robert E. Lewis. "The Complement System." *Atlas of Immunology*. Boca Raton, Florida, London, and New York: Taylor & Francis Group, 1999, pp. 207–224.

Cunningham, David Douglas, and Timothy R. Lewis. "Scientific Investigation into the Causes of Cholera. I. Report of Interviews with Professor Ernst Hallier, at Jena, Oct. 1868." *The Lancet*. Edited by James G. Wakley. Volume **1**. London: John James Croft, 2 January 1869, pp. 3–4. https://www.books.google.com / accessed 1 July 2015.

_____. "Scientific Investigation into the Causes of Cholera. II." *The Lancet*. Edited by James G. Wakley. Volume **1**. London: John James Croft, 9 January 1869, pp. 38–41. https://www.books.google.com / accessed 1 July 2015.

_____. "Scientific Investigation into the Causes of Cholera. III." *The Lancet*. Edited by James G. Wakley. Volume **1**. London: John James Croft, 16 January 1869, pp. 76–78. http://www.books.google.com / accessed 1 July 2015.

Dakin, Henry Drysdale. "The Antiseptic Action of Hypochlorites: The Ancient History of the 'New Antiseptic.'" *The British Medical Journal*. 4 December 1915; **2**(2866): 809–810. http://www.ncbi.nlm.nih.gov / accessed 16 May 2014.

_____. "Biochemistry and War Problems." *The British Medical Journal*. 23 June 1917; **1**(2947): 833–837. In *British Medicine in the War, 1914–1917*, pp. 11–14.

_____. "On the Use of Certain Antiseptic Substances in the Treatment of Infected Wounds." *The British Medical Journal*. 28 August 1915; **2**(2852): 318–320. http://www.ncbi.nlm.nih.gov / accessed 25 September 2014.

_____, and Edward Kellogg Dunham. *A Handbook on Antiseptics*. New York: The Macmillan Company, 1918. http://www.books.google.com / accessed 16 May 2014.

Daniel, Thomas M. "The History of Tuberculosis: Historical Review." *Respiratory Medicine*. November, 2006; **100**(11): 1862–1870. http://www.sciencedirect.com / accessed 24 August 2014.

Darwin, Charles. *Charles Darwin's Notebooks, 1836–1844: Geology, Transmutation of Species, Metaphysical Enquiries*. Transcribed and Edited by Paul H. Barrett, Peter J. Gautrey, Sandra Herbert, David Kohn, and Sydney Smith. Ithaca, New York, and London: British Museum (Natural History) and Cornell University Press, 1987.

De, Sambhu N., J. K. Sakar, and B. P. Tribedi. "An Experimental Study of the Action of Cholera Toxin." *Journal of Pathology and Bacteriology.* 1951; **63**: 707–717.

De, Sambhu N. *Cholera: Its Pathology and Pathogenesis.* London: Oliver and Boyd, 1961.

_____. "Enterotoxicity of Bacteria-free Culture-filtrate of Vibrio cholera." *Nature.* 1959; **183**: 1533–1534.

"Death of an Eminent Surgeon: Sir William Watson Cheyne, former Associate of Lister." *The Glasgow Herald.* 21 April 1932; **96**:11. https://www.news.google.com / accessed 7 July 2015.

Debré, Patrice. *Louis Pasteur.* Translated by Elborg Forster. 2nd edition. Baltimore: The Johns Hopkins University Press, 1998.

"Declaration on the Use of Bullets Which Expand or Flatten Easily in the Human Body." The Hague Conference. Declaration IV.3 (29 July 1899). *The Avalon Project: Documents in Law, History and Diplomacy.* Yale Law School. http://avalon.law.yale.edu / accessed 4 October 2014.

Déclat, Gilbert. *Nouvelle Applications de l'Acide Phénique en Medicine et en Chirurgie.* Paris and London: Hippolyte Baillière, 1865.

DeGroote, M.A., et al. "Comparative Studies Evaluating Mouse Models Used for Efficacy Testing of Experimental Drugs against Mycobacterium Tuberculosis." *Antimicrobial Agents & Chemotherapy.* March, 2011; **55**(3): 1237–1247. http://www.ncbi.nlm.nih.gov / accessed 9 September 2014.

Denys, J., and J. Havet. "Sur la part des globules blanc dans le Pouvoir bactericide du sange de chien." *La Cellule.* 1894; Volume **10**: 7–35.

DePage, Antoine. "General Considerations as to the Treatment of War Wounds." *Annals of Surgery.* June, 1919; **69**(6): 575–588. http://www.ncbi.nlm.nih.gov / accessed 16 May 2014.

DePaolo, Charles. "Pasteur and Lister: A Chronicle of Scientific Influence." *The Victorian Web: Literature, History, & Culture in the Age of Victoria.* Editor-in-Chief and Webmaster, George P. Landow. http://www.victorianweb.org.

_____. "Periodization." *Encyclopedia of Time.* Edited by Samuel L. Macey. Garland Reference Library of Social Science. Volume **810**. New York and London: Garland Publishing, Inc., 1994, pp. 446–447.

DePaolo, Patrick A. "Sir William Watson Cheyne (1852–1932): Bacteriologist and Surgeon." *The Victorian Web: Literature, History, & Culture in the Age of Victoria.* Editor-in-Chief and Webmaster, George P. Landow. http://www.victorianweb.org / accessed 19 October 2014.

De Villiers, J. C. "The Medical Aspect of the Anglo-Boer War, 1899–1902: Part I." *Military History Journal.* December, 1983; 6(2). *South African Military History Society.* http://samilitaryhistory.com / accessed 20 March 2016.

_____. "The Medical Aspect of the Anglo-Boer War, 1899–1902: Part II." *Military History Journal.* June, 1984; 6(3). *South African Military History Society.* http://samilitaryhistory.com / accessed 20 March 2016.

Dimond, Lyn, and Robert McQueen. "The Carrel-Dakin Treatment and a Method for Its Application on an Extensive Scale." *The British Medical Journal.* 22 September 1917; **2**(2960): 387–388. http://www.ncbi.nlm.nih.gov / accessed 25 September 1914.

Doerr, Stephen. "Abscess." *eMedicinehealth.* Edited by Melissa Conrad Stöppler. http://www.emedicinehealth.com / accessed 22 January 2015.

"Dogs' Protection Bill." *The British Medical Journal.* 3 May 1919; **1**(3044): 552–553. http://www.ncbi.nlm.nih.gov / accessed 21 October 2014.

Doyle, Arthur Conan. *The Great Boer War, with Maps.* London: Smith, Elder, & Company, 1900. https://www.books.google.com / accessed 11 March 2016.

Drews, Gerhart. "Ferdinand Cohn, a Founder of Modern Microbiology." *ASM News.* August, 1999; **65**(8). http://201.114.65.51 / accessed 19 May 2015.

Dubos, René. *Louis Pasteur: Free Lance of Science.* New York: Da Capo, 1950.
_____. *Pasteur and Modern Science.* Garden City, New York: Anchor Books, Doubleday and Company, 1960.
Dutta, A. C. *Botany for Degree Students.* 2nd edition. Bombay, Calcutta, Madras: Oxford University Press, 1968.
Eardley, W. G. P., K. V. Brown, T. J. Bonner, A. D. Green, and J. S. Clasper. "Infection in Conflict Wounded." *Philosophical Transactions of the Royal Society. B: Biological Sciences.* 13 December 2010; **366**: 204–218. http://rstb.royalsocietypublishing.org / accessed 1 May 2014.
Edmunds, Arthur. "The Late Sir Watson Cheyne." Obituary. *The Lancet.* 7 May 1932; (originally: **1**[5691]); **219**(5691): 1013.
Ehrlich, Paul. "On Immunity with Especial Reference to the Relations Existing between the Distribution and the Action Antigens." *Harben Lectures,* Lecture I. *A Journal of the Royal Institute of Public Health: A Journal of Preventive Medicine.* Edited by William R. Smith. January-December 1907. **15**(6): 321–340. https://www.books.google.com / accessed 11 January 2015.
Eisler, Ronald. "Mercury Uses and Sources" [Corrosive Sublimate]." *Eisler's Encyclopedia of Environmentally Hazardous Priority Chemicals.* Amsterdam and London: Elsevier, 2007, pp. 409–421.
Esmarch, Erwin von. "Fifth Congress of the Society of German Surgeons." *The London Medical Record: A Review of the Progress of Medicine, Obstetrics, and the Allied Sciences.* 15 June 1876; Volume **IV**. London: Smith, Elder, & Company, 1876, pp. 233–281. https://www.books.google.com / accessed 15 December 2014.
Esmarch, J. Friedrich von. *Resection in Gunshot Injuries.* Translated by S. F. Statham. Philadelphia: J. B. Lippincott & Company, 1862. https://www.archive.org / accessed 8 March 2015.
"Eusol." *The Compact Oxford English Dictionary.* Volume **II**. P-Z; Supplement: P. 3958.
Euzeby, J. P. *Vibrio Cholerae. List of Prokaryotic Names with Standing in Nomenclature.* https://www.google.com / accessed 16 August 2014.
Evans, Alfred S. "Causation and Disease: The Henle-Koch Postulates Revisited." *The Yale Journal of Biology and Medicine.* May, 1976; **49**(2): 175–195. http://www.ncbi.nlm.nih.gov / accessed 19 August 2014.
_____. "Pettenkofer Revisited: The Life and Contributions of Max von Pettenkofer (1818–1890)." *The Yale Journal of Biology and Medicine.* June, 1973; **46**(3): 161–176. http://www.ncbi.nlm.nih.gov / accessed 19 August 2014.
Evans, Alice C. "Studies on Hemolytic Streptococci. II: *Streptococcus Pyogenes.*" *Journal of Bacteriology.* 1936; **31**(6): 611. http://jb.asm.org / accessed 21 July 2014.
Ewart, James Cossar. "The Life History of Bacterium termo and Micrococcus, with further Observations on Bacillus." *Proceeding of the Royal Society of London.* 1 January 1878. Volume **27**: 474–485. https://www.archive.org / accessed 29 June 2014.
"Experimental Pathology—I." *The Zoophilist.* 1 November 1883; **2**(4): 191–193. https://www.books.google.com / accessed 1 September 2014.
Eyler, John M. "The Changing Assessments of John Snow's and William Farr's Cholera Studies." *Series: History of Epidemiology/Soz.-Präventivmed.* 2001; **46**: 225–232. http://www.epidemiology.ch / accessed 3 July 2015.
_____. *Victorian Social Medicine: The Ideals and Methods of William Farr.* Baltimore: The Johns Hopkins University Press, 1979.
_____. "William Farr on the Cholera: The Sanitarian's Disease Theory and the Statistician's Method." *Journal of the History of Medicine.* April, 1973; **28**: 79–100. www.medicine.mcgill.ca / accessed 16 September 2015.
Facklam, Richard. "What Happened to the Streptococci: Overview of Taxonomic and Nomenclature Changes." *Clinical Microbiology Reviews.* October, 2002; **15**(4): 613–630. https://www.ncbi.nlm.nih.gov / accessed 15 January 2015.

Farley, John. "Philosophical and Historical Aspects of the Origin of Life." *Works of the Catalan Society of Biology*. 1986; **39**: 37–47. http://www.raco.cat / accessed 28 November 2014.
_____. "The Spontaneous Generation Controversy." *Journal of the History of Biology*. 1972; **5**(2): 285–319.
_____. *The Spontaneous Generation Controversy from Descartes to Oparin*. Baltimore: The Johns Hopkins University Press, 1977.
_____, and G. L. Geison. "Science, Politics and Spontaneous Generation in Nineteenth Century France: The Pasteur-Pouchet Debate." 1974. *Bulletin of the History of Medicine*. **48**: 161–198.
Farmer, III, J. J., and J. Michael Janda. "Family 1. *Vibrionaceae*." In *Bergey's Manual of Systematic Bacteriology*. Volume **2**: The Proteobacteria, Part B: The Gammaproteobacteria. Edited by Don J. Brenner, Noel R. Krieg, and James R. Staley. Berlin, Heidelberg, Dordrecht, New York: Springer Science & Business Media, 2004, pp. 491–494.
Farr, William. *Report on the Mortality of Cholera in England, 1848–1849*. London: W. Clowes, 1852.
Fauntleroy, A. M. "The Surgical Lessons of the European War." *Annals of Surgery*. August, 1916; **64**(2): 136–150. http://www.ncbi.nlm.nih.gov / accessed 22 May 2014.
Fehleisen, Friedrich. "On Erysipelas [The Etiology of Erysipelas]." Translated by Leslie Ogilvie. *Recent Researches on Micro-Organisms in Relation to Suppuration and Septic Diseases*. Preface, Translated, and Abstracted by W. Watson Cheyne. Volume **115**. London: The New Sydenham Society, 1886, pp. 261–286. https://www.books.google.com / accessed 22 July 2014.
Fetrow, Charles W., Juan R. Avila, et al. "Wormwood." *The Complete Guide to Herbal Medicines*. Foreword by Simeon Margolis. Preface by Charles W. Fetrow and Juan R. Avila. New York and London: Pocket Books/Simon & Schuster, 2000.
"Fifth Congress of the Society of German Surgeons." *The London Medical Record: A Review of the Progress of Medicine, Surgery, Obstetrics, and the Allied Sciences*. Volume **IV**. London: Smith, Elder & Company, 1876. https://www.books.google.com / accessed 15 December 2014.
Fisher, Richard B. *Joseph Lister: 1827–1912*. New York: Stein and Day Publishers, 1977.
Fleming, Alexander. "The Action of Chemical and Physiological Antiseptics in a Septic Wound." *The British Journal of Surgery*. July 1919-April 1920; 7(25–28): 99–129. https://www.google.books.com / accessed 15 May 2014.
Flügge, Carl. *Micro-organisms: with Special Reference to the Etiology of the Infective Diseases*. 1886. Translated by W. Watson Cheyne. Volume **132**. London: The New Sydenham Society, 1890. https://www.biodiversitylibrary.org / accessed 15 July 2014.
Foster, Michael. "Physiology." In *Handbook for the Physiological Laboratory*. Edited by John Burdon Sanderson. 1873. Philadelphia: P. Blakiston, Son & Company, 1884, pp. 347–420. https://www.archive.org / accessed 16 August 2015.
Foster, W. D. "The Early History of Clinical Pathology in Great Britain." *Medical History*. July, 1959; **3**(3): 173–187. http://www.ncbi.nlm.nih.gov / accessed 24 May 2014.
_____. *A History of Medical Bacteriology and Immunology*. London, Fakenham and Reading: William Heinemann Medical Books, 1970.
Fraenkel, Carl. *Text-Book of Bacteriology*. Translated and Edited by J. H. Linsey. 3rd edition. New York: William Ward & Company, 1891. https://biodiversitylibrary.org / accessed 19 April 2015.
French, Richard D. *Antivivisection and Medical Science in Victorian Society*. Princeton and London: Princeton University Press, 1975.
"*fresszellen* [phagocytosis]." *The New Cassell's German Dictionary: German-English/English-German*. Based on the Editions of Karl Breul. Revised and re-edited by Harold T. Betteridge. Foreword by Gerhard Cordes. 1958; New York: Funk & Wagnalls, 1971, pp. 165–166.

Gabriel, Richard A., and Karen S. Metz. *A History of Military Medicine.* Contributions in Military Studies, 124. 2 volumes. New York: Greenwood Press, 1992.

Garrison, Fielding Hudson. *An Introduction to the History of Medicine.* Philadelphia and London: W. B. Saunders Company, 1914.

Gaw, Jerry L. *A Time to Heal: The Diffusion of Listerism in Victorian Britain.* Transactions of the American Philosophical Society. Vol. **89**.1. Philadelphia: American Philosophical Society, 1999.

Gaynes, Robert P. *Medical Pioneers in Infectious Diseases.* Washington, D.C.: American Society for Microbiology Press, 2011.

Geison, Gerald L. *The Private Science of Louis Pasteur.* Princeton: Princeton University Press, 1995.

Gerhard, Dietrich. "Periodization in History." *Dictionary of the History of Ideas: Studies of Selected Pivotal Ideas.* Edited and Preface by Philip P. Wiener. 4 volumes and Index. New York: Charles Scribner's Sons, 1973. Vol. III: 476–481.

Gibbes, Heneage. "Bacteria and Micrococci: Bacilli in Tuberculosis." *Proceedings of the Medical Society of London.* Volume **6**. Edited by Isambard Owen and Alfred Pearce Gould. London: J. E. Adlard, 1884. 314–320. https://www.books.google.com / accessed 18 November 2014.

Gilroy, Derek W. "Resolution of Acute Inflammation and Wound Healing." *Fundamentals of Inflammation.* Edited by Charles N. Serhan, Peter A. Vard, and Derek W. Gilroy. Cambridge and New York: Cambridge University Press, 2010, pp. 17–27.

Godlee, Rickman John. *Lord Lister.* 2nd edition. 1917; London: The Macmillan Company, 1918.

Götz, Friedrich, Tammy Bannerman, and Karl-Heinz Schleifer. "The Genera Staphylococcus and Macrococcus." *Prokaryotes: A Handbook on the Biology of Bacteria.* In *Prokaryotes.* Volume **4**: Bacteria. 2006, pp. 5–75. http://link.springer.com / accessed 1 June 2014.

Grain/léger. Larousse Modern Dictionary: French-English/English-French. By Marguerite-Marie Dubois, et al. Edited by William Maxwell Landers, et al. Paris: Librairie Larousse, 1960, pp. 353–354.

Grano/ligero. The University of Chicago Spanish–English/English-Spanish Dictionary. Diccionario: Español-Inglés/Inglés-Español. Compiled and Edited by Carlos Castillo, Otto Bond, and Barbara Garcia. New York: Washington Square Press, 1948. P. 106.

Gray, Henry. *Anatomy, Descriptive and Surgical.* Drawings by H. V. Carter. Edited by T. Pickering Pick and Robert Howden. 15th edition. New York: Barnes & Noble, 2010.

Gray, Henry M. W. *The Early Treatment of War Wounds.* London: Henry Frowde, 1919.

Greenwood, John T., and F. Clifton Berry, Jr. *Medics at War: Military Medicine from Colonial Times to the 21st Century.* Annapolis, Maryland: Naval Institute, 2005.

Griswold, E. "What Conditions on the Field Justify Amputation in Gunshot Wounds?" *Transactions of the International Medical Congress.* Ninth Session. Edited by John B. Hamilton. Volume **2**. Washington, D.C.: William F. Fell & Company, 1887. 284–289.

Guiry, M. D., and G. M. Guiry. *AlgaeBase. World-wide electronic publication.* Galway: National University of Ireland. http://www.algaebase.org / accessed 9 February 2015.

Guthrie, Douglas. *Lord Lister: His Life and Doctrine.* Edinburgh: E & S. Livingstone, 1949.

Haeckel, Ernst. *The Riddle of the Universe.* Translated by Joseph McCabe. Introduction by H. James Birx. Great Mind Series. Buffalo and Amherst, New York: Prometheus Books, 1992.

Hager, Thomas. *The Demon under the Microscope: From Battlefield Hospitals to Nazi Labs: One Doctor's Heroic Search for the World's First Miracle Drug.* New York: Three Rivers Press/Crown Publishing Group, 2007.

Haller, John S., Jr. *Battlefield Medicine: A History of the Military Ambulance from the Napoleonic Wars through World War I.* Carbondale, Illinois: Southern Illinois University Press, 1992.

_____. "Treatment of Infected Wounds during the Great War, 1914 to 1918." *Southern Medical Journal.* March, 1992; **85**(3): 303–315.

Hallet, Christine E. *Veiled Warriors: Allied Nurses of the First World War.* Oxford: Oxford University Press, 2014.

Hallier, Ernst. *Das Cholera-Contagium: Botanische Untersuchungen, Aerzten und Naturforschern.* Liepzig: Wilhelm Endelmann Verlag, 1867.

Hamilton, Susan. "On the Cruelty to Animals Act, 15 August 1876." *BRANCH Collective.* http://www.branchcollective.org / accessed 24 September 2014.

Hamlin, Christopher. *A Science of Impurity: Water Analysis in Nineteenth Century Britain.* Berkeley and Los Angeles: The University of California Press, 1990.

Hankin, E. H. "On Immunity (read before the Congress of Hygiene and Demography)." *The Lancet.* 15 August 1891; **138**(3546): 339–340. In *The Lancet.* 2 volumes. Edited by Thomas H. Wakley and Thomas Wakley, Jr. London: *The Lancet*, 1891, pp. 339–340.

_____. "Report on the Conflict between the Organism and the Microbe." *The British Medical Journal.* 12 July 1890; **2**(1541): 65–68. http://www.ncbi.nlm.nih.gov / accessed 28 December 2014.

Hannigan, William C. "Neurological Surgery during the Great War: The Influence of Colonel Cushing." *Neurosurgery.* September, 1998; **23**(3): 283–410. http://journals.lww.com / accessed 17 April 2014.

Hardy, Anne. "Methods of Outbreak Investigation in the 'Era of Bacteriology' 1880- 1920." *Präventivmed.* 2001; **48**: 355–360. http://www.epidemiology.ch / accessed 4 July 2014.

Harrison, Mark. *The Medical War: British Military Medicine in the First World War.* New York and Oxford: Oxford University Press, 2010.

_____. *Public Health in British India: Anglo-Indian Preventive Medicine, 1859–1914.* Cambridge: Cambridge University Press, 1994.

Hart, D. Berry. "An Address on Pasteur and Lister." *The British Medical Journal.* 13 December 1902; **2**(2189): 1838–1840. http://www.ncbi.nlm.nih.gov / accessed 7 July 2014.

Hays, J. N. *The Burdens of Disease: Epidemics and Human Response in Western History.* New Brunswick, New Jersey: Rutgers University Press, 2000.

Helling, Thomas S., and Emmanuel Daon. "In Flanders Field: The Great War, Antoine Depage, and the Resurgence of Debridement." *Annals of Surgery.* August, 1998; **228**(2): 173–181. http://www.ncbi.nlm.nih.gov / accessed 6 May 2014.

"hematoxylin." *The Random House College Dictionary.* Edited and Preface by Jess Stein. New York: Random House, 1973. P. 616.

Henle, Jacob F. G. "Concerning Miasmatic, Contagious, and Miasmatic-Contagious Diseases" (1840). *Milestones in Microbiology*, pp. 76–79.

Henry, Herbert. "On Some Anaërobes Found in Wounds and Their Mode of Action in the Tissues." *British Medicine in the War, 1914–1917*, pp. 75–77.

Herrick, Claire E. J. "Casualty Care during the First World War: The Experience of the Royal Navy." *War in History.* April, 2000; **7**(2): 154–179. http://wih.sagepub.com / accessed 15 May 2014.

_____. "Cheyne, Sir (William) Watson, first baronet (1852–1932), bacteriologist and surgeon." *Oxford Dictionary of National Biography.* Oxford University Press. Online edition: May, 2006. http://www.oxforddnb.com / accessed 24 May 2014.

_____. "Of War and Wounds: The Propaganda, Politics, and Experience of Medicine in World War I." University of Manchester: Unpublished Ph.D. Thesis, 1996.

Herzenberg, John E. "Johann Friedrich August von Esmarch: His Life and Contributions to Orthopaedic Surgery." *Iowa Orthopaedic Journal.* 1988; Volume. **8**: 85–91. http://www.forensicgenealogy.info http://www.forensicgenealogy.info / accessed 25 March 2014.

"Heterogenesis." *The Oxford English Dictionary. Compact Edition.* Volume I, A-O. Oxford: At the Clarendon Press, 1971. P. 1298.

Hill, R. L. "Taxonomy of the Staphylococci." CNS and the Epidemiological Typing of *S. Epidermidis* coccus species with the API-Staph-Ident. System. *Journal of Clinical Microbiology.* 1981; **6**: 509–516.

"Historical Perspectives Centennial: Koch's Discovery of the Tubercle Bacillus." *Morbidity and Mortality Weekly*. 19 March 1982; **31**(10): 121–123. *Centers for Disease Control and Prevention*. http://www.cdc.gov / accessed 3 August 2015.

Howard, Edward R. "Joseph Lister: His Contributions to Early Experimental Physiology." *Notes & Records of the Royal Society*. Published online: 29 May 2013; **67**:191–198 / http://rsnr.royalsocietypublishing.org / accessed 2 September 2015.

Howard-Jones, Norman. "Robert Koch and the Cholera Vibrio: A Centenary." *British Medical History*. 4 February 1984; Volume **288**: 379–381. http://www.google.com / accessed 6 August 2014.

Howell, E. V. "Alcohol as an Antidote for Carbolic Acid." Proceedings of the Annual Meeting (Charlotte), 18–19 May 1898. *North Carolina Pharmaceutical Association*. Volumes 19–26. Raleigh, North Carolina: Edwards & Broughton, 1898, pp. 26–30. https://www.books.google.com / accessed 5 October 2015.

Hunter, William. "On the Nature, Action, and Therapeutic Value of the Active Principles of Tuberculin." *The British Medical Journal*. 25 July 1891; **2**(1595): 169–176. http://www.ncbi.nlm.nih.gov / accessed 2 March 2015.

Huxley, Thomas Henry. "Biogenesis and Abiogenesis." The Presidential Address to the British Association for the Advancement of Science for 1870. *Discourses: Biological and Geological Essays*. New York: D. Appleton and Company, Publishers, 1897, pp. 229–271.

_____. "On the Physical Basis of Life" (1868). *Autobiography and Selected Essays by Thomas Henry Huxley*. Edited with Introduction and Notes by Ada L. F. Snell. The Riverside Literature Series. Boston and New York: Houghton Mifflin Company, 1909, pp. 95–114.

Isaacs, Jeremy, D. "D. D. Cunningham and the Aetiology of Cholera in British India, 1869–1897." *Medical History*. July, 1998; **42**(3): 297–405. http://www.ncbi.nlm.nih.gov / accessed 1 July 2015.

The Journal of the American Medical Association. Editorial: "The Treatment of War Wounds." 28 October 1916; **67**(18): 1304–1305. https://www.books.google.com / accessed 19 August 2015.

Kaufmann, Stefan H. E. "Elie Metchnikoff's and Paul Ehrlich's Impact on Infection Biology." *Microbes and Infection*. November–December, 2008; **10**(14–15): 1417–1419. http://www.sciencedirect.com / accessed 29 December 2014.

_____. "Immune Response to Tuberculosis: Experimental Animal Models." *Tuberculosis*. February, 2003; **83**(1–3): 107–111. http://www.sciencedirect.com / accessed 9 September 2014.

_____. "Robert Koch's Highs and Lows in Search for a Remedy for Tuberculosis." *Nature Medicine*. Special Web Focus: Tuberculosis (2000). http://www.nature.com / accessed 4 March 2015.

Keen, William W. *The Treatment of War Wounds*. Philadelphia and London: W. B. Saunders Company, 1917.

Keilty, Robert, and Jesse E. Packer. "Experimental Studies of Various Antiseptic Substances for Use in Treatment of Wounds (Based on the Work of Sir W. Watson Cheyne)." *The Journal of the American Medical Association*. 26 June 1915; **64**(26): 2123–2125. www.jama.jamanetwork.com / accessed 26 February 2014.

Kelly, Howard A. "Jules Lemaire: The First to Recognize the True Nature of Wound Infection and Inflammation, and the First to Use Carbolic Acid in Medicine and Surgery." *The Journal of the American Medical Association*. 20 April 1901; **36**(16): 1083–1088. www.jama.jamanetwork.com / accessed 22 September 2015.

Kent, William Saville. *A Manual of Infusoria: Including a Description of all known Flagellate, Ciliate, and Tentaculiferous Protozoa, British and Foreign, and an Account of the Organization and Affinities of the Sponges*. Volume **1**. London: David Bogue, 1880–1881. https://www.books.google.com / accessed 9 February 2015.

Kenwood, Henry Richard. *Public Health Laboratory Work*. Philadelphia: P. Blakiston, Son & Company, 1893. https://www.archive.org / accessed 31 March 2015.

_____. *Public Health Laboratory Work*. 2nd edition. London: H. K. Lewis, 1896. https://www.archive.org / accessed 9 April 2015.

_____. *Public Health Laboratory Work*. 5th edition. London: H. K. Lewis, 1911. https://babel.hathitrust.org / accessed 9 April 2015.

Kern, Lilly. *Deutsche Bakteriologie im Spiegel Englischer medizinischer Zeitschriften, 1785–1885*. Zurich: Jurls Druck und Verlag, 1972.

Klein, Emanuel Edward. "The Anti-Cholera Vaccination: An Experimental Critique." *British Medical Journal*. 25 March 1893; **1**(1682): 632–634. http://www.ncbi.nlm.nih.gov / accessed 3 August 2014.

_____. *The Bacteria in Asiatic Cholera*. 1886/1887. London and New York: Macmillan and Company, 1889.

_____. "Histology." In *A Handbook for the Physiological Laboratory*. Edited by John Burdon Sanderson. 1873. Philadelphia: P. Blakiston, Son & Company, 1884, pp. 17–173. https://www.archive.org / accessed 16 August 2015.

_____. *Micro-Organisms and Disease: An Introduction to the Study of Specific Organisms*. 3rd edition, revised. 1884. London and New York: Macmillan, 1896. https://www.books.google.com / accessed 18 August 2015.

_____. "The Organisms of Cholera." *British Medical Journal*. 24 January 1885; **1**(1256): 200. http://www.ncbi.nlm.nih.gov / accessed 3 August 2014.

_____. "Remarks on the Etiology of Asiatic Cholera." *British Medical Journal*. 28 March 1885; **1**(1265): 650–652. http://www.ncbi.nlm.nih.gov / accessed 1 August 1885.

_____. "Some Remarks on the Present State of Our Knowledge of the Comma-Bacilli of Koch." *British Medical Journal*. 4 April 1885; **1**(1266): 693–695. http://www.ncbi.nlm.nih.gov / accessed 1 August 2014.

Klein, Emanuel Edward, and Heneage Gibbes. *An Inquiry into the Etiology of Asiatic Cholera*. London: India Office, 1885. http://collections.nlm.gov / accessed 22 August 2014.

Koch, Robert. "Abstract of an Address on the Value of Tuberculin in the Treatment of Surgical Tuberculosis." *The British Medical Journal*. 2 May 1891; **1**(1583): 951–961. http://www.ncbi.nlm.nih.gov / accessed 2 March 2015.

_____. "An Address on Bacteriological Research." *British Medical Journal*. 16 August 1890; **2**(1546): 380–383. http://www.ncbi.nlm.nih.gov / accessed 2 September 2014.

_____. "An Address on Cholera and Its Bacillus." *British Medical Journal*. 30 August 1884; **2**(1235): 403–407. http://www.ncbi.nlm.nih.gov / accessed 8 August 2014.

_____. "An Address on Cholera and Its Bacillus." *British Medical Journal*. 6 September 1884; **2**(1236): 453–459. http://www.ncbi.nlm.nih.gov / accessed 8 August 2014.

_____. "An Address on the Fight against Tuberculosis in the Light of the Experience that Has been Gained in the Successful Combat of Other Infectious Diseases." *The British Medical Journal*. 27 July 1901; **2**(2117): 189–193. http://www.ncbi.nlm.nih.gov / accessed 2 September 2014 .

_____. "The Current State of the Struggle against Tuberculosis." *Nobel Lectures: Physiology or Medicine 1901–1921*. Amsterdam: Elsevier Publishing Company, 12 December 1905. http://www.nobelprize.org / accessed 3 March 2015.

_____. "The Etiology of Anthrax, Founded on the Course of Development of the Bacillus Anthracis" (1876). Translated by K. Codell Carter. *Essays of Robert Koch*, pp. 1–17.

_____. "The Etiology of Tuberculosis" (1882). Translated by Thomas D. Brock. *Milestones in Microbiology*, pp. 109–115.

_____. "The Etiology of Tuberculosis [Koch's Postulates]" (1884). Translated by Thomas D. Brock. *Milestones in Microbiology*, pp. 116–118.

_____. "A Further Communication on a Remedy for Tuberculosis." *British Medical Journal*.

22 November 1890; **2**(1560): 1193–1199. http://www.ncbi.nlm.nih.gov / accessed 2 September 2014.

_____. "A Further Communication on a Remedy for Tuberculosis." *British Medical Journal.* 17 January 1891; **1**(1568): 125–127. http://www.ncbi.nlm.nih.gov / accessed 2 September 2014.

_____. *Investigation into the Etiology of Traumatic Infective Diseases.* Translated by W. Watson Cheyne. Volume **88**. London: The New Sydenham Society, 1880. https://www.books.google.com / accessed 11 July 2014.

_____. "Investigations of the Etiology of Wound Infections" (1878). Translated by K. Codell Carter. *Essays of Robert Koch*, pp. 19–56.

_____. "Lecture at the First Conference for Discussion of the Cholera Question" (1884). Translated by K. Codell Carter. *Essays of Robert Koch*, pp. 151–170.

_____. "Methods for the Study of Pathogenic Organisms" (1881). Translated by Thomas D. Brock. *Milestones in Microbiology*, pp. 101–108.

_____. "On the Anthrax Inoculation" (1882). Translated by K. Codell Carter. *Essays of Robert Koch*, pp. 98–115.

_____. "On Bacteriological Research" (1890). Translated by K. Codell Carter. *Essays of Robert Koch*, pp. 179–186.

_____. "On Cholera Bacteria" (1884). Translated by K. Codell Carter. *Essays of Robert Koch*, pp. 171–177.

"Koch's Methods of Investigating Pathogenic Bacteria." *British Medical Journal.* 29 April 1882; **1**(1113): 624–625. http://www.ncbi.nlm.nih.gov / accessed 2 September 2014.

Kohn, George Childs, editor. "Asiatic Cholera Pandemic of 1881–1896." *Encyclopedia of Plague & Pestilence from Ancient Times to the Present.* Foreword by Mary-Louise Scully. Revised Edition. New York: Facts on File/Checkmark Books, 2001, pp. 11–16.

Koob, Derry D., and William Boggs. *The Nature of Life.* Addison-Wesley Series in Life Science. Consulting Editor, John W. Hopkins, III. Reading, Massachusetts, and London: Addison-Wesley Publishing Company, 1972.

Krause, Fedor. *Die Tuberkulose der Knocken und Gelenke.* Leipzig: Verlag Von F. C. Vogel, 1891, pp. 152–155. https://www.books.google.com / accessed 2 August 2015.

Krüger, Christine K. "German Suffering in the Franco-German War, 1870/71." *German History.* 2011; **29**(3): 404–422. http://gh.oxfordjournals.org / accessed 31 July 2012.

La Garde, Louis A. *Gunshot Injuries: How They Are Inflicted, Their Complications and Treatment.* 2nd revised edition. New York: William Wood and Company, 1916, pp. 136–164.

Larson, E. M. "Review of the New Surgical Methods Taught by the World War." *The Journal-Lancet.* 15 August 1920; **40**(16): 449–453. Republished in *The Journal-Lancet.* Edited by W. A. Jones. Minneapolis, Minnesota: W. L. Klein, 1920. https://www.books.google.com / accessed 31 October 2014.

LaTour, Bruno. *The Pasteurization of France.* Translated by Alan Sheridan and John Law. Cambridge, Massachusetts: Harvard University Press, 1993.

Lawrence, Christopher, and Richard Dixey. "Practicing on Principle: Joseph Lister and the Germ Theories of Disease." *Medical Theory, Surgical Practice: Studies in the History of Surgery.* Edited by Christopher Lawrence. Clio Medica/Wellcome Institute Series in the History of Medicine. London: Routledge, 1992, pp. 153–215.

Lawrence, Christopher, and Anna K. Mayer, editors. *Regenerating England's Science, Medicine and Culture in Inter-war Britain.* Clio Medica/Wellcome Institute Series in the History of Medicine. Amsterdam & Atlanta: Edition Rodopi, 2000.

Lawrence, Christopher. "Lister, Joseph, Baron Lister (1827–1912)." In *Oxford Dictionary of National Biography.* Oxford: Oxford University Press, 2004. http://www.oxforddnb.com / accessed 10 November 2014.

Leeson, John Rudd. *Lister as I Knew Him.* New York: William Wood and Company, 1927.

Lehnhardt, M., et al. "Carl von Reyher's Studies of Wound Therapy." *Chirurgie.* 2002; **73**: 721–724. http://www.journals.elsevierhealth.com / accessed 17 January 2011.

Lemaire, François-Jules. *De L'Acide Phénique: De Son Action Sur Les Végétaux, Les Animaux, Les Ferments, Les Venins, Les Virus, Les Miasmes et de des Applications … et à la Thérapeutique.* 1863; Charleston, South Carolina: Nabu Press, 2011.

Lewis, T. R. "A Memorandum on the 'Comma-Shaped Bacillus' Alleged to be the Cause of Cholera." *Physiological and Pathological Researches: Being a Reprint of the Principal Scientific Writings.* Edited by Sir William Aitken, G. E. Dobson, and A. E. Brown. London: Lewis Memorial Committee, 1888. https://www.books.google.com / accessed 9 August 2014.

Licitra, Giancarlo. *Staphylococcus: Etymologia. Centers for Disease Control and Prevention.* September, 2013; **19**(9). http://www.nc.cdc.gov / accessed 21 July 2014.

Liebig, Justis. "Concerning the Phenomena of Fermentation, Putrefaction and Decay, and Their Causes" (1839). *Milestones in Microbiology*, pp. 24–27.

Lippi, D., and E. Gotuzzo. "The Greatest Steps towards the Discovery of *Vibrio cholerae.*" *Clinical Microbiology and Infection.* 2013: 1–5.

Lister, Joseph Baron. "Address [M. Pasteur's 'Jubilee'] (1893)." In *Lord Lister,* by Rickman J. Godlee, pp. 518–522.

_____. "The Address in Surgery: Delivered on August 10, 1871, to the Thirty-Ninth Annual Meeting of the British Medical Association Held in Plymouth" (1871). Vol. **II** of *The Collected Papers*, pp. 172–198.

_____. "An Address on the Antiseptic Management of Wounds" (1893)." Volume **II** of *The Collected Papers*, pp. 349–364.

_____. "An Address on the Antiseptic System of Treatment in Surgery" (1868/1869). Vol. **II** of *The Collected Papers*, pp. 51–85.

_____. "An Address on Corrosive Sublimate as a Surgical Dressing" (1884). Vol. **II** of *The Collected Papers*, pp. 293–308.

_____. "An Address on the Effect of the Antiseptic Treatment upon the General Salubrity of Surgical Hospitals" (1875). Vol. **II** of *The Collected Papers*, pp. 247–255.

_____. "An Address on the Relations of Minute Organisms to Inflammation" (1881). Volume **I** of *The Collected Papers, pp.* 399–410.

_____. "An Address on the Treatment of Wounds" (1881). Vol. **II** of *The Collected Papers*, pp. 275–292.

_____. *The Collected Papers.* Oxford: At the Clarendon Press, 1909.

_____. "A Contribution to the Germ Theory of Putrefaction and Other Fermentative Changes, and to the Natural History of Torulae and Bacteria" (1875). Vol. **I** of *The Collected Papers*, pp. 275–308.

_____. "Demonstrations of Antiseptic Surgery before Members of the British Medical Association" (1875–1876). Vol. **II** of *The Collected Papers*, pp. 256–274.

_____. "Edinburgh University Graduation Ceremony [1 August 1876]." *The Edinburgh Medical Journal.* January–June, 1876; **22**(pt. 1). Edinburgh: Oliver and Boyd; London: Simpkin, Marshall, and Company, 1876, pp. 282–284. https://www.babel.hathitrust.org / accessed 23 September 2015.

_____. "A Further Contribution to the Natural History of Bacteria and the Germ Theory of Fermentative Changes" (1873). Vol. **I** of *The Collected Papers*, pp. 309–334.

_____. "Illustrations of the Antiseptic System of Treatment in Surgery" (1867). Vol. **II** of *The Collected Papers*, pp. 46–50.

_____. "An Introductory Lecture (On the Causation of Putrefaction and Fermentation)" (1869). Vol. **II** of *The Collected Papers*, pp. 477–488.

_____. Letter: Lister to Pasteur, 10 February 1874. Godlee. *Lord Lister*, pp. 275–77.

_____. "Letters from Joseph Lister to William Watson Cheyne, 1880–1911." *Joseph Lister: Surgeon and Founder of a System of Antiseptic Surgery.* The Royal College of Surgeons of Edinburgh. http://www.library.rcsed.ac.uk / accessed 18 February 2015.

_____. "Merciful Advantages Due to Animal Experimentation: Hitherto Unpublished Letter

from Lord Lister to Dr. W. W. Keen, Exclaiming at Attempts to Restrict Medical Research"
(4 April 1898). [*Philadelphia*] *Public Ledger*. 7 January 1917. https://www.archive.org /
accessed 8 July 2012.

_____. "A Method of Antiseptic Treatment Applicable to Wounded Soldiers in the Present
War" (1870). Vol. **II** of *The Collected Papers*, pp. 161–164.

_____. "Note on the Double Cyanide of Mercury and Zinc as an Antiseptic Dressing (1907)."
Volume **II** of *The Collected Papers*, pp. 329–331.

_____. "Notice of Further Researches on the Coagulation of the Blood" (December, 1859).
Volume **I** of *The Collected Papers*, pp. 105–108.

_____. "Obituary Notice of the Late Joseph Jackson Lister, F.R.S., Z.S., With Special Reference
to His Labours in the Improvement of the Achromatic Microscope" (1 March 1870).
Vol. **II** of *The Collected Papers*, pp. 543–552.

_____. "On the Antiseptic Principle in the Practice of Surgery" (1867). Vol. **II** of *The Collected
Papers*, pp. 37–45.

_____. "On a Case Illustrating the Present Aspect of the Antiseptic Treatment in Surgery"
(1871). Vol. **II** of *The Collected Papers*, pp. 165–171.

_____. "On the Coagulation of the Blood. The Croonian Lecture delivered before the Royal
Society of London" (11 June 1863). Volume **I** of *The Collected Papers*, pp. 109–134.

_____ "On the Early Stages of Inflammation" (1857/1858). Volume **I** of *The Collected Papers
of Joseph, Baron Lister*. Hector C. Cameron, W. Watson Cheyne, Rickman J. Godlee, C.
J. Martin, and Dawson Williams. 2 volumes. Honolulu, Hawaii: University Press of the
Pacific, 2003, pp. 208–275. Textual citations to *The Collected Papers* abbreviated as: *CP*.

_____. "On the Effects of the Antiseptic System of Treatment upon the Salubrity of a Surgical
Hospital" (1870). Vol. **II** of *The Collected Papers*, pp.123–136.

_____. "On the Interdependence of Science and the Healing Art, Being the Presidential
Address to the British Association for the Advancement of Science" (1896). Vol. **II** of
The Collected Papers, pp. 489–514.

_____. "On the Lactic Fermentation and Its Bearings on Pathology" (1878). Vol. **I** of *The
Collected Papers*, pp. 353–86.

_____. "On the Nature of Fermentation" (1878). Vol. **I** of *The Collected Papers*, pp. 335–352.

_____. "On the Nature of Fermentation" (1 October 1877). Vol. **I** of *The Collected Papers*,
pp. 335–352.

_____. "On a New Method of Treating Compound Fracture, Abscess, Etc." (1867). Vol. **II** of
The Collected Papers, pp. 387–398.

_____. "On a New Method of Treating Compound Fracture, Abscess, Etc., with Observations
on the Conditions of Suppuration" (1867). Vol. **II** of *The Collected Papers, pp.* 1–36.

_____. "On the Principles of Antiseptic Surgery" (1891). Vol. **II** of *The Collected Papers*, pp.
340–348.

_____. "On Recent Improvements in the Details of Antiseptic Surgery" (1875). Vol. **II** of
The Collected Papers, pp. 206–246.

_____. "On the Relations of Micro-Organisms to Disease" (1880/1881). Vol. **I** of *The Collected
Papers*, pp. 387–98.

_____. "On Spontaneous Gangrene from Arteritis and the Causes of Coagulation of the Blood
in Diseases of the Blood-Vessels" (1858). Volume **I** of *The Collected Papers*, pp. 69–84.

_____. "Remarks on a Case of Compound Dislocation of the Ankle with Other Injuries;
Illustrating the Antiseptic System of Treatment" (1870). Vol. **II** of *The Collected Papers*,
pp. 137–155.

_____. "The Third Huxley Lecture" (1900). Vol. **II** of *The Collected Papers*, pp. 515–542.

Lister, Joseph, Baron: Centenary Volume, 1827–1927. Edited by Arthur Logan Turner. For the
Lister Centenary Committee of the British Medical Association. Edinburgh: Oliver &
Boyd, 1927.

Lockwood, C. B. "Erysipelas." *A System of Surgery*. Volume **1**: 157–170.

_____. "Pyæmia." *A System of Surgery.* Volume 1: 171–186.

_____. "Tetanus and Tetany." *A System of Surgery.* Volume 1: 128–153.

Logan, Niall A. *Bacterial Systematics.* Blackwell Scientific Publications. London, Edinburgh, Boston: Oxford at the Alden Press, 1994.

Lyell, Alan. "Alexander Ogston, Micrococci, and Joseph Lister." *The Journal of the American Academy of Dermatology.* February, 1989; **20**(2 Pt.1): 302–310. http://www.ncbi.nlm.nih. gov / accessed 8 October 2014.

_____. "Alexander Ogston (1844–1929)-Staphylococci." *The Scottish Medical Journal.* October, 1977; **22**(4): 277–278. http://www.ncbi.nlm.nih.gov / accessed 8 October 2014.

MacCormac, William. *Antiseptic Surgery: An Address Delivered at St. Thomas's Hospital, with the Subsequent Debate.* London: Simon, Elder, & Company, 1880. https://books.google. com / accessed 12 December 2014.

_____. *Notes and Recollections of an Ambulance Surgeon, Being an Account during the Campaign of 1870.* London: J. A. Churchill, 1871. https://www.books.google.com / accessed 16 December 2014.

_____. "On the Wounded in the Transvaal War: An Address Delivered Before the Royal Medical and Chirurgical Society, May 22nd, 1900." *Medico-Chirurgical Transactions. Proceedings of the Royal Society of Medicine.* 1900; Volume 83: 315–333. https://www.ncbi.nlm. nih.gov / accessed 11 December 2014.

"MacCormac, Sir William (1836–1901)." Biographical Entry. *Plarr's Lives of the Fellows Online.* The Royal College Surgeons of England. http://livesonline.rcseng.ac.uk / accessed 28 September 2014.

Magnan, Valentin. *On Alcoholism; On the Different Forms of Alcoholic Delirium.* [1872 edition]. *Reviews and Notices. The British Medical Journal.* 13 March 1875; **1**(741): 346–347. http:// www.ncbi.nlm.nih.gov / accessed 24 February 2015.

"Dr. Magnan of Paris, the results obtained." *The British Medical Journal.* 13 February 1875; **1**(737): 211–212. http://ncbi.nlm.nih.gov / accessed 9 February 2015.

_____. *On Alcoholism; On the Various Forms of Alcoholic Delirium.* Translated by W. S. Greenfield. London: H. K. Lewis, 1876. Medical Heritage Library. https://www.archive.org / www.openlibrary.org / accessed 24 February 2015.

Makins, George Henry. "The Development of British Surgery in the Hospitals on the Lines of Communication [in] France." *British Medicine in the War, 1914–1917,* pp. 58–75.

_____. "Surgical Diseases Due to Microbic Infection and Parasites." *A System of Surgery.* Volume 1: 294–338.

Manring, M. M., Alan Hawk, Jason H. Calhoun, Romney C. Andersen. "Treatment of War Wounds: A Historical Review." *Clinical Orthopaedics and Related Research.* 2009; **467**: 2168–2191. http://www.ncbi.nlm.nih.gov / accessed 27 September 2014.

Martin, Walton. "Sir William Watson Cheyne, 14 Dec.1852–19 April 1932." *Bulletin of the New York Academy of Medicine.* May, 1952; **8**(5): 336–337. http://link.springer / accessed 11 March 2015.

Masters, Barry R. "Ernst Abbé and the Foundation of Scientific Microscopes." *Optics & Photonics News.* http://www.osa-opn.org / accessed 29 May 2014.

McCallum, Jack Edward. *Military Medicine: From Ancient Times to the 21st Century.* Santa Barbara, California: ABC-CLIO/Greenwood, Incorporated, 2008.

McCarthy, E. F. "The Toxins of William B. Coley and the Treatment of Bone and Soft-Tissue Sarcomas." 2006; **26**: 154–158. http://www.ncbi.nlm.nih.gov / accessed 7 March 2016.

McCartney, G. E., and F. H. H. Mewburn. "The Technique of the Carrel-Dakin Treatment." *The British Medical Journal.* 9 February 1918; **1**(2980): 170–171. http://www.ncbi.nlm. nih.gov / accessed 25 September 2014.

McDowall, Jennifer. "Cholera Toxin." *Interpro: Protein sequence analysis & classification. The European Bioinformatics Institute.* http://www.ebi.ac.uk / accessed 21 August 2014.

"Medical News [William Watson Cheyne awarded *Syme Surgical Fellowship*]." *The Medical*

Times & Gazette: A Journal of Medical Science, Literature, Criticism and News. 18 August 1877. Volume **2**. London: J & A. Churchill, 1877. P. 187.

Meselson, M., et al. "The Sverdlovsk Anthrax Outbreak of 1979." *Science*. 1994; **266**: 1202–1208. http://www.anthrax.osd.mil / accessed 22 July 2013.

Metchnikoff, Élie. "A Disease of Daphnia Caused by a Yeast. A Contribution to the Theory of Phagocytes as Agents for Attack on Disease-Causing Organisms" (1884). *Milestones in Microbiology*, pp. 132–138.

_____. *The Founders of Modern Medicine: Pasteur, Koch, Lister*. Translated by D. Berger. 1912. New York: Walden Publications, 1939.

_____. *Immunity in Infectious Diseases*. Translated by Francis G. Binnie. 1901. Cambridge: Cambridge Univ. Press, 1905.

_____. "Lecture on Phagocytosis and Immunity. Delivered at the Institut Pasteur, December 29th, 1890." Translated by J. G. Adami. *The British Medical Journal*. 31 January 1891; **1**(1570): 213–217. http://www.ncbi.nlm.nih.gov / accessed 29 December 2014.

_____. *Lectures on the Comparative Pathology of Inflammation, Delivered at the Pasteur Institute in 1891*. Translated by F. A Starling and E. H. Starling. London: Kegan Paul, Trench, Trübner & Company, Incorporated, 1893.

_____. "Untersuchungen über die Intracellulaire Verdauung bei Wirbellosen Tieren" ["Investigation into the Intracellular Digestion of Vertebrate Animals"]. *Arbeit Zoologischen Instituten, Universitat Wien*. 1883; **5**: 141–168.

"Microbe." *The Oxford English Dictionary. Compact Edition*. Volume **I**. P. 1788.

"Microbiology." *The Oxford English Dictionary. Compact Edition*. Volume **I**. P. 155.

Micrococcus, Genus. *LSPN: List of Prokaryotic Names with Standing in Nomenclature*. http:// www.bacterio.net/micrococcus.html / accessed 27 May 2014.

Miller, Kenneth R., and Joseph Levine. *Biology*. Upper Saddle River, New Jersey: Pearson Education, Incorporated, 2002.

Milligan, E. T. C. "The Early Treatment of Projectile Wounds by Excision of the Damaged Tissues." *The British Medical Journal*. 26 June 1915; **2**(2843): 1081; http://www.ncbi. nlm.nih.gov / accessed 20 May 2014.

Moore, James W. "Toxic Particulates in the Air." In *The Changing Environment*. Springer Series on Environmental Management. New York: Springer-Verlag, 1986, pp. 112–142. www. link.springer.com / accessed 17 August 2015.

Morgan, W. Parry. "The Treatment of Wound Infections." *The British Medical Journal*. 13 May 1916; **1**(2889): 685–688. http://www.ncbi.nlm.nih.gov / accessed 22 May 2014.

Moriarity, T. B. "Listerism in Gun-shot Wounds." *Half-Yearly Compendium of Medical Science: A Synopsis*. Edited by D. G. Brinton. Part **XXIX**. January, 1882. Third Series. No. 5. Philadelphia: Medical Publication Office, pp. 297–301.

_____. "Listerism—Simple Application of, to Recent Injuries." *The Lancet*. 3 September 1881; **118**(3027): 411–413.

Morison, Rutherford. "Remarks on the Treatment of Infected, Especially, War, Wounds." *The British Medical Journal*. 20 October 1917; **2**(2964): 503–506. http://www.ncbi.nlm.nih. gov / accessed 19 May 2014.

Mortimer, Philip P. "The Bacteria Craze of the 1880s." *The Lancet*. 13 February 1999; **353**(9152): 581–584. http://www.thelancet.com / accessed 7 March 2015.

Moynihan, Berkeley. "The Institutes of Surgery: An Historical Review." *British Medicine in the War, 1914–1917*, pp. 1–14.

Müller, Volker. "Bacterial Fermentation." *Encyclopedia of Life Sciences*. 2001. Nature Publishing Group. www.els.net / http://web.oranim.ac.il / accessed 19 July 2014.

Murray, Clinton, Mary K. Hinkle, and Heather C. Yun. "History of Infections Associated with Combat-Related Injuries." *The Journal of Trauma, Injury, Infection, and Critical Care*. 2008; **64**(3): S221–231. http://afids,org / accessed 20 March 2014.

Murray, John. "Lecture on the Action of the Cholera-Poison on the Body; and Its Nature and

History Outside the Body." *The British Medical Journal.* 9 April 1870; **1**(484): 355–357. http://www.ncbi.nlm.nih.gov / accessed 1 July 1870.

_____. "Lecture on the Action of the Cholera-Poison on the Body; and Its Nature and History Outside the Body." *The British Medical Journal.* 16 April 1870; **1**(485): 383–384. http://www.ncbi.nlm.nih.gov / accessed 1 July 1870.

_____. *Observations on the Pathology and Treatment of Cholera.* London: Smith, Elder, & Company, 1874. https://www.archive.org / accessed 4 August 2015.

_____. *Report on the Treatment of Epidemic Cholera.* Calcutta: Office of Superintendent of Government Printing, 1 June 1869. http://collections.nlm.nih.gov / accessed 4 August 2015.

Nevins, J. Birkbeck. "On the New French Codex." *The British Medical Journal.* 29 June 1867; **1**(339): 766–768. http://www.ncbi.nlm.nih.gov / accessed 9 February 2015.

Newsom, S.W. "Ogston's Coccus." *The Journal of Hospital Infections.* December, 2008; **70**(4): 369–372. http://www.ncbi.nlm.nih.gov / accessed 7 July 2014.

Nicati, W., and M. Reitsch. "Expériences sur la vitalité du bacilli virgule cholérique." *Revue d'Hygiene et Police Sanitaire.* 1885; **7**: 353–378.

_____. "Recherches sur le Choléra: Experiences d'inoculation." *Revue de Médicine* (Paris). 1885; **5**:449.

O'Connor, W. J. "Edgar March Crookshank (1859–1928)." *British Physiologists, 1885–1914: A Biographical Dictionary.* New York: St. Martin's Press; Manchester: Manchester University Press, 1991. P. 192.

Ogawa, M. "Uneasy Bedfellows: Science and Politics in the Refutation of Koch's Bacterial Theory of Cholera." *Bulletin of the History Medicine.* 2000; **74**: 671–707.

"Ogston, Sir Alexander (1844–1929)." *Oxford Dictionary of National Biography.* http://www.oxforddnb.com / accessed 7 July 2014.

Ogston, Alexander. "Continental Criticism of English Rifle Bullets." *The British Medical Journal.* 25 March 1899; **1**(1995): 752–757. http://www.ncbi.nlm.nih.gov / accessed 7 March 2015.

_____. "The Effects of the Dum-Dum Bullet from a Surgical Point of View." Correspondence. *The British Medical Journal.* 28 May 1898; **1**(1952): 1425. http://www.ncbi.nlm.nih.gov / accessed 7 October 2014.

_____. "The Influence of Lister upon Military Surgery." *The British Medical Journal.* 13 December 1902; **2**(2189): 1837–1838. http://www.ncbi.nlm.nih.gov / accessed 7 July 2014.

_____. "Micrococcus Poisoning." *Journal of Anatomy & Physiology.* July, 1882. **16**(4): 526–567. http://www.ncbi.nlm.nih.gov / accessed 11 May 2014.

_____. "Micrococcus Poisoning." *Journal of Anatomy and Physiology.* October, 1882. **17**(1): 24–58. http://www.ncbi.nlm.nih.gov / accessed 1 October 2014.

_____. "On Abscesses" (1880). *The Staphylococci: Proceedings of the Alexander Ogston Centennial Conference.* Translated by W. Witte. Edited by Alexander MacDonald and George Smith. Aberdeen, Scotland: Aberdeen University Press, 1981, pp. 277–286. Reprinted in *Reviews of Infectious Diseases/Classics in Infectious Diseases.* January-February, 1984; **6**(1): 122–128. Oxford University Press. http://www.jstor.org / accessed 7 July 2014.

_____. "Report upon Micro-Organisms in Surgical Diseases." *The British Medical Journal.* 12 March 1881; **1**(1054): 369–375. http://www.ncbi.nlm.nih.gov / 9 May 2014.

_____. "Veld Sores." *The British Medical Journal.* 20 April 1901; **1**(2103): 951–952. http://www.ncbi.nlm.nih.org / accessed 7 October 2014.

_____. "The Wounds Produced by Small-Bore Bullets: The Dum-Dum Bullet and the Soft-Nosed Mauser." *The British Medical Journal.* 17 September 1898; **2**(1968): 813–815. http://www.ncbi.nlm.nih.gov / accessed 4 October 2014.

"Ogston, Sir Alexander." *The British Medical Journal.* 16 February 1929; **1**(3554): 325–327 / accessed 8 October 2014.

Ogston, Alexander George. "Ogston, Sir Alexander (1844–1929), Surgeon and Bacteriologist."

Oxford Dictionary of National Biography. http://www.oxforddnb.com / accessed 7 July 2014.

Orenstein, Abigail. "The Discovery and Naming of Staphylococcus aureus." 2011. http://www.antimicrobe.org / accessed 26 May 2014.

Orme, I. M. "Immune Responses in Animal Models." *Current Topics in Microbiology and Immunology.* In *Tuberculosis: Current Topics in Microbiology and Immunology.* Edited by T. M. Shinnick. New York, Berlin, and Heidelberg: Springer, 1996; **215**: 181–196. http://link.springer.com / accessed 10 September 2014.

The Oxford English Dictionary. Compact Edition: Complete Text Reproduced Micrographically. 2 volumes. Oxford: At the Clarendon Press, 1971.

Pacini, Filippo. *Osservazioni Microscopiche e Deduzioni Patologiche.* Firenze: Federigo Bencini, 1854. https://www.books.google.com / accessed 9 February 2015.

_____. "Osservazioni microscopiche e deduzioni patologiche sul Cholera Asiatica." *Gazetta Medica Italiana: Toscana.* 2nd Series. 19 Decimbre 1854; **4**(50): 397–401; **4**(51): 405–412.

Paget, Stephen. *Pasteur and After Pasteur.* Edited by John D. Comrie. London: Adam and Charles Black, 1914.

"Pain in Research Animals: General Principles and Considerations." National Research Council (US) Committee on Recognition and Alleviation of Pain in Laboratory Animals. *Recognition and Alleviation of Pain in Laboratory Animals.* Washington, D.C.: National Academies Press, 2009. http://www.ncbi.nlm.nih.gov / accessed 9 August 2014.

Paneth, N., P. Vinten-Johansen, H. Brody, and M. Rip. "A Rivalry of Foulness: Official and Unofficial Investigations of the London Cholera Epidemic of 1854." *American Journal of Public Health.* October, 1998; **88**(10): 1545–1553. http://ajph.aphapublications.org / accessed 7 August 2014.

Parkes, Edmond Alexander. *A Manual of Hygiene.* Edited by F. S. B. François de Chaumont. With an Appendix. .. by Frederick N. Owen. 2 vols. 6th edition. New York: William Wood & Company 1883. Volume **I**, pp. 134–139. https://www.archive.org / accessed 4 July 2015.

Passet, J. *Untersuchungen über die Aetiologie der eiterigen Phlegmone des Menschen.* Weisbaden: Verlag von J. F. Bergmann, 1884.

Pasteur, Louis.

_____. "An Address delivered by Louis Pasteur at the 'Sorbonne Scientific Soirée' of April 7, 1864." Volume **I** (1863–1864): 257–264. Translated by Alex Levine. In *Scientific Progress.* By Alex Levine. Dubuque, Iowa: Kendall and Hunt, 2010. www.pasteurbrewing.com / accessed 8 November 2014.

_____. "An Address: Vaccination in Relation to Chicken Cholera and Splenic Fever." *The British Medical Journal.* 13 August 1881; **2**(1076): 283–284. www.ncbi.nlm.nih.gov / accessed 16 August 2015.

_____. "Animal Infusoria Living in the Absence of Free Oxygen, and the Fermentation They Bring About" (1861). *Milestones in Microbiology*, pp. 39–41.

_____. "The Attenuation of the Causal Agent of Fowl Cholera" (1880). *Milestones in Microbiology*, pp. 126–131.

_____. "The Germ Theory and Its Application to Medicine (Revised)" (1878). Translated by H. C. Ernst. *Germ Theory and Its Applications to Medicine & On the Antiseptic Principle of* [sic. *in*] *the Practice of Surgery: Louis Pasteur and Joseph Lister.* Amherst, New York: Prometheus Books, 1996; pp. 110–117.

_____. "Influence of Oxygen on the Development of Yeast and on the Alcoholic Fermentation" (Abstract) (1861). *Milestones in Microbiology*, pp. 41–43.

_____. "Investigation into the Role Attributed to Atmospheric Oxygen in the Destruction of Animal and Vegetable Substances after Death" (1861). *A Documentary History of Biochemistry, 1770–1940.* Edited by Mikulas Teich and Dorothy N. Needham. Rutherford, New Jersey: Fairleigh Dickinson University Press, 1992.

_____. "Letter: Pasteur to Lister, 29 June 1876." Godlee. *Lord Lister, pp.* 307–310.

_____. Letter: Pasteur to Lister, 2 January 1880. Godlee. *Lord Lister,* pp. 436–437.

_____. "Memoir on the Alcoholic Fermentation" (1861). *Milestones in Microbiology,* pp. 31–38.

_____. "On the Extension of the Germ Theory to the Etiology of Certain Common Diseases" (1880). Translated by H. C. Ernst. *Germ Theory and Its Applications to Medicine,* pp. 118–130.

_____. "On the Organized Bodies Which Exist in the Atmosphere; Examination of the Doctrine of Spontaneous Generation" (1861). *Milestones in Microbiology,* pp. 43–48.

_____. "On a Vaccine for Fowl Cholera and Anthrax" (1881). *Milestones in Microbiology,* pp. 131–132.

_____. "The Physiological Theory of Fermentation" (1879). Translated by F. Faulkner and D. C. Robb. *Germ Theory and Its Applications to Medicine,* pp. 15–109.

_____. "Prevention of Rabies: A Method by Which the Development of Rabies after a Bite May be Prevented." *The Founders of Modern Medicine,* pp. 379–387.

_____ "Report on the Lactic Acid Fermentation (Abstract)" (1857). *Milestones in Microbiology,* pp. 27–30.

Paterson, Robert. *Memorials of the Life of James Syme, Professor of Clinic Surgery in the University of Edinburgh, Etc.* Edinburgh: Edmonston & Douglas, 1877.

"Pavlov, Ivan Petrovich (1849–1936)." *Random House Webster's Dictionary of Scientists.* Preface by David Abbott and Roy Porter. New York: Random House, 1997. P. 376.

Pelling, Margaret. *Cholera, Fever, and English Medicine: 1825–1865.* Oxford Historical Monographs. Oxford: Oxford University Press, 1978.

Peterson, Johnny W. "Bacterial Pathogenesis." *Medical Bacteriology.* Edited by S. Baron. 4th edition. Galveston, Texas: University of Texas Medical Branch at Galveston, 1996. *NCBI Bookshelf.* http://www.ncbi.nlm.nih.gov / accessed 4 March 2014.

Pettenkofer, Max von. *Cholera: How to Prevent and Resist It.* Translated with an Introduction and Appendix by Thomas Whiteside Hime. Revised edition. London, Paris, Madrid: Balliére, Tindall, and Cox: 1875. https://www.books.google.com / accessed 19 August 2014.

"Physiological Fallacies" [Antivivisectionist papers by various authors]. First Series. London: Williams and Norgate, 1882. https://www.books.google.com / accessed 31 August 2014.

Porter, I. A. "Alexander Ogston—Bacteriologist." *The British Medical Journal.* 7 August 1954; **2**(4883): 355–356. http://www.ncbi.nlm.nih.org / accessed 8 October 2014.

Pretorius, Fransjohan. "The Second Anglo-Boer War: An Overview." *Scientia Militaria: South African Journal of Military Studies.* 2000; **30**(2): 111–125. http://www.scientsiamilitaria-journals.ac.za / accessed 2 October 2014.

"Prosecution at Norwich: Experiments on Animals." *The British Medical Journal.* 12 December 1874; **2**(728): 751–754. http://www.ncbi.nlm.nih.con / accessed 23 February 2015.

Prüll, Cay-Rüdiger. "Pathology at War 1914–1918: Germany and Britain in Comparison." *Medicine and Modern Warfare.* Edited by R. Cooter, M. Harrison, and S. Sturdy. Clio Medica/The Wellcome Series in the History of Medicine. Amsterdam and Atlanta: Rodopi, 1999, pp. 131–161.

Ray, C. Claiborne. "Don't Eat the Pits." Questions/ Science Times. *The New York Times.* 16 February 1993. www.nytimes.com / accessed 16 September 2015.

Redi, Francesco. "Experiments on the Generation of Insects" (1688). *Great Experiments in Biology.* Translated by Mab Bigelow. Edited by Mordecai Gabriel and Seymour Fogel. The Prentice-Hall Animal Science Series. Edited by H. Burr Steinbach. Englewood Cliffs, NJ: Prentice-Hall, 1955, pp. 187–189.

Richards, Stewart. "Anesthetics, Ethics, and Aesthetics: Vivisection in the Late Nineteenth Century British Laboratory." *The Laboratory Revolution in Medicine.* Edited by Andrew

Cunningham and Perry Williams. New York and Cambridge, UK: Cambridge University Press, 1992, pp. 142–169.

Richardson, Benjamin Ward. *The Cause of Coagulation: Being the Astley Cooper Prize Essay for 1856.* London: John Churchill: London, 1858. https://babel.hathitrust.org / accessed 9 August 2015.

Richardson, Ruth. "Inflammation, Suppuration, Putrefaction, Fermentation: Joseph Lister's Microbiology." *Royal Society Publishing.* 29 May 2013. http://royalsocietypublishing.org / accessed 10 July 2014.

Rickard, J. "Battle of Poplar Grove, 7 March 1900." 7 March 2007. http://www.historyofwar. org / accessed 28 September 2014.

Roe, S. "John Turbeville Needham and the Generation of Living Organisms." *Isis.* 1983; **74**: 159–184.

Romano, Terrie. *Making Medicine Scientific: John Burdon Sanderson and the Culture of Victorian Science.* Baltimore: The Johns Hopkins University Press, 2003.

Rosenbach, A. J. F. "Micro-organisms in Human Traumatic Infective Diseases." *Recent Researches on Micro-organisms in Relation to Suppuration and Septic Diseases.* Preface, Translated, and Abstracted by W. Watson Cheyne. Volume **115**. London: The New Syndenham Society, 1886, pp. 397–438. https://www.books.google.com / accessed 22 July 2014.

_____. *Mikro-organismen bei den Wund-infections-krankheiten des Menschen.* Weisbaden: Verlag von J. F. Bergmann, 1884.

_____. "Process of Preparing a Preventive of Tuberculosis: The Use of Rosenbach's Tuberculin in Surgical Tuberculosis." *The Lancet.* December, 1911; **178**(4606): 1621–1624. https://www.google.com / accessed 21 July 2014.

Rowe, R.M. "A Note on the Carrel-Dakin-Daufresne Treatment." *The British Medical Journal.* 22 September 1917; **2**(2960): 387. http://www.ncbi.nih.nlm.gov / accessed 25 December 2014.

Ruffer, M. Armand. "Remarks Made at the Discussion on Phagocytosis and Immunity." *The British Medical Journal.* 19 March 1892; **1**(162): 591–596. http://www.ncbi.nlm.nih.gov / accessed 29 December 2014.

Sakula, Alex. "Robert Koch: Centenary of the Discovery of the Tubercle Bacillus, 1882." *Thorax.* 1982; **37**: 246–251. http://thorax.bmj.com / accessed 24 August 2014.

Sanderson, John Burdon. "An Address on the Relation of Science to Experience in Medicine." *The British Medical Journal.* 11 November 1899; **2**(2028): 1333–1335. http://www.ncbi. nlm.nih.gov / accessed 21 October 2014.

_____. "The Croonian Lectures on the Progress of Discovery Relating to the Origin and Nature of Infectious Diseases." *The British Medical Journal.* 28 November 1891; **2**(1613): 1135–1139. http://www.ncbi.nih.nlm.gov / accessed 10 September 2014.

_____. "Introduction to a Discussion on Tuberculosis in All Its Relations." *British Medical Journal.* 22 August 1891; **2**(1599): 403–406. http://www.ncbi.nlm.nih.gov / accessed 29 August 2014.

_____. "Physiology." In *Handbook for the Physiological Laboratory,* pp. 175–348.

_____, ed. *Handbook for the Physiological Laboratory.* 1873. Philadelphia: P. Blakiston, Son & Company, 1884. https://www.archive.org / accessed 16 August 2015.

Santer, M. "Joseph Lister: First Use of a Bacterium as a 'Model Organism' to Illustrate the Cause of Infectious Disease of Humans." *Notes and Records of the Royal Society.* London, 2010; **64**: 59–65. http://www.ncbi.nlm.nih.gov / accessed 16 March 2016.

Savage, W. G. *Milk and the Public Health.* London: Macmillan, 1912.

_____. "Part VII.: Bacteriological Examinations." In *Public Health Laboratory Work.* By Henry Richard Kenwood. 5th edition. London: H. K. Lewis, 1911, pp. 379–440. https://www. babel.hathitrust.org / accessed 9 April 2015.

Schechter, D. C., and H. Swan. "Jules Lemaire: A Forgotten Hero of Surgery." *Surgery.* June, 1961; **49**: 817–826. www.ncbi.nlm.nih.gov / accessed 22 September 2015.

Schlich, Thomas. "Farmer to Industrialist: Lister's Antisepsis and the Making of Modern Surgery in Germany." *Notes & Records: The Royal Society Journal of the History of Science.* 2013. http://rsn.royalsocietypublishing.org / accessed 14 December 2014.

Schwann, Theodore. "Preliminary Reports on Experiments Concerning Alcoholic Fermentation and Putrefaction" (1837). *Milestones in Microbiology*, pp. 16–19.

Semmelweis, Ignaz. "Lecture on the Genesis of Puerperal Fever (childbed fever)" (1850). Translated by Thomas D. Brock. *Milestones in Microbiology*, pp. 80–82.

Silverstein, Arthur. *A History of Immunology.* 2nd edition. 1989. New York and London: Academic Press/Elsevier, Incorporated, 2009.

Simpson, James Young. *Hospitalism: Its Effects on the Results of Surgical Operations, Etc.* Part I. Edinburgh: Oliver and Boyd, 1869.

Singhal, A., et al. "Experimental Tuberculosis in the Wistar Rat: A Model for Protective Immunity and Control of Infection." *PLOS/One.* 12 April 2011; **6**(4):18632. http://www.plosone.org / accessed 9 September 2014.

Smith, D., E. Wiegeshaus, and V. Balasubramanian. "Animal Models for Experimental Tuberculosis." *Clinical Infectious Diseases.* 2000; **31**(Supplement 3): S68–70. http://cid.oxfordjournals.org / accessed 9 September 2014.

Smith, George. "Ogston the Bacteriologist." Aberdeen, Scotland: Aberdeen University Press, 1982, pp. 9–21. http://www.scholar.google.com / accessed 31 December 2014.

Smith, J. Lorrain, A. Murray Drennan, Theodore Rettie, and William Campbell "Antiseptic Action of Hypochlorous Acid and Its Application to Wound Treatment."

The British Medical Journal. 24 July 1915; **2**(2847): 129–136. http://www.ncbi.nlm.nih.gov / accessed 17 July 2015.

Snodgrass, Mary Ellen. *World Epidemics: A Cultural Chronology of Disease from Prehistory to the Era of SARS.* Jefferson, North Carolina, and London: McFarland, 2003.

Snow, John. "Cholera and the Water Supply in the South Districts of London, 1854." http://www.ph.ucla.edu / accessed on 9 February 2015.

_____. *On the Mode of Communication of Cholera.* London: John Churchill, 1855. http://www.ph.ucla.edu / accessed on 9 February 2015.

Solliday, Jim. "History of Oil Immersion Lenses." *Southwest Museum of Engineering, Communications and Computation History.* www.smecc.org / accessed 22 September 2015.

"Some of Last Year's Vivisections." *The Zoophilist and Animals' Defender.* 2 May 1892; **12**: 7–8; London: Victoria Street Society for the Protection of Animals from Vivisection, 1892–1893. https://www.books.google.com / accessed 31 August 2014.

Sorenson, R. A. "The Inducible System: History of Development of Immunology as a Component of Host-Parasite Interactions." *Infection, Resistance, and Immunity.* Edited and Preface by Julius P. Kreier. Second Edition. New York: Taylor & Francis, 2002, pp. 61–82.

Spallanzani, Lazaro. "Tracts on the Nature of Animals and Vegetable; Observations and Experiments upon the Animalcula of Infusions" (1799). *Milestones in Microbiology*, pp. 13–16.

"Spontaneous Generation." *The Oxford English Dictionary. Compact Edition.* Volume II. P-Z, Supplement and Bibliography. Oxford: At the Clarendon Press, 1971. P. 2977.

"Staphylococcus." *The Compact Oxford English Dictionary.* Volume II. P-Z. P. 3019.

"Staphylococcus erysipelas." *Webster's Seventh New College Dictionary*, pp. 159, 283, 854.

"Staphylococcus pyogenes aureus." *Webster's Seventh New Collegiate Dictionary*, pp. 348, 854, 696.

Sternberg, George Miller. *A Manual of Bacteriology.* New York: William Wood & Company, 1893.

_____. *A Text-Book of Bacteriology.* 2nd edition. New York: William Wood and Company, 1901. https://www.books.google.com / accessed 17 February 2015.

Stevenson, William Flack. *Report on the Surgical Cases Noted in the South African War, 1899–1902.* London: His Majesty's Stationery Office: Harrison and Sons, 1905. https://www.books.google.com / accessed 12 March 2016.

_____. *Wounds in War: The Mechanism of Their Production and Their Treatment.* 3rd edition.

New York: William Wood and Company, 1910. https://www.archive.org / accessed 26 February 2014.

Stimson, L. A. "Bacteria and Their Effects." *Popular Science Monthly*. February 1975; Volume **6**. http://en.wikisource.org / accessed 23 December 2014.

Stossel, Thomas P. "The Early History of Phagocytosis." *Phagocytosis: The Host*. Edited by Siamon Gordon. Volume **5** of *Advances in Cell and Molecular Biology of Membranes and Organelles*. Series editor, Alan M. Tartakoff. Stamford, Connecticut: JAI Press, Incorporated, 1999, pp. 3–18.

Stover, Robert. "Great Man Theory of History." *The Encyclopedia of Philosophy*. Edited by Paul Edwards. 8 volumes in 4. 1967; New York: Macmillan Publishing Company, Inc., & The Free Press; London: Collier Macmillan Publishers, 1972. Volumes **3 & 4**: 378–382.

Strahler, Arthur N. *Introduction to Physical Geography*. 2nd edition. 1965. New York and London: John Wiley and Sons, 1970.

"Streptococcus." *The Compact Oxford English Dictionary*. Volume **II**. P-Z. P. 3090.

Strick, James E. "Darwinism and the Origin of Life: The Role of H. C. Bastian in the British Spontaneous Generation Debates, 1868–1873." *Journal of the History of Biology*. 1999; **32**: 51–92.

_____. "New Details Add to Our Understanding of Spontaneous Generation Controversies: Reexamining Several Episodes Reveals the Complexity and Human Richness of Science in the Making." *American Society for Microbiology*. 1999; **63**(4): 192–198.

_____. *Sparks of Life: Darwinism and the Victorian Debate over Spontaneous Generation*. 2nd edition. Cambridge, Massachusetts: Harvard University Press, 2002.

Swain, Valentine A. J. "Franco-Prussian War 1870–1871: Voluntary Aid for the Wounded and Sick." *The British Medical Journal*. 29 August 1970; **3**(5721): 514–517. http://www.ncbi.nlm.nih.gov / accessed 11 December 2014.

Takeda, Yoshifumi. "*Vibrio Cholerae* and Cholera Toxin: From Calcutta to Kolkata." *Science and Culture*. May-June, 2010; **76**(5–6): 149–152. http://www.scienceandculture-isna.org / accessed 6 August 2014.

Tanner, H. H. "The Treatment of Wounds by Saline Solution." *Surgery in War*. Edited by Alfred John Hull. Preface by Sir Alfred Keogh. Philadelphia: P. Blakiston's Sons & Company, 1916, pp. 70–81.

Tansey, E. M. "'The Queen Has Been Dreadfully Shocked': Aspects of Teaching Experimental Physiology Using Animals in Britain, 1876–1986. *American Journal of Physiology*. June 1998; **274**(6, Pt. 2): S18–33, www.ncbi.nlm.nih.gov / accessed 26 November 2015.

Tauber, Alfred I. *Metchnikoff and the Origins of Immunology*. New York and Oxford: Oxford University Press, 1991. https://www.books.google.com / accessed 3 August 2015.

_____. "Metchnikoff and the Phagocytosis Theory." *Perspectives/Molecular Cell Biology*. November, 2003; Volume **4**: 897–401. http://blog.bu.edu / accessed 27 July 2014.

Tee, Garry J. "Andrew Cheyne, Father of Sir William Watson Cheyne." *Auckland-Waikato Historical Journal*. April 1998; No. **71**: 25–26.

Todor, Kenneth. "Pathogenic Clostridia, including Botulism and Tetanus." *Todor's Online Textbooks of Bacteriology*. http://textbookbacteriology.net / accessed 21 August 2014.

Tortora, Gerald J., and Bryan Derrickson. *Introduction to the Human Body: The Essentials of Anatomy and Physiology*. New York: John Wiley & Sons, 2007.

"The Treatment of War Wounds." *The Journal of the American Medical Association*. 28 October, 1916; **67**(18): 1304–1305). https://www.books.google.com / accessed 16 April 2014.

Treves, Frederick, editor. *A System of Surgery*. 2 vols. Philadelphia: Lea Brothers & Company, 1895–1896.

Trucksis, Michele, Timothy L. Conn, Alessio A. Fasano, and James B. Kaper. "Production of *Vibrio cholerae* Accessory to Cholera Enterotoxin (Ace) in the Yeast *Pichia pastoris*." *Infection and Immunity*. December, 1997; **65**(12): 4984–4988. http://www.ncbi.nlm.nih.gov / accessed 19 July 2015.

Tuberculosis. *Physiological Fallacies [anti-vivisection papers by various authors]*. Preface by Frances Power Cobbe. First Series. London: Williams and Norgate, 1882, pp. 99–106. https://www.google.books / accessed 31 August 2014.

Turner, Gerard L'Etrange. *Nineteenth-Century Scientific Instruments*. Foreword by Dame Margaret Weston. London, Berkeley, and Los Angeles: Sotheby Publications/ The University of California Press, 1983.

Tyndall, John. "Dust and Disease: Being Part of a Lecture delivered at the Royal Institution of Great Britain." *The British Medical Journal*. 24 June 1871; **1**(547): 661–662. http://www.ncbi.nlm.nih.gov / accessed 1 August 2015.

United States Department of Agriculture. APHIS Policy #11, Painful Procedures (Issue dated: 14 April 1997), 1997a. www.aphis.usda.gov / accessed 9 August 2014.

Vallery-Radot, René. *The Life of Pasteur*. Translated by R. L. Devonshire. 2nd edition. New York: Doubleday, Page, & Company, 1915.

Van Arsdale, W. W. "On the Present State of Knowledge in Bacterial Science in Its Surgical Relations." *Annals of Surgery*. January/June, 1886; **3**(3): 221–226. http://www.ncbi.nlm.nih.gov / accessed 2 August 2015.

Vernon, Keith. "Pus, Sewage, Beer and Milk: Microbiology in Britain, 1870–1940." *History of Science*. September, 1990; **28**(81 pt. 3): 289–325.

Vibrio. *The Compact Oxford English Dictionary*. Volume **II**: P-Z, Supplement and Bibliography. P. 3623.

Vibrio cholerae Pacini. List of Prokaryotic Names with Standing Nomenclature. http://www.google.com / accessed on 16 August 2014.

Vincoli, Jeffrey W. *Risk Management for Hazardous Chemicals*. Volume **1**. Boca Raton, Florida: CRC Press, 1997, pp. 703–726. https://www.books.google.com / accessed 20 August 2015.

"Virus." *Pocket Oxford Latin Dictionary*. Edited by James Morwood. 1994. Oxford and New York, 2005. P. 206.

Vivisection. *The Compact Oxford English Dictionary*. Volume **II**. P-Z. P. 3648.

Vivisection Again! *The Medical Times and Gazette: A Journal of Medical Science, Literature, Criticism and News*. 30 January 1875. Volume **I**: 116. London: J & A. Churchill, 1875. https://www.books.google.com / accessed 23 February 2015.

The Vivisectors' Directory, Being a List of the Licensed Vivisectors in the United Kingdom, Together with the Leading Physiologists in Foreign Laboratories. Edited by Benjamin Bryan. Preface by Frances Power Cobbe. London: Victoria Street Society for the Protection of Animals from Vivisection, 1884.

The Vivisectors' Directory. *The Zoophilist*. 1 December 1882; Vol. **III**. New Series 13: 215–218; 1 January 1883; Vol. **III**. New Series 14: 240–242. https://www.books.google.com / accessed 22 September 2014.

Vivus/sectio. *Pocket Oxford Latin Dictionary*, pp. 207, 411.

Volkmann, Richard von, [of] Halle: Obituary." *The Medical Record*. 7 December 1889. Edited by George Frederick Shrady and Thomas Lathrop Stedman. Volume **36**. 6 July-28 December, 1889. New York: William Wood & Company, 1889. P. 636. https://www.books.google.com / accessed 8 March 2015.

Wainwright, Milton, and Joshua Lederberg. "History of Microbiology." *Encyclopedia of Microbiology*. Volume **2**. Academic Press, 1992, pp. 419–437. http://www.profiles.nlm.nih.gov / accessed 19 May 2015.

Waldren, Mike. "Dum Dum Bullets: 'It Will Rip the Arm from a Normal Healthy Human.'" *Police Firearms Officers Association: Police History Series*. 2012. http://www.pfoa.co.uk / accessed 4 October 2014.

Walker, Emma E. "The Effects of Absinthe." *Medical Record*. 13 October 1906; Volume **70**. *The Wormwood Society*. http://wormwoodsociety.org / accessed 23 February 2015.

Wall, Rosemary. *Bacteria and Britain, 1880–1939*. Studies for the Society of the Social History of Medicine: **17**. London: Pickering & Chatto, 2013.

Warner, John Harley. "The History of Science and the Sciences of Medicine." *Osiris*, 2nd Series (1995), Vol. **10**, Constructing Knowledge in the History of Science: 164–93.

Waterson, A. P., and Lise Wilkinson. *An Introduction to the History of Virology*. Cambridge and New York: Cambridge University Press, 1978.

Watson, Caroline C., and Christoph J. Greissenauer, Marious Loukas, Jeffrey P. Blount, R. Shane Tubbs. "William Watson Cheyne (1852–1932): A Life in Medicine and his Innovative Surgical Treatment of Congenital Hydrocephalus." *Child's Nervous System*. November, 2013; **29**(11): 1961–1965. http://link.springer.com / accessed 15 May 2014.

"Watson Cheyne on Tubercle-Bacilli." *Medical Times and Gazette: A Journal of Medical Science, Literature, Criticism, and News*. 31 March 1883; Vol. **1**: 354–355; London: J. & A. Churchill, 1883. https://www.books.google.com / accessed 31 August 2014.

Welch, William H. "Conditions Underlying the Infection of Wounds." *American Journal of the Medical Sciences*. November, 1891; **102**(5): 439–464. http://journals.lww.com / accessed 7 May 2014.

_____. "Conditions Underlying the Infection of Wounds." *Papers and Addresses: Bacteriology*. Volume **2** of *Papers and Addresses [of] William Henry Welch*. 3 vols. Edited by Walter Cleveland Burket. Baltimore: The Johns Hopkins University Press, 1920. https://www. books.google.com / accessed 7 May 2014, pp. 392–419.

West, James F. "On the Surgical Cases in which Lister's Plan of Treating Wounds is Preferable to any other Method of Dressing." *Saint Thomas's Hospital Reports*. Volume **10**. Edited by Robert Cory and Francis Mason. London: J. & A. Churchill, 1880, pp. 73–83. https:// www.books.google.com / accessed 10 December 2014.

Whitehead, Ian. "The British Medical Officer on the Western Front: The Training of Doctors for War." *Medicine and Modern Warfare*. Edited by R. Cooter, M. Harrison, and S. Sturdy. Clio Medica/The Wellcome Series in the History of Medicine. Volume **55**. Amsterdam and Atlanta: Rodopi, 1999, pp. 163–184.

Williams, C. Theodore. "Harveian Oration on Old and New Views on the Treatment of Consumption." *The British Medical Journal*. 21 October 1911; **2**(2651): 961–968. http://www. ncbi.nlm.nih.gov / accessed 31 August 2014.

Willy, Christian, Peter Schneider, Michael Engelhardt, Alan R. Hargens, and Scott J. Mubarack. "Richard von Volkmann: Surgeon and Renaissance Man." *Clinical Orthopaedics and Related Research*. February, 2008; **466**(2): 500–506. http://www.ncbi.nlm.nih.gov / accessed 10 December 2014.

Wilson, Leonard G. "The Early Recognition of Streptococci as Causes of Disease." *Medical History*. 1987; **31**: 403–414. http://www.ncbi.nlm.nih.gov / accessed 9 May 2014.

Winslow, Charles-Edward Amory. *The Conquest of Epidemic Disease: A Chapter in the History of Ideas*. 1943. Madison: University of Wisconsin Press, 1980.

Wolf, Anthony S. *Endangered Lives: Public Health in Victorian Britain*. West Sussex, U.K.: Littlehampton Book Services, 1983.

Wong, Thiang Yian, Robert Schwarzenbacher, and Robert C. Liddington. "Towards Understanding Anthrax: Structural Basis of Target Recognition by Anthrax Lethal Factor." http://www-ssrlac.stanford.edu / accessed 21 December 2014.

Woodhead, G. Sims, J. Burdon Sanderson, E. E. Klein, and A. A. Kanthack. "A Discussion on Phagocytosis and Immunity." *The British Medical Journal*. 20 February 1892; **1**(1625): 373–380. http://www.ncbi.nlm.nih.gov / accessed 29 December 2014.

Woodhead, German Sims. "Surgical Bacteriology." *A System of Surgery*. Edited by Frederick Treves. Volume **1**, pp. 1–52.

Worboys, Michael. "Almroth Wright at Netley: Modern Medicine and the Military in Britain, 1892–1902." *Medicine and Modern Warfare*. Edited by R. Cooter, M. Harrison, and S. Sturdy. Clio Medica/The Wellcome Institute Series in the History of Medicine. Volume **55**. Amsterdam and Atlanta: Rodopi, 1999, pp. 77–97.

_____. "History of Bacteriology." *Encyclopedia of Life Sciences.* John Wiley & Sons, 2001: Pp. 1–5. www.els.net / accessed 31 December 2014.

_____. "Joseph Lister and the Performance of Antiseptic Surgery." *Notes & Records of the Royal Society of London.* 20 September 2013; **67**(3): 199–209. http://www.escholar. manchester.ac.uk / accessed 31 December 2014.

_____. *Spreading Germs: Diseases, Theories, and Medical Practices in Britain, 1865–1900.* Cambridge History of Medicine. Edited by Charles Rosenberg. Cambridge: Cambridge University Press, 2000.

_____. "Was There a Bacteriological Revolution in Late Nineteenth-Century Medicine?" *Studies in the History, Philosophy of Biology, and Biomedical Science.* March, 2007; **38**(1): 20–42.

_____. "Wright, Sir Almroth Edward (1861–1947), medical scientist." *Oxford Dictionary of National Biography.* Oxford University Press, 2004; Online edition: October, 2006. http://www.oxforddnb.com / accessed 24 September 2014.

"The Wounded about Sedan." *The Lancet.* Edited by James G. Wakley. Volume **2**. London: *The Lancet,* 8 October 1870, pp. 511–512. https://www.google.books.org / accessed 12 January 2015.

Wright, Almroth E. "An Address on Wound Infections; and on Some New Methods for the Study of the Various Factors Which Come into Consideration in Their Treatment." *The British Medical Journal.* 10 April 1915; **1**(2832): 625–628. http://www.ncbi.nlm.nih.gov / accessed 16 May 2014.

_____. "An Address on Wound Infections; and on Some New Methods for the Study of the Various Factors Which Come into Consideration in Their Treatment." *The British Medical Journal.* 17 April 1915; **1**(2833): 665–668. http://www.ncbi.nlm.nih.gov / accessed 16 May 2014.

_____. "An Address on Wound Infections; and on Some New Methods for the Study of the Various Factors Which Come into Consideration in Their Treatment." *The British Medical Journal.* 24 April 1915; **1**(2834): 720–723. http://www.ncbi.nlm.nih.gov / accessed 16 May 2014.

_____. "An Address on Wound Infections; and on Some New Methods for the Study of the Various Factors Which Come into Consideration in Their Treatment." *The British Medical Journal.* 1 May 1915; **1**(2835): 762–764. http://www.ncbi.nlm.nih.gov / accessed 16 May 2014.

_____. "Conditions which Govern the Growth of the Bacillus of 'Gas Gangrene' in Artificial Culture Media, in the Blood Fluids *in vitro,* and in the Dead and Living Organism." *Proceedings of the Royal Society of Medicine.* Delivered 18 December 1916; published 1917; **10**(Gen. Rep.), 1: 1–32. http://www.ncbi.nlm.nih.gov / accessed 20 May 2014.

_____. "A Lecture on Wound Infections and Their Treatment." *The British Medical Journal.* 30 October 1915; **2**(2861): 629–635. http://www.ncbi.nlm.nih.gov / accessed 17 April 2014.

_____. "A Lecture on Wound Infections and Their Treatment." *The British Medical Journal.* 13 November 1915; **2**(2863): 717–721. http://www.ncbi.nlm.nih.gov / accessed 16 May 2014.

_____. "A Lecture on Wound Infections and their Treatment." *Proceedings of the Royal Society of Medicine.* Delivered 8 October 1915; Published 1916; **9**(Gen. Rep.): 1–72. http:// www.ncbi.nlm.nih.gov / accessed 7 October 2014.

_____. "Memorandum on the Employment of Bandages for the Irrigation of Wound-Surfaces with Therapeutic Solutions and the Draining of Wounds." *The British Medical Journal.* 16 October 1915; **2**(2859): 564–567. http://www.ncbi.nlm.nih.gov / accessed 8 October 2014.

_____. "Memorandum on the Treatment of Infected Wounds by Physiological Methods (Drainage of Infected Tissues by Hypertonic Salt Solution, and Utilization of the Antibacterial Powers of the Blood Fluids and White Corpuscles)." *The British Medical Journal.*

3 June 1916; **1**(2892): 793–797. http://www.ncbi.nlm.nih.org / accessed 10 November 1916.

_____. "The Question as to How Septic War Wounds Should Be Treated (Being a Reply to Polemical Criticism published by Sir W. Watson Cheyne in the '*British Journal of Surgery*')." *The Lancet.* 16 September 1916; **188**(4855): 503–516.

_____. *Wound Infections and Some New Methods for the Study of the Various Factors Which Come into Consideration in Their Treatment.* Address delivered before the Royal Society of Medicine. London: University of London Press, 1915, pp. 49–92. https://babel. hathitrust.org / accessed 22 May 2014.

_____, and Stewart R. Douglas. "An Experimental Investigation of the Role of the Blood Fluids in Connection with Phagocytosis." *Proceedings of the Royal Society of London.* 3 January 1903; **72**: 357–370. http://rspl.royalsocietypublisjing.org / accessed 8 January 2015.

Yeager, Robert G. "Protozoa: Structure, Classification, Growth, and Development." *Medical Microbiology.* Edited by Samuel Baron. 4th edition. Galveston, Texas: University of Texas Medical Branch at Galveston, 1996. http://www.ncbi.nlm.nih.gov / accessed on 11 February 2015.

Yeo, Richard. *Defining Science: William Whewell, Natural Knowledge and Public Debate in Early Victorian Britain.* Cambridge: Cambridge University Press, 1993.

Zabecki, David T. "German Artillery in the First World War." In *King of Battle: Artillery in World War I.* Edited by Sanders Marble. Series: History of Warfare. Volume 108. Boston: Brill, 2015, pp. 101–125.

Zimmermann, M. "Life and Work of the Surgeon Ernst von Bergmann (1836–1907), Long-Term Editor of the '*Zentralblatt für Chirurgie*.'" *Zentralblatt für Chirurgie.* 2000; **125**(6): 552–560. http://www.ncbi.nlm.nih.gov / accessed 14 December 2014.

Zooglœa(s). *Webster's Seventh New Collegiate Dictionary.* 1040; and *The Oxford English Dictionary. Compact Edition.* Volume **II**. P-Z. P. 3871.

The Zoophilist. 2 April 1883; Volume II. New series; **4**: 52. https://www.google.books.com / accessed 16 September 2014.

_____. "Notes and Notices." 1 July 1891; Volume **XI**. No. 3. New Series: 37. In *The Animal's Defender and Zoophilist.* Volume **XI**. London: The Victoria Street Society for the Protection of Animals from Vivisection, 1891–1892. https://www.books.google.com / accessed 22 September 2014.

Zymotic. *The Compact Oxford English Dictionary.* Volume **II**. P-Z. P. 3872.

Index

Numbers in **bold italics** refer to pages with photographs.